143-

Nutrition and Cardiovascular Disease

Progress in
Biochemical Pharmacology

Vol. 19

Series Editor
R. Paoletti, Milan

Basel · München · Paris · London · New York · Tokyo · Sydney

US-Italy Symposium on Nutrition and Cardiovascular Disease, Rome, December 5, 1980

Nutrition and Cardiovascular Disease

Volume Editor
R.J. Hegyeli

Assistant Director for International Programs, Office of the Director
National Heart, Lung, and Blood Institute, National Institutes of Health
Bethesda, Md.

37 figures and 80 tables, 1983

Basel · München · Paris · London · New York · Tokyo · Sydney

Progress in Biochemical Pharmacology

Vol. 17: Hormones and the Kidney
  6th Kanematsu Conference on the Kidney, Sydney 1980.
  Stokes, G. S. and Mahony, J. F., Sydney (eds.)
  VIII+268 p., 64 fig., 20 tab., 1980. ISBN 3-8055-1090-X
Vol. 18: Endocrinological Aspects of Alcoholism
  4th Annual Conference on Alcoholism, El Paso, Tex., 1980
  Messiha, F. S. and Tyner, G. S., Lubbock, Tex. (eds.)
  XII+232 p., 65 fig., 27 tab., 1981. ISBN 3-8055-2689-X

National Library of Medicine, Cataloging in Publication
  US-Italy Symposium on Nutrition and Cardiovascular Disease (1980: Rome, Italy).
  Nutrition and cardiovascular disease.
  Volume editor, R. J. Hegyeli. – Basel, New York: Karger, 1983.
  (Progress in biochemical pharmacology; v. 19).
  1. Cardiovascular Diseases – etiology – congresses. 2. Cardiovascular Diseases – prevention & control – congresses. 3. Diet – adverse effects – congresses. I. Johnsson-Hegyeli, Ruth, 1931 – II. Title. III. Series.
  WI PR666H v. 19 / WG 100 U84n 1980
  ISBN 3-8055-3571-6

Drug Dosage
  The authors and the publisher have exerted every effort to ensure that drug selection and dosage set forth in this text are in accord with current recommendations and practice at the time of publication. However, in view of ongoing research, changes in government regulations, and the constant flow of information relating to drug therapy and drug reactions, the reader is urged to check the package insert for each drug for any change in indications and dosage and for added warnings and precautions. This is particularly important when the recommended agent is a new and/or infrequently employed drug.

All rights reserved
  No part of this publication may be translated into other languages, reproduced or utilized in any form or by any means, electronic or mechanical, including photocopying, recording, microcopying, or by any information storage and retrieval system, without permission in writing from the publisher.

© Copyright 1983 by S. Karger AG, P.O. Box, CH-4009 Basel (Switzerland)
  Printed in Switzerland by Reinhardt AG, Basel
  ISBN 3-8055-3571-6

Contents

List of Contributors VII
Editor's Note VIII
Preface IX

Epidemiological Observations

Keys, A. (Minneapolis, Minn.): From Naples to Seven Countries – A Sentimental Journey 1
Blackburn, H.; Prineas, R. (Minneapolis, Minn.): Diet and Hypertension: Anthropology, Epidemiology, and Public Health Implications 31
Lanzola, E.; Turconi, G.; Allegrini, M.; Marco, R. de; Marinoni, A.; Miracca, P. (Pavia): Relation between Coronary Heart Disease and Certain Elements in Water and Diet 80

Nutrition and Lipoproteins

Rifkind, B.M. (Bethesda, Md.): Nutrient – High-Density Lipoprotein Relationships: An Overview 89
Havel, R.J. (San Francisco, Calif.): Dietary Regulation of Plasma Lipoprotein Metabolism in Humans 110
Farinaro, E.; Rubba, P.; Postiglione, A.; Riccardi, G.; Mancini, M. (Naples): Studies on Diet and Plasma Lipids in Naples 123
Baggio, G.; Gabelli, C.; Fellin, R.; Baiocchi, M.R.; Martini, S.; Baldo, G.; Manzato, E.; Crepaldi, G. (Padova): Metabolism of Lipoproteins in the Postprandial Phase . 129
Avogaro, P.; Bittolo Bon, G.; Belussi, F.; Cazzolato, G. (Venice): Relationships of Body Weight and Fatness to Lipoprotein Components. Variations following Hypocaloric Diet 141
Kritchevsky, D. (Philadelphia, Pa.): Dietary Influences on Lipids and Lipoprotein Levels in Animals and Atherosclerosis 151

Rabbi, A. (Rome): Nutrients and Lipoproteins: Dietary Profiles of Population Groups and Experimental Models 166

Diet and Hypertension

Dustan, H. P. (Birmingham, Ala.): Role of Nutrition in Hypertension and Its Control – Experimental Aspects 177
Sirtori, C. R.; Lovati, M. R.; Gianfranceschi, G.; Farina, R.; Franceschini, G. (Milan): Experimental Studies on Nutrition, Hypertension, and Cardiovascular Diseases 192
Tobian, L. J. (Minneapolis, Minn.): Interrelationship of Sodium, Volume, CNS, and Hypertension 208

Risk Factors and Prevention of CHD

Menotti, A. (Rome): An Italian Preventive Trial of Coronary Heart Disease: The Rome Project of Coronary Heart Disease Prevention 230
Stamler, J. (Chicago, Ill.): Nutrition-Related Risk Factors for the Atherosclerotic Diseases – Present Status 245

Subject Index 309

List of Contributors

Avogaro, P., M.D., Ph.D., General Regional Hospital, National Council for Research, Preventive Medicine, Sub-Project Atherosclerosis Unit of Venice, I-30100 Venice (Italy)

Baggio, G., M.D., Department of Internal Medicine, Division of Gerontology and Metabolic Diseases, Policlinico, Via Giustiniani 2, I-35100 Padova (Italy)

Blackburn, H., M.D., Professor and Director, Laboratory of Physiological Hygiene, School of Public Health, Professor of Medicine, Medical School, University of Minneapolis, Minneapolis, MN 55455 (USA)

Dustan, H. P., Director, Cardiovascular Research and Training Center, The University of Alabama in Birmingham, School of Medicine, 1002 Zeigler Building, University Station, Birmingham, AL 35294 (USA)

Farinaro, E., M.D., Center for Arteriosclerosis and Metabolic Diseases, Semeiotica Medica, 2nd Medical School, University of Naples, I-80131 Naples (Italy)

Havel, R. J., Cardiovascular Research Institute and Department of Medicine, University of California, San Francisco, CA 94143 (USA)

Keys, A., University of Minnesota, Minneapolis, MN 55455 (USA)

Kritchevsky, D., Ph.D., Associate Director, Wistar Institute of Anatomy and Biology, 36th Street at Spruce, Philadelphia, PA 19104 (USA)

Lanzola, E., M.D., Istituto di Scienze Sanitarie Applicate, University of Pavia, Via Taramelli 1, I-27100 Pavia (Italy)

Menotti, A., M.D., Research Group of the Rome Project of Coronary Heart Disease Prevention (PPCC), Laboratory of Epidemiology and Biostatistics, Istituto Superiore di Sanità, Viale Regina Elena 299, I-00161 Rome (Italy)

Rabbi, A., Istituto Nazionale della Nutrizione, Via Ardeatina 546, I-00179 Rome (Italy)

Rifkind, B. M., M.D., FRCP, Deputy Associate Director for Etiology of Arteriosclerosis and Hypertension, Chief, Lipid Metabolism and Atherogenesis Branch, Division of Heart and Vascular Diseases, National Heart, Lung, and Blood Institute, National Institutes of Health, Room 4A14A – Federal Building, Bethesda, MD 20205 (USA)

Sirtori, C. R., Center E. Grossi Paoletti, Institute of Pharmacology and Pharmacognosy and Chemotherapy Chair, University of Milan, I-20129 Milan (Italy)

Stamler, J., M.D., Professor and Chairman, Department of Community Health and Preventive Medicine, and Dingman Professor of Cardiology, Northwestern University Medical School, Chicago, IL 60611 (USA)

Tobian, L. J., Professor of Internal Medicine, Chief, Hypertension Section, Department of Medicine, Box 285, Mayo Building, University of Minnesota, Minneapolis, MN 55455 (USA)

Editor's Note

The scientific opinions expressed in these papers are those of the individual authors and do not necessarily represent the positions of either the U.S. or Italian government organizations sponsoring the joint symposium, the cochairmen of the symposium, or the editor of this volume.

Preface

This volume contains the papers developed during 1981 in follow-up of the US-Italy Joint Symposium on Nutrition and Cardiovascular Diseases in Rome, Italy, on December 5 and 6, 1980. The symposium was held in honor of Dr. *Ancel Keys* for his contributions to the field of international epidemiology. This was the third in a series of joint cardiovascular symposia conducted under the US-Italy Joint Agreement for cooperation in the field of Health and Medicine. The US-Italy cooperation in the cardiovascular area focuses on prevention of cardiovascular disease through the control of risk factors such as hyperlipidemia, hypertension, and smoking. The two earlier symposia dealt with the measurement and control of cardiovascular risk factors, and prostaglandins and heart disease.

The contributors to this volume are the Italian and US participants in the symposium. The symposium was cochaired by Prof. *Rodolfo Paoletti,* Director of the Institute of Pharmacology and Pharmacognosy, University of Milan, Italy, and Dr. *Antonio M. Gotto,* Jr., Chairman of the Department of Medicine, Baylor College of Medicine and the Methodist Hospital, Houston, Tex., USA.

Bilateral activity under the US-Italy program focuses on the exchange of information and data on research topics of mutual interest and need to both countries, and the exchange of scientists for short periods of time to work on collaborative studies. Progress is reviewed annually and opportunities for meaningful exchange and cooperation are identified and evaluated jointly. Through such a review, the relationship between nutrition and cardiovascular disease was identified as a particularly timely subject for joint consideration. There is intense international interest in this area as evidenced by the wealth of international data accumulated in recent years, the number of ongoing studies, and the increasing demands for public action on prevention of cardiovascular disease. Changes in lifestyles, including modifications in diet, are believed to be factors in the changing trends in mortality and morbidity rates from cardiovascular diseases that have been observed in various countries. Ongoing debate centers on the relative importance of diet in the development of cardiovascular disease and the feasibility of modifying nutritional patterns for total populations.

The relationship between nutrition and cardiovascular diseases is one of long-standing international research interest, with reports of population differences dating back to the early 1900s. Active documentation of the influences of dietary change came after World War II and led to a number of epidemiological studies with participants from all over the world. These studies have been combined with clinical and experimental work examining specific interactions in both humans and animals. The present volume provides extensive critical reviews of this work from different perspectives and explores major controversial issues and questions. The papers in this volume cover epidemiological observations over the past 40 years, the role of lipoproteins in cardiovascular diseases, the relationship of diet to hypertension specifically, and programs aimed at prevention of coronary heart disease (CHD) through changes in diet.

Ancel Keys opens the volume with an historic overview of the development of the well-known Seven Countries Study and some of the methodological problems of international epidemiological investigations. *Blackburn* provides an anthropological perspective on changes in diet since early man; he also reviews recent epidemiological findings and suggests preventive measures for public health action. *Lanzola* presents data on a specific survey of minerals in water and diet and their relation to CHD.

Much of the recent interest in the relationship of nutrition to cardiovascular diseases is focused on the role of the various lipoproteins. *Rifkind* provides an overview of the aims and progress of the North American Lipid Research Clinics Program, while *Havel* considers the concepts of lipid metabolism. A series of Italian perspectives follow from *Farinaro, Baggio,* and *Avogaro* on diet and lipoproteins; *Kritchevsky* reviews the experimental data, and *Rabbi* shares his findings on a specific study of the hypocholesterolemic effects of dietary changes in experimental animals.

Much work is also going on concerning the role of nutrition in hypertension and its control. *Dustan* provides a broad overview of experimental studies on obesity and salt and potassium intake, and *Sirtori* specifically reviews animal data, while *Tobian* examines the clinical aspects of nutritional control of hypertension.

The final aim of all these studies is the elucidation of ways to prevent both the onset and progression of cardiovascular diseases in human beings. *Menotti* brings us up to date on the Rome Project of Coronary Heart Disease Prevention conducted as part of the World Health Organization European Multifactor Preventive Trial of CHD, and the volume concludes with *Stamler's* review of current knowledge of nutrition-related risk factors and a discussion on the controversial aspects of the diet-heart relationship.

In summary, the papers present a diversity of findings from the many studies carried on over the past 40 years. The field continues to move quite rapidly. It is our hope that this volume will encourage further investigation of the many questions that still remain.

Ruth Johnsson Hegyeli, Bethesda, Md., USA
Rodolfo Paoletti, Milan, Italy
Antonio M. Gotto, Jr., Houston, Tex., USA

Epidemiological Observations

From Naples to Seven Countries – A Sentimental Journey

Ancel Keys

University of Minnesota, Minneapolis, Minn., USA

The US–Italy Joint Symposium on Nutrition and Cardiovascular Diseases inevitably calls to mind the start of my international studies on this subject. In the late 1940s we had given extremely low-fat diets to patients with familial hypercholesterolemia and produced spectacular lowering of the serum cholesterol level, but changes in cholesterol in the diet had little effect [23]. At the same time, reading *Snapper's* [67] 'Chinese Lessons to Western Medicine' led to digging into obscure Dutch language reports on differences between Java and the Netherlands in the diet and disease. Then, in 1950, *Malmros* [53] published his summary of the wartime experience in the Scandinavian countries. We were reminded that *Ludwig Aschoff* [5] told, without giving data, of a decrease in the severity of coronary atherosclerosis seen at autopsy in Germany during the period of the end of World War I and after when the diet was greatly reduced in fats. Data on coronary atherosclerosis recorded at autopsy indicated a substantial change in Finland associated with the dietary privations especially in meats and butterfat, caused by the Russian invasion [73].

All of this seemed to fit into one picture. In 1916 *De Langen* [13] had reported that the Javanese were much less prone to arteriosclerosis and its complications than the people in the Netherlands and that this difference was associated with differences in the diet – much lower in fat in Java – and in the average concentration of cholesterol in the blood serum. In regard to cholesterol at least, this was not a racial peculiarity; Javanese stewards on Dutch steamships who ate the usual Dutch diet had serum cholesterol concentrations similar to the Dutchmen. Dutch surgeons who worked in Java confirmed the difference in disease [14] and *Snapper* [67] said his experience in China was in agreement.

The Scandinavian data summarized by *Malmros* [53] showed a remarkable fall in mortality from arteriosclerotic heart disease as the diet became more and more restricted in fats during the German occupation; with the return of less restricted diets the mortality rose. In connection with my position as an authority on starvation I was privy to data from the Netherlands showing the same picture for changes in coronary mortality associated with changes in the diet caused by the German occupation. Data on the story in the Netherlands were published later [63]. In 1951 I was Chairman of the Joint FAO–WHO United Nations Committee on Nutrition meeting 10 days in Rome. *Gino Bergami,* Director of the Istituto Filippo Bottazzi, the University of Naples, was attached to that Committee and I told him something of my interest in this subject. He, in turn, told me about the typical low-fat diet of Naples, insisted that coronary heart disease was not a big problem in that city, and invited me to Naples to see for myself.

At the end of 1951 this came to mind. In January, 1952, I had to lecture at the Scottish Universities but afterward would be free from duties at Oxford University until Easter. The children were in boarding schools and the thought of southern Italy and escape from coal and food rationing was too alluring. My wife and I loaded the car with equipment, including the Evelyn colorimeter for cholesterol measurements, and headed south.

Flaminio Fidanza, then a recent medical graduate, was waiting for us at the Hotel Santa Lucia in Naples. He had little English then and we had no Italian but we managed to agree on how to get on with the job. The Istituto Bottazzi was colder than the Churchill Hospital Laboratory at Oxford but *Margaret* got the cholesterol method – Bloor – going and soon *Flaminio* was bringing in subjects. No tests of probability were needed to show a vast difference between these workers in Naples and their counterparts in Minnesota. The average serum cholesterol was about 170 mg/dl by the Bloor method, meaning more like 160 by the Abell-Kendall method that later became standard [42]. The dietary survey was crude but there was no mistaking the general picture. A little lean meat once or twice a week was the rule, butter was almost unknown, milk was never a beverage except in coffee or for infants, 'colazione' on the job often meant half a loaf of French bread crammed with boiled lettuce or spinach. Pasta was eaten every day, usually also with bread (no spread), and something like a fourth of the calories in the diet were provided by olive oil and wine. There was no evidence of nutritional deficiency but in the working class only the women were fat.

As to coronary heart disease, we had to rely on what the local internists told us and what we saw when they took us through the wards. Certainly the disease was rare in the hospitals. Among the cardiac patients inspection of the charts showed that the murmurs of valvular disease were far more common than complaints of angina pectoris and major Q-waves or S-T depressions in the electrocardiogram. During the time we worked in Naples we learned of no acute myocardial infarction patients in the 60-bed medicine service in the University Hospital next door to the Istituto Bottazzi.

A few months later Madrid revealed much the same picture [50] – no coronary patients in the public hospitals, but our host, Prof. *Carlos Jimenez Diaz,* showed us the records of a fair number of cases among his wealthy private patients. As in Naples, no one in the small class of rich people ever went to a public hospital. When we were entertained by members of that privileged class the tables were loaded with meats and rich foods, a world apart from the dietary picture that Dr. *F. Vivanco* and his team were obtaining from their elaborate dietary surveys in the worker's quarter, Quattros Caminos, where we made our surveys.

In Madrid, *Francisco Grande,* who later joined us in Minneapolis and stayed 20 years, persuaded men of the top economic class to allow blood sampling. The average concentration of cholesterol in the blood was much closer to that in Minnesota than to that of their fellow countrymen of the working class in Madrid.

The experiences in Italy and Spain were consistent with the comparisons of the Dutch and Javanese and the changes associated with dietary restrictions imposed by war in Europe. Meanwhile, in the metabolic unit of the Hastings State Hospital in Minnesota we had been finding that an exchange of fat for carbohydrate calories, or the reverse, produced a predictable change in the serum cholesterol concentration. But we had not yet started to make changes in the kind of fat in the diet and so had an oversimplified picture. The pattern could be: Fat in the diet raises the cholesterol in the blood which promotes atherosclerosis and the end is clinical disaster. It all made an impressive story for an invitational lecture, 'The Cholesterol Problem', presented in Amsterdam later in 1952 at the Joint Session of the International Congresses of Diabetes and Nutrition [24] and amplified in New York [25].

Armed with all this we were able to persuade some colleagues to join in more systematic work in Naples. The team included *Paul Dudley White, Haqvin Malmros, Gunnar Biörck, Joseph T. Doyle, Brian Bronte-*

Stewart, Ratko Buzina, and *Bengt Swahn* as well as *Flaminio Fidanza* and other members of the Naples Medical School. Our previous impressions were fully confirmed by surveys covering the Naples Fire Department, clerical workers, men doing heavy work at the Ilva steel mill, and members of the local Rotary Club. Some of the findings were reported later that year at the Second World Congress of Cardiology where *Paul White* as the president had commissioned me to organize the first Symposium on Cardiovascular Epidemiology [51]. An overflow crowd in the biggest auditorium went away with many converts to the view that here was a fascinating approach to the coronary problem, complementary to clinical studies and experiments on animals. Personally, I went away determined to go further afield with better methods applied to more carefully defined samples of men.

South Africa and Italy Again

In 1955 Prof. *John Brock* invited us to study Bantu, Cape Coloured, and 'Europeans' (whites) in Cape Town. *Brian Bronte-Stewart,* who had been indoctrinated with us at Naples the year before, did a magnificent job of organizing the program. The report of that work [7] aroused much interest, but about the virtual absence of coronary heart disease among the Bantu the cynical response was, 'Who wants to be a Bantu?' It was suggested that the elimination of coronary heart disease would only mean an exchange for worse ailments. Recently, we showed that this is by no means necessarily so [41].

Later in 1955 *Paul White, Henry Taylor,* and *Reuben Berman* joined at Cagliari, Sardinia, with *Arrigo Poppi* and *Teodoro Posteli* of Bologna helping there and later at the University of Bologna Medical School. The Sardinians proved to be much like the Neapolitans in the diet, the average concentration of cholesterol in the blood serum, and in the relative rarity of coronary heart disease. When we were there it seemed that most of the families kept chickens and, being springtime, eggs were abundant; most of our subjects, men in the Police and Fire Departments of Cagliari, employees of the Medical School, and coal miners, were eating more eggs than the average in Minnesota. The serum cholesterol concentrations were low, however, and this seemed to be linked with the very small consumption of meats and dairy products.

In Bologna the picture was very different. Bologna is in the heart of what the Italians call 'Emilia la grassa', the region of Emilia, the fat land, and the Bolognese kitchen is proclaimed to be the richest in Italy. In any case, the serum cholesterol levels were much higher than measured a few weeks earlier in Cagliari; many of the senior police officers, long in administrative jobs, were obese; and the wards of the Department of Medicine had a good share of coronary patients as reported by the American cardiologists who checked them. All of these data from Sardinia and Bologna remain in the notebooks, publication put aside in the face of the pressure of other work in Japan and Finland.

Japan

The year 1956 produced the greatest contrast on opposite sides of the world. *Noboru Kimura* had spent half a year in Minneapolis and had responded to the invitation to present something at the 1954 Symposium on Epidemiology by assembling 10,000 autopsy records from Fukuoka [51]. Asked about organizing a survey in Japan like those made in Italy and South Africa he responded with enthusiasm. In Fukuoka he enlisted dozens of physicians, dieticians, and clerks to help in recruiting and examining samples of men in various occupations – farmers, clerks, coal miners, members of the Rotary Club, city firemen. Except in the Rotarians, serum cholesterol levels proved to be as low as in the Bantu. *Margaret* separated the lipoproteins by paper electrophoresis and found the concentration of cholesterol in the alpha fraction, now called high-density lipoprotein (HDL), to be little different from that in the Bantu and in Minneapolis.

Meanwhile, *Paul White* spent weeks trying to find a case of myocardial infarction in the big Kyushu Medical School Hospital, in district hospitals, and in the private clinics patronized by the upper class. In the Department of Pathology the best they could do was to produce the heart of a physician who had a fatal infarction a few months after returning home from 30 years of practicing medicine in Honolulu. Fukuoka then had a population of over a million.

In Honolulu, as a way station to Japan, the late Dr. *Nils Larsen* got the help of the Japanese Kuakini Hospital so we could make a survey of Japanese as well as white residents of Honolulu. The Japanese in Hawaii had migrated from the area around the Inland Sea of Japan and so could be considered to be of the same general Japanese stock as the people of Fukuoka and northern Kyushu. The Japanese in California, who were examined later, also came from the same stock, mostly as a further migration from Hawaii. The Japanese in the three areas, Fukuoka, Honolulu, and Los Angeles, differed in the mode of life, notably in the diet. In 1956 the Japanese in Honolulu were partly Americanized, the Nisei in Los Angeles almost completely so.

Racial differences could scarcely be invoked to explain the fact that the men of Japanese ancestry in these three areas were characterized by three dietary habits, three levels of serum cholesterol, three contrasts in the frequency of coronary heart disease. The patient population of the Japanese hospital in Los Angeles had a representation of coronary heart disease like a hospital in Minnesota or a general hospital in Los Angeles catering to white people.

White men in Fukuoka, members of the large US Army establishment there at the time, were also sampled. Queries about their diets indicated that the PX assured no change from what they were accustomed to eat 'back home'. The average concentration of cholesterol in the serum was close to that found for men of their ages in Minneapolis.

Many of these facts were reported in detail under the title 'Lessons from Serum Cholesterol Studies in Japan, Hawaii and Los Angeles' [47]. Long afterward similar studies by others on a much more elaborate scale were confirmatory.

During the first stay in Fukuoka we made a dietary experiment on Japanese coal miners, substituting butterfat, isocalorically, for some of the rice in their customary diet. Several men complained of nausea and diarrhea and refused to continue, but in those who completed the trial the serum cholesterol response to the diet change was quantitatively like that found in experiments in the Hastings State Hospital metabolic ward [46]. Japanese and white Americans seem to be alike in cholesterol metabolism.

Finland

In the early 1950s *Martti Karvonen* of Helsinki and I had several times talked about the coronary problem in Finland, reputed to be even more serious than in the United States. We agreed that a survey there would be rewarding and, finally, in the fall of 1956, it was possible to go ahead.

North Karelia, near the (new) frontier with Russia, was thought to have the highest incidence of coronary heart disease so the work started in the villages not far from Joensuu. The first village had an infirmary with six beds for male patients. One patient was a young man who had been bitten by a bear. A second patient had cancer of the lung and a third bed was occupied by an old man wheezing with asthma. The other 3 patients had coronary heart disease. The little infirmary of Ilomantsi seemed to be north Karelia in microcosm.

From Ilomantsi we went into the woods to have sauna with some lumberjacks. Two of them confessed to being slowed up by angina pectoris but more interesting was a glimpse into local eating habits. A favorite after-sauna snack was a slab of full-fat cheese the size of a slice of bread on which was smeared a thick layer of 'that good Finnish butter'. A later detailed dietary survey by *Maija Pekkarinen* and the late *Paavo Roine* found that butter, milk, and cheese accounted for 40% of the average total dietary calories of the middle-aged men in that area [62]. We were not surprised when Minneapolis reported serum cholesterol values over 300 mg/dl in more than 15% of the samples we sent from east Finland.

During that work in 1956, besides total cholesterol, that in the alpha lipoprotein (HDL) fraction was also measured, using paper electrophoresis for the separation as had been done in Fukuoka. *Flaminio Fidanza* helped with this tedious work in Finland, running two strips for each sample, one to be stained to show where the other should be cut for elution and analysis. The values were little different from those recorded in Japan earlier in the year. Some of the data from that 1956 exploration in Finland were published [44], but we lost interest in HDL cholesterol; it seemed unlikely that a variable so similar in the Japanese and Finns could have much to do with coronary heart disease. However, in view of the recent enthusiasm about HDL cholesterol, we remembered the men studied in 1956. *Sven Punsar* writes that the 24-year mortality data covering nearly a thousand men will soon be complete. We eagerly await comparison with the findings in the 25-year follow-up in Minnesota [37].

Seven Countries

Back in 1947 we had realized that cross-sectional surveys could never suffice for the identification and evaluation of factors promoting or hindering the development of coronary heart disease and, accordingly, had started the prospective study on the men of the Twin Cities of Minnesota. Obviously, that approach should also be applied in international studies but the difficulties of organization, finance, and staffing were alarming to contemplate. Still, determination to pursue that vision was developed with *Flaminio Fidanza* of Naples and *Ratko Buzina* of Zagreb in a long day in the gardens of the Aventine in Rome. Courage was given by the conviction that support would come from *Noboru Kimura* of Japan, *Martti Karvonen* of Finland, *Vittorio Puddu* of Rome, and *Henry Taylor* of Minnesota. The decision to proceed was sealed at a memorable dinner in the open air that night in Trastevere on the other side of the Tiber.

Alerted to the idea, our colleagues in Athens, *Andy Dontas* and *Christ Aravanis,* had argued for Greece where they assured low costs, excellent cooperation, and a chance to learn about serum cholesterol and coronary heart disease in a population always on a diet high in total fats but low in saturated fatty acids, namely a diet with lots of olive oil and very little other fat. But, before being committed to a study to cover 10 years or more, we had to see whether an international team could be assembled to work harmoniously with a fixed protocol and standardized methods to assure comparability between regions, and to determine the logistics and costs of such work.

To these ends, full-scale field trials were made in 1957, first in the large village of Nicotera near the Straits of Messina in Italy, later on the island of Crete in a series of villages within striking distance from Heraklion. A remarkable staff came to take part, all without pay and, mostly, travelling at their own expense, representing nine countries: England, Finland, France, Greece, Italy, Japan, the Netherlands, the United States, and Yugoslavia.

Alfonso Del Vecchio, then an assistant at the Naples Medical School, later to be in the pharmaceutical industry in Milan, had arranged for the work at his native village of Nicotera and enlisted the help of the local physicians to check the death records of men over 40 years of age who died during the previous 2 years. We learned nothing about the frequency of coronary heart disease from that source. 'Coronary heart disease', 'Myocardial infarction', 'arteriosclerotic heart disease', did not appear on the death certificates but six deaths were attributed to 'collapse' (collasso) and others to 'old age'. The most enigmatic was the cause of death specified only by the single word, 'jealousy'. Further inquiry elicited the explanation that the man had indeed been a victim of jealousy; the term 'gelosia' was explained as being more specific than 'omicidio'.

The men of Nicotera were very cooperative, pleased to come for examination, though not always at the right time, and the only outright refusals were a few men who, we learned later, had been under treatment for tuberculosis at one time; they were fearful of being sent to the sanatorium again. The only problem with the subjects was reluctance to submit to venesection. An unexpected technical problem arose in connection with the preparation of serum samples dried on filter paper for cholesterol measurement in Minneapolis. It was September, the season of flies in southern Italy as elsewhere, and there were no screens until we improvised mosquito netting protection for the filter papers being air dried. Fly specks contain cholesterol and with 0.1-ml serum samples a couple of fly specks will give an erroneously high value in the ultimate analysis. As it was, even with an occasional fly speck, median concentration of serum for the men 55–59 years old was only 166 mg/dl. Very few of our men in Nicotera could be classed as obese by any criterion; the median relative weight was 92% of the average for men of given age and height in the American insurance company tables, but hypertension was not unknown. 10% of the men aged 55–59 had systolic pressures of 170 mmHg or more. Among 438 men aged 45–59, only 5 had major Q-waves (Minnesota code I.1) and only 2 had negative T-waves of 5 mm or more in rest. A detailed dietary survey of the men of Nicotera, made later in three seasons by *Flaminio* and *Adalberta Fidanza, G. Ferro-Luzzi,* and *M. Proia,* showed an average of only 23% of total calories from all fats and oils with olive oil accounting for most of this [15].

Nicotera, then, was very informative in 1957. *Paul White* surveyed the hospitals of nearby Reggio Calabria and Vibo Valentia and reported a remarkable lack of coronary patients but many cases of valvular heart disease. Finally, it was a personal pleasure to go away with a large parchment scroll, hand-lettered and embellished with multi-colored decorations, attesting to my honorary citizenship in the Commune of Gioia Tauro, the nearest town to Nicotera. But Nicotera was not included in the eventual Seven Countries Study for practical reasons of finance, logistics, and lack of local professional help.

In 1957 tourism had not yet overwhelmed Crete. In Heraklion, the largest city on the island, we were housed in the still unfinished Hotel Astir, the only hostelry boasting private baths. Our preselected villages were chiefly on the main road east of Heraklion but most of that main road was innocent of cement or asphalt and so narrow that meeting the occasional bus created a major traffic problem, usually solved by a little extemporary road work and balancing the Plymouth station wagon on the edge of a slope, half a dozen men straining to keep it upright. Lack of electricity in the villages meant using storage batteries for the electrocardiograph machines. At the end of each day in the field the batteries had to be taken to a power station for overnight recharging.

The men of Crete were at least as cooperative as the men of Nicotera. Their qualms about venesection were overcome by the insistence of the local authorities – village leaders, police, and the bearded, black-robed priests – who themselves demonstrated their bravery by being subjects extending their arms for the needle. Outside the entry door of each of our village headquarters would gather old men garbed in great baggy bloomers, high boots, and the traditional black turban-like cap of Crete. They were the 'papoos', grandfathers, we were told, so when regular subjects were late for appointments a number of them aged 70–90 went through the examination procedure. They, like the other Greek subjects, immediately understood the rhythm of the step test and most finished the 3 min without panting and with only a small increase in pulse rate.

In one small mountain village of Crete we saw an old man with a hoe over his shoulder walking out to the fields holding the arm of a companion. He was almost blind, we were told, which was explained by his age, 104 years, and he was going to do his usual half day of cultivating work. We never got to examine him because the next time we tried to get to the village the road was impassible; the fall rains had started. I was reminded that at the turn of the century *Ilya Metchnikoff, Louis Pasteur's* successor as Director of the Pasteur Institute, had attempted a kind of census of centenarians and concluded their number in Greece was proportionally many times that in France. That recollection came back long afterwards when the 10-year follow-up of the Cretans examined in 1960 was completed; they had the lowest death rate of the 16 cohorts. Follow-up of the Cretans in 1979 again confirmed their longevity.

Those 1957 trials showed that the basic scheme for international prospective studies could work. A roster of all men of specified age permanently in the area was compiled from tax, voting, and church records and over 90% of the men invited to participate came to be examined. The various nationalities in the professional staff were quickly united in the common interest, differences in usual customs of examination and interpretation gave way without friction to the agreed procedures and standards of diagnosis, and friendships developed that have persisted over the years. At least part of the success was due to living as well as working together in the field without other distractions during the weeks needed for the job.

The definitive long-range program began in Croatia in Yugoslavia in 1958. At the same time *Noboru Kimura* and staff examined a cohort of men at Tanushimaru on the island of Kyushu near Fukuoka and *Henry Taylor* started the examination of the American railroad men in the northwest part of the United States.

Initially the plan called for studying men in six countries but the Ministry of Health of the Netherlands, impressed by the 1957 trials, authorized a study at Zutphen so there would be seven countries. And Dr. *B. S. Djordjevic,* Dean of the Medical School of the University of Belgrade, observed the work on Corfu in Greece and returned to Belgrade to add three cohorts in Serbia to counter the two organized by *Ratko Buzina* in Croatia. The final result, as noted earlier, was 16 cohorts in seven countries, comprising 12,763 men aged 40 through 59 at entry.

Reexaminations were made after 5 years and, except for the US railroad men where financial support was not forthcoming, again after 10 years. The coverage of the Italian railroad men at the 10-year reexamination was hampered by lack of funds but, of 722 men alive at the time, 484 were examined at the San Camillo Hospital in Rome and another 20 men were examined at their homes. Among those not examined after 10 years, 113 were still in full-time service with the railroad with no indication of work-limiting disorder and 71 were retired for age without disability, and the other men missed, either refused examination (11 men) or were temporarily lost because of change of address. Accordingly, the 10-year incidence of non-fatal coronary heart disease was probably somewhat underestimated for the Italians, more seriously for the Americans. In all of the cohorts some men did not respond to the invitation for the 10-year examination, an experience similar to that in other prospective studies. This was one reason why in the analysis of incidence the emphasis was put on 'hard CHD' (death from coronary heart disease or definite but non-fatal myocardial infarction), and death in general. Yearly visits were made to the areas of each of the cohorts for the purpose of ascertaining interim deaths and their causes, the latter not only from death certificates but also from information elicited from local physicians, family, friends, and hospital records. Those materials from each of the several areas were scrutinized by *Alessandro Menotti* in Rome and by the group in Minneapolis; differences of opinion were resolved by discussion.

The habitual diet was a major variable for study and it proved to be the most difficult, expensive, and troublesome. At the entry examination each man was asked questions about the diet to discover peculiarities and differences from the usual dietary pattern of the area. Though expressed preferences or dislikes of particular food items or seasoning were frequent, major individual departures from the general local custom were rare. The answers to these questions had minimal scientific value except for their indication that within each local area there seemed to be marked uniformity in the general picture of nutrients and their sources in the diets of the individuals.

An attempt to get quantitative specification of the nutrients, including the several fatty acids, in the habitual diets of the individuals would demand impossible expense so recourse was had to characterizing the dietary pattern of the separate cohorts by repeated 7-day surveys of men representing statistical subsamples of the cohorts. All food items were weighed, plate wastes were allowed for, and special tables of food composition were used for the different areas in the calculation of the nutrients. In addition, replicate meals were homogenized, freeze-dried, and sent to Minneapolis for detailed chemical analysis. Results of these dietary studies have been reported [15, 45]. Some further work is in progress or planned with the voluminous dietary records, particularly those from Croatia, Finland, rural Italy, and Greece. In the Netherlands, where very detailed dietary data were repeatedly obtained for every man, *Daan Kromhout* is starting extensive analyses.

The word 'troublesome' was used in regard to the diet as a variable, not only because of the labor and expense of the survey work but for two additional reasons. First, within these relatively homogeneous areas the intraindividual variation in the diet was similar to the interindividual variation so it is impossible to rank individuals reliably. Second, the surveys show the average diets when they were made but in many areas the diets are changing with profound implications for the future.

The 10-year experience of the study has been given in some detail in the Commonwealth Fund Book recently published by the Harvard University Press [41]. In 10 years there were 1,507 deaths, 410 from coronary heart disease. Except for Serbia, the 15-year mortality follow-up is now complete, thanks to the devoted efforts of *Alessandro Menotti* in Rome and the cooperation of the several responsible investigators. The problem now is how to analyze all this 15-year material; nothing can be done nowadays without money.

Other Prospective Studies

Soon after the entrance examinations in Minnesota were finished the National Heart Institute embarked on a similar but far larger program at Framingham, Massachusetts, covering both sexes and a much wider age range. Devised and organized by a committee, the responsibility for the operations of the project was soon given to *Roy Dawber* who directed the work with skill and devotion for many years. Shortly after the start at Framingham, *John Chapman* began the prospective study of civil servants in Los Angeles to be followed by similar programs with civil servants in Albany, New York, directed by *Joseph Doyle;* long-time studies on industrial employees in Chicago, directed by *Jeremiah Stamler* and by *Oglesby Paul;* and so on to reach Tecumseh, Michigan, Evans County, Georgia, San Francisco and, in more recent years, spilling over into many parts of the world.

Here it is enough to comment on the remarkable consistency of the findings in these prospective studies and to note a sequel, the attempt to utilize the new knowledge in prevention programs. Still, there is much to be learned from prospective studies without intervention. Several examples may be mentioned.

End Points

All of these prospective studies were originally aimed to discover and evaluate characteristics associated with increased risk of developing coronary heart disease but the material can also be relevant to other end points. Some efforts have been made, especially at Framingham, to consider the incidence of other cardiovascular diseases but in these studies the numbers are generally too small to hope to find statistically significant relationships between pre-disease characteristics and any but the most frequent disorders. As the cohorts grow older, however, cancer mortality becomes substantial and the number of all-causes and non-coronary deaths calls for analysis.

Currently, some attention is being given to what these follow-up studies can offer about the significance of pre-disease characteristics for the development of neoplasms. The suggestion that exceptionally low levels of total cholesterol in the blood serum may be a risk factor for cancer has some evidence to support it but most of the data so far reported seem to be negative. More material and analyses are needed.

In the concentration on coronary heart disease in these prospective studies almost no attention has been paid to total mortality, all-causes deaths. I have already mentioned the fact that there is no evidence that a reduction in the incidence of coronary heart disease necessarily involves a trade-off for an increase in other lethal disorders but we ask what may be the relevance of the coronary risk factors for non-coronary mortality. Cigarette smoking is clearly a risk factor for lung cancer and probably other neoplasms, including cancer of the bladder. In the Seven Countries Study elevated arterial blood pressure proved to be a risk factor for all-causes and for non-coronary deaths. On the other hand, the serum cholesterol concentration was a risk factor for all-causes deaths only in some populations suffering a very high incidence of coronary heart disease or, to look at the relationship in another way, only in certain populations characterized by high serum cholesterol levels was this variable important for all-causes deaths [41]. In the 18-year experience at Framingham the serum cholesterol concentration tended to be a negative risk factor for all-causes death, statistically significant in both men and women at ages 65–74 [66]. That analysis, however, concerns subjects who were not specified as to state of health at the last examination when cholesterol was measured.

In the examinations in prospective studies on coronary heart disease the characteristics selected for recording and measurement at entry were primarily chosen with the idea they might have relevance to the circulatory system. But, should different items be considered in a search for risk factors for other causes of death, say cancer? What items?

The Mathematical Model

In the early days of the prospective studies the variables explored as being possibly related to the likelihood of coronary heart disease in the future were looked at one at a time or the subjects were classified as being high or low in each of several characteristics. The introduction of the multiple regression model, and its more elegant form in the multiple logistic equation, and the availability of computers and programs, allowed graduation from those earlier elementary analytical methods and the large loss of information they involved. That was a great step forward but it is not always appreciated that the analysis easily becomes a prisoner of the model.

These multiple regressions presuppose that the frequency distributions are multivariate normal, with different means but with the same variances and covariances [71]. It was noted, however, that it is not necessary for these stipulations to be met in the material under analysis; it is enough if 'the linear compound of risk factors... be multivariate normal' [71]. It is stated that this condition 'holds approximately' for the seven-factor trial with 12-year Framingham data but the practical test is how well the solution of the equation discriminates between cases and non-cases.

Applied to the incidence of coronary heart disease, this test showed that good discrimination was achieved with the solution to the multiple equation using seven independent variables at Framingham [71] and with as few as four variables (age, systolic pressure, serum cholesterol, and cigarette smoking) in a 20-year follow-up of our business and professional men in Minnesota [49], and in 5-year and 10-year follow-up experience in the Seven Countries Study [40, 41]. But it would be a serious mistake to conclude that such results establish true cause and effect between the several independent variables and the end point and that the form of the relationship in the model is the 'best fit' that can be achieved and expresses accurately underlying biological relationships.

In regard to cause and effect, the most that can be said is that the results are, or are not, consistent with an hypothesis based on sound physico-chemical and pathogenetic considerations. In regard to atherosclerotic heart disease it is easy to defend the ingredients in the hypothesis regarding age, blood pressure, and serum cholesterol and, less certainly, cigarette smoking. Most of the other variables proposed from time to time to be risk factors are less firmly supported by consideration of basic mechanisms. But, to repeat, if they statistically contribute to discrimination, at least they have practical value.

Comparison of the 'importance' of the several independent risk factors is another matter. Solution of the multiple regression or the multiple logistic equation with the modern computer programs yields coefficients and their standard errors so we have t-values, the latter very useful in showing something about the contribution of each variable to the total discrimination. But it must be emphasized that unless the t-values of two coefficients are substantially different from each other it is rarely justifiable to say much about the relative importance of the two variables.

Consider, for example, the 18-year experience at Framingham of sudden death from coronary heart disease among men aged 45–74 and free of coronary heart disease at the examination when the independent variables were measured [66]. In the multivariate analysis the coefficient found for diastolic blood pressure has a t-value of 1.61 (table 7–2) while that for serum cholesterol is 1.99 (table 7–3) [66]. So the conventional conclusion would be that the diastolic blood pressure was not significantly related to later sudden death while serum cholesterol would be rated as significant because a t-value of 1.99 indicates a probability of less than 5% chance explanation, while a t-value of 1.61 indicates 11% chance explanation. However, are these two probabilities significantly different from each other? The 95% confidence limits of the diastolic pressure coefficient, standardized by multiplying by the population standard deviation, are -0.045 and $+0.404$, while the corresponding limits for serum cholesterol are $+0.004$ and $+0.497$. The great degree of overlap makes it obvious that these two variables do not, in fact, differ significantly in discriminating power in regard to sudden death from coronary heart disease. The conventional conclusion would be wrong because it is based on a completely arbitrary definition of 'significance'.

This argument against using arbitrary classifications of t-values in evaluating the relative importance of variables may seem unnecessary but the insistence on a very special meaning of $p = 0.05$ is all too common. Perhaps it may also seem unwarranted to call attention to the fact that the form of a relation obtained from the statistical analysis is fixed by the model used and a basic relationship may be missed because an inappropriate model is used.

A good example is the relation of the serum cholesterol concentration to age. In the early 1930s men ranging in age from the late teens to over 90 were surveyed at the Rockefeller Institute. The correlation between cholesterol and age proved to be statistically non-significant and this was the basis for the teaching in all the textbooks for many years that the serum cholesterol is not related to age. In 1949 we surveyed a large sample of men in Minnesota and found a highly significant curvilinear relationship, the average cholesterol concentration rising steadily from the teens to middle age, a tendency to a plateau into the sixties, and thereafter a progressive decline [48]. The individual data were given in the Rockefeller Institute paper and it was easy to show that they too described the same curvilinear relationship. The error had been in assuming that if there were a relationship it must be linear. The regression-correlation calculation simply showed that there was no significant linear relation over the whole age span of nearly 80 years. The curvilinear relationship missed in this way has been found in many populations since the report from Minnesota in 1950.

An unwarranted assumption of linearity is frequently made in analyses of such data as are obtained in epidemiological studies though the assumption is seldom stated explicitly. The usual regression analysis, including such elaborations as the use of the multiple logistic equation, looks for a linear relation and yields the coefficients that best describe a straight line. Unfortunately, the search for and analysis of curvilinear relationships immediately poses the question: What kind of curve?

One kind of curve is described by the quadratic, using the value of a measurement as one independent variable and its square as another term in a multiple regression equation. This simple transformation has been applied with notable success in examining the relationship between mortality and relative body weight in men followed for 10 years or more in the United States and in Europe [16, 36, 39, 41]. In these samples a U-shaped relationship is described in which the lowest death rate is found for men slightly above the middle of the relative body weight distribution, the death rate rising progressively on both sides of the distribution so that the risk of premature death is markedly increased in the most extremely under- and overweight men. This U-shaped curve obtained by using the quadratic of the relative body weight fits the data better than the linear equation in a number of samples of middle-aged men but in some other samples no significant relation between relative weight and mortality is found with either linear or quadratic approaches. Examples are the northern European men in the Seven Countries Study and the men employed by the Western Electric Company in Chicago [41, 69].

The quadratic is only one kind of curve, one way to look at curvilinear relationships, and we have not found this approach to be useful in relating Seven Countries data on mortality to either serum cholesterol or to blood pressure, though for both of those independent variables a curvilinear relationship is indicated [41]. This whole matter of looking for and evaluating non-linear relationships greatly needs attention.

The results of a dozen prospective studies in America and Europe have been remarkably consistent in finding that relative body weight has nothing like the medical significance so long attributed to it by some insurance company actuaries [39]. All of the scientific prospective studies agree that being somewhat above the average weight for height and age in the population does not entail undue risk of a heart attack or premature death. None of these data indicates any advantage of matching the 'recommended' weight shown in the tables of the Metropolitan Life Insurance Company.

The serious weakness of the insurance company materials and the methods applied to them was pointed out beginning more than a quarter of a century ago [26, 27, 65]. Those publications had no effect whatever on the professional and general public acceptance of the view, repeated over and over, that overweight is the greatest health hazard today and that a few pounds over what was first called 'ideal weight' is a step towards disaster. Negative arguments are seldom effective against repeated positive propaganda, as salesmen and demagogues well know. Now, however, we are no longer dependent on insurance company material as a substitute for proper prospective studies.

Age and the Duration of Follow-Up

The great majority of prospective studies have concentrated on relations between incidence and characteristics recorded at the start of follow-up. However, as the follow-up continues and the time span increases, it is necessary to ask whether the characteristics recorded at entry have the same significance over the years. In regard to coronary heart disease, the classic risk factors, arterial blood pressure, serum cholesterol, cigarette smoking, have many times been shown to be highly significant for periods up to a dozen years or so but very little has been reported for later follow-up. There is much indication that the discriminating power of these characteristics is not the same at different ages, better discrimination in regard to future incidence of the disease being found in younger than in older persons.

There are two questions. Does an entry measurement cease to be properly representative of the individuals concerned as the years pass and if so when? Does the relation between the characteristics and the eventual end point change with age? I have said that for at least some of the risk factors there is a change in prognostic significance with age so the question really is: How much does it change with age? A current task is to examine these questions with the data on the Minnesota men first seen in 1947 and soon it will be possible to start on similar analyses of the mortality of the men examined in Finland in 1956 and traced through 1980.

Conceivably this question of age and the passage of time is important in our finding that the HDL cholesterol concentration in the serum is not a significant risk factor (negative) as reported by others [37]. With the Minnesota men the end point was death; other investigators have used the incidence of coronary heart disease as the end point. Perhaps more important is the fact that the follow-up period for the Minnesota men was 25 years while in other studies the period covered has generally been 10 years or less.

The Rocky Road of Progress

We are privileged to work in a time of great advances, of an explosive growth of the epidemiological approach to the discovery and evaluation of risk factors – and anti-risk factors. There is growing insistence on the application of the new knowledge to a public health attack on the risk factors. But this raises large issues when it touches on the diet of the population. In the medical profession there are conservatives who hold to the belief that Americans have the world's best diet, who resent a threatened intrusion into the domain of the private physician, and per se dislike change. Certain food industries see great commercial implications and marshall their forces of propaganda for or against dietary recommendations according to their view of possible effects on profits. And there are the protagonists of personal theories who belittle all but their own views.

The result is a vast amount of argument that gives pause to members of the health professions and confuses and misleads the general public. This is abetted by editors of medical journals who say their readers delight in controversy. The mass media are ever alert to publicize claims of medical discoveries, especially if they challenge previous ideas, but consider that subsequent disproof is not 'news'. The fuel for most of this argument is provided by writings of persons who are victims of their own vanity, ill-informed about the facts, unused to rigorous logic, ignorant of statistical theory, or seduced by commercial interests. Unfortunately, some of these disseminators of misinformation are prominent physicians or biologists, so they readily find speaking platforms and publishers. Their negative views have particular appeal to colleagues in the medical sciences who are disturbed by what they consider over-zealous evangelism for new ideas of prevention. But let us go from generalities to specifics.

The Persistence of Claims Long Since Disproved

It is a common truism that it is easier to introduce a new idea into medicine than to remove an old one proved wrong. Many examples can be cited in regard to the diet and coronary heart disease. It has been said that the primitive Eskimo has 'clean' arteries in spite of a diet high in animal fat. The fact is that, averaged over the year, the diet of the primitive Eskimo was [sic] low in fat and much of that highly unsaturated. There are almost no data on the arteries of primitive Eskimos and few of them ever reached the age when coronary heart disease is common in more civilized populations. The past tense is proper because the primitive diet of the Eskimo was disappearing 50 years ago with white flour and sugar becoming the major source of calories.

And there is the Navajo Indian, once claimed to have no coronary heart disease in spite of living on a 'typical American diet' which strangely was associated with low levels of cholesterol in the blood serum. Actually, even in the mid 1950s when those claims were made, cases of coronary heart disease were not unknown at the Reservation Hospital but more to the point is that the diet was very far from being 'typical American'; it was very low in meats and all kinds of dairy products. Those claims about the Eskimo and the Navajo Indian were disposed of many years ago [28, 31, 35, 64]. Still, that nonsense surfaces again now and then in the popular literature.

A good example of the persistence of fallacy is the marvellous town of Roseto, Pennsylvania, where the people were said to subsist on the richest of fatty diets and were often obese but still were untouched by the epidemic of coronary heart disease that affected neighboring towns. The reason, so it was said, lay in the warm mutual support among the members of this Italian-American community. The evidence for the lack of coronary heart disease at Roseto was the absence of death certificates attributing the cause of death to myocardial infarction. But a check on the situation at Roseto quickly deflated the claims. One physician signed almost all of the death certificates for Roseto and apparently myocardial infarction was not in his lexicon; he preferred the term arteriosclerotic heart disease. In death rate from arteriosclerotic heart disease, and in all-causes death rate, Roseto was not significantly different from the neighboring towns that did not have the benefit of that 'psychosocial community support' [29, 30, 35]. But the demonstration that the claim for Roseto was fallacious did not end the matter. From time to time the wonderful case of Roseto reappears in the world popular press and a few months before writing this article a book on the subject was advertized!

An idea that has persisted most tenaciously in spite of overwhelming evidence against it concerns dietary sucrose. In 1957 *John Yudkin* sought to counter the theory of the importance of fats in the diet in the development of coronary heart disease by pointing to other characteristics of populations that also tended to be correlated with the coronary mortality rates – the numbers of bath tubs, radios, etc., and the consumption of sugar [77]. Later he insisted that sugar in the diet is the major cause of coronary heart disease as 'proved' by a close correlation between the coronary death rates of 15 countries with the use of sugar in those countries, by an historical increase in coronary mortality in England parallel to a rise in sugar consumption [78, 79], and by his claim that coronary patients use more sugar than 'controls' [80]. *Yudkin's* indictment of sugar in the diet was enthusiastically embraced by a retired naval surgeon [8]. Besides being the cause of coronary heart disease, *Cleave* [8] added that sugar in the diet promotes obesity, dental caries, varicose veins, diabetes and gall stones, and very likely causes colon infections, acne, eczema, and toxemia of pregnancy.

Yudkin's evidence for his theory that sucrose in the diet is the principal cause of coronary heart disease has been repeatedly refuted in detail on every point he attempted to make [4, 6, 10, 18, 20, 21, 33, 34, 52, 55, 59, 60, 70, 72, 74, 76]. The correlation he claimed between sugar use and coronary death rates of populations holds only when countries with low coronary mortality and very high sugar consumption are excluded – Colombia, Costa Rica, Cuba, Honduras, Venezuela [20, 33]. The historical parallel between the rise in coronary mortality and the rise in sugar consumption in Great Britain and in the United States is chronologically wrong by some 50 years [33]. The claim that coronary patients use more sugar than others has been disproved many times in larger and more precise surveys [6, 18, 52, 59, 76]. A large prospective study found no relation between the habitual consumption of sugar and the incidence of coronary heart disease [60]. *Yudkin* claimed that sucrose in the diet has an adverse effect on the plasma lipids [70], but many other experiments are negative [4, 10, 20, 21, 55, 72]. The idea that isocaloric exchange of sucrose for starch elevates the serum cholesterol is specifically denied by the results of independent experiments.

Why then does *Yudkin's* sucrose theory persist in periodical revivals in the world popular press? The answer is simple; *Yudkin's* claims have been repeatedly fed to the press by commercial interests that find this a way to distract attention from the picture of foods from animal sources as villains in the coronary story. An enormous mass of negative evidence may not suffice to bury fallacious claims when propaganda continues to revive what should be a dead issue.

Bias and Distortion in Reviews

Professional and public understanding of the coronary problem is hampered by publication of such false claims and mistaken interpretations on a single issue as noted in the previous section. More damage to progress can result from 'critical' reviews written by persons of professional stature whose personal prejudice leads them to adopt the tricks of some lawyers – distort, misquote, omit so as to gain the argument they favor. Only the informed specialist will fail to give at least some credence to a lengthy critique by a 'Distinguished Professor of Biochemistry and Biophysics', or by a Past Chairman of the Scientific Council of the International Society of Cardiology.

Reiser's [61] 'Perspective in Nutrition' has been answered in extenso [43]. No rebuttal has been forthcoming and since then we have seen no favorable references to *Reiser's* 'Perspective' in reputable medical journals but the paper remains in the literature as a source for exploitation by unscrupulous interests.

An equally distorted polemical review attacking the hypothesis of a relationship of the diet and plasma lipids to the incidence of coronary heart disease may cause more trouble because the author is a prominent cardiologist who only lately left the academic world for the pharmaceutical industry [75]. That recent contribution by *Werkö* [75] to confusion has been subjected to three independent critical examinations which agree it is replete with misstatements, claims supported by no evidence, demonstrations of ignorance of statistical theory, and careful omission of any mention of the mass of well-documented evidence that would contradict his thesis [38, 54, 68].

Both *Reiser* [61] and *Werkö* [75] express concern about the harm to the meat and dairy industries from adoption of the recommendation to the populations in the prosperous western countries to reduce the amount of saturated fatty acids and cholesterol in the diet. Recommendations to that effect were first made officially by the Medical Directors of the Scandinavian countries [32], followed by many other official and responsible organizations [1, 2, 9, 11, 12, 19, 56–58]. Similar recommendations but limited to persons considered to be at high risk because of elevated levels of cholesterol in the plasma, bad family history, etc., have been made by national medical societies [3, 22].

This is a time when many persons in the health professions and in the general public are wrestling with the question of what, if anything, should be done about the diet. In the meantime dietary patterns are changing in the United States and elsewhere in response to economic factors, changes in food technology, a relentless barrage of advertizing and clever propaganda. A greater role of noncommercial education is resisted but it is hardly possible to deny that decisions about efforts to prevent coronary heart disease, either by individuals or by organizations, must be based on full information and understanding of the issues. A biased selection of facts, filled out with misinformation and distortion, is a disservice to both medicine and the general public. Unfortunately, at this juncture a special committee of the National Research Council Food and Nutrition Board issued a remarkable report warning that there is no definite proof that dietary adjustment will prevent coronary heart disease. That committee, financed by the food industry, chose to ignore the enormous amount of evidence on the subject from experiments on animals and from epidemiological researches covering more than 60 years. A detailed response to that report is made in this volume by *Stamler*. Here it may be noted that the Chicago Heart Association expressed a frequent concern: 'The Chicago Heart Association believes a great disservice has been done to the American public by the Food and Nutrition Board of the National Research Council.'

Minnelea

After so many periods of work in Italy, arriving at Naples by ship in the good old days with a station wagon and gear for field work, we had a special attachment to the Mediterranean. Most of a Sabbatical year (1963–1964) was spent at Naples as a Special Research Fellow of the US Public Health Service and we often explored the coast to the south, the Amalfi drive, the Gulf of Salerno and its miles of sandy beach where the landings suffered in 1943, to Paestum and its splendid Doric temples, and on to Agropoli where the mountains come down to the sea again. Beyond Paestum the road was bad in those days – it still is very crooked and slow – but 35 km beyond Agropoli we found the place where we could pursue our dream of a 'guaranteed picnic spot'.

With friends we bought land stretching from the road down to the sea 70 m below, the slope of Monte della Stella rising 1,131 m (almost 4,000 feet) behind us. Across the sea 220 km due west is the Costa Smeralda (Emerald Coast) of Sardinia; the same distance south is the eastern tip of

Sicily and the straits of Messina, the passage between the Tyrrhenian and the Ionian Seas where the whirlpools and racing currents of Scylla and Charybdis terrified sailors in the days before steamships. A beautiful 5-hour drive takes us to the Adriatic and the car ferries to Greece and Yugoslavia, convenient for the continuation of work on the subjects of the Seven Countries Study. This is far from the rigors of winter in Minnesota; year 'round we enjoy the outdoors almost every day'.

We are 3 km by road from the fishing village of Pioppi, half that far by foot along the shore. The village has a permanent population of 500 but in recent summers it overflows with 3,000 vacationers, mostly Italian but many from all Europe to the north, especially West Germans. An old farmer and his wife live across the road on the slope above but our immediate neighbors are *Rose* and *Jerry Stamler* of Chicago, *Annikki* and *Martti Karvonen* of Helsinki, and the brothers *Flaminio* and *Alberto Fidanza,* professors at the Universities of Perugia and Rome, respectively. 6 km down the coast beyond Pioppi are the ruins of the Greek city Elea, called Velia in Italian, about which more later.

Our little settlement is called Minnelea, a combination of Sioux Indian and Greek words following the example of Minneapolis, 'minne' meaning waters and 'poli', people or population. Thus was invented 'Minnelea', the waters of Elea, a coinage admired by the Italians for its euphony.

Elea

Until recently Elea was known only to a few scholars interested in the migrations of the Greeks in early classical times and in the Eleatic School of Philosophy that developed there. Though it seems Elea once had 20,000 inhabitants and was a choice winter resort of rich and cultured Romans, it declined with the Roman Empire and was forgotten for more than a thousand years. Serious excavations began only after World War II and it is still relatively unknown to tourists and most Italians. Hundreds of tour busses come as far south as Paestum but Elea, in spite of its archeological importance and the beauty of the site, makes too long a day.

Recent discoveries, only beginning to reach the world literature, indicate an exciting importance for Elea in medical history, so instead of closing this 'sentimental journey' with the remarks in the last section on recent problems in cardiovascular epidemiology, it is appropriate to go back 25 centuries in time to the founding of Elea and evidence that it became a medical teaching center, the precursor of the Medical School of Salerno, heretofore long recognized as the first in Europe.

Elea was founded about 535 BC by the people of Phocaea, a Greek city on the coast of what is now Turkey, near present-day Izmir, and a little south of ancient Troy. The Phocaeans were leaders in the naval commerce in all the Mediterranean and beyond to the 'tin island' of Britain. A colony of Phocaeans settled on the island of Corsica about 565 BC and from there founded Marseilles in present-day France. The colony on Corsica was joined there by the rest of the population of Phocaea who abandoned their homes rather than submit to the rule of the Persians of Cyrus the Great. When *Harpagus,* the general of the Persian army, arrived at the gates of Phocaea and demanded submission, the city fathers said they would consult the people and answer the next day. At midnight they quietly took their ships and slipped away.

The Phocaeans were not long united at Alalia on Corsica when they were attacked by the combined fleets of the Etruscans and the Carthaginians. Though the battle was in favor of the Phocaeans, they decided they must look elsewhere for a home. In the remaining ships they sailed to what is now Reggio Calabria at the tip of the Italian boot where a friendly Greek colony had settled earlier. From Reggio they scouted the coast to the north and finally chose the site of Elea 210 km by sea from Reggio.

The full story of Elea is far too long to attempt here; besides, it is still being unfolded. Evidence that it was a medical center was first presented by a local physician, *Pietro Ebner* [17], who is just now following this with a major work on Elea and its place in medical history. The conclusion of *Ebner* [17] is that there was a medical school (facultà Medica) at Elea, the first in Europe, antedating Salerno by many centuries.

Elea's greatest personage in history was *Parmenides,* the philosopher whose reasoning was put into a dialogue by *Plato* who taxed *Parmenides* for inconsistencies but recognized him as the precursor of idealism. The connection of *Parmenides* with medicine is indicated by a commemorative marble describing him as 'son of Apollo the Healer' found at Elea in what seems to have been a sanctuary of the god of medicine where there was also a cult statue of *Aesculapius,* the Greek god of healing who was slain by a thunderbolt hurled by *Zeus* who feared he would make men immortal. Other Greek commemorative marbles recently unearthed in the ruins of what *Ebner* [17] suggested was the infirmary are dedicated to various physicians (Iatros) with indications they could have been successive chiefs of the 'medical School'. From the excavations also came a full length statue of *Eyxino,* considered to have been 'capo' (dean?) of the 'Collegio Medico di Velia', by *Ebner* [17]. A headless statue of a woman found there is thought to represent *Hygieia,* the Greek goddess of health from whom we derive the modern word 'hygiene'. Excavations at Elea have also uncovered some superb coins of the 3rd and 4th centuries BC; on the reverse side they bear the serpent, emblem of *Aesculapius.*

In 387 BC Elea became allied with Rome and later the people of Elea had all the rights and duties of Roman citizens. By the 1st century BC Elea was a favorite vacation spot for rich and illustrious Romans, such as *Cicero* and *Horace* from whom we know that Roman physicians often advised patients to have a long sojourn in the mild climate of Elea. Vacationers and patients returning to Rome would be sure to carry water from a well in the middle of Elea, reputed to have health-giving properties. The well is still there, always full of water and in the springtime full of frogs.

With the decline of the Roman Empire Elea decayed and by the 10th century AD the remaining population, clustered around the Acropolis on the edge of a spur of the mountain that ends in a sea-cliff, used the stones of the Acropolis to build a great watch tower and fortress as protection against the Saracens who raided all the coasts of the Mediterranean, robbing, murdering, raping and carrying away people to hold for ransom or sell as slaves. Finally, even the tower was abandoned and in the 14th century the people of Elea moved 25 km back into the mountains to found Novi Velia. But the great tower, now a national monument, still stands, clearly visible from our Minnelea.

In the meantime the Medical School of Salerno developed, distant only 100 km to the north. In the Codices (846 AD) the School was mentioned as 'ancient'. Salerno, too, was of Greek origin. For centuries it was called the 'urba Graeca' or Greek City and undoubtedly attracted Greeks from nearby Paestum and Elca. It is difficult to reject the thought that the School of Salerno had its origin in the School of Elea.

The School of Salerno is justly renowned for its emphasis on preventive medicine as set forth in the 'Regimen Sanitatis Salerni'. The first stanza of the poem contains the famous advice that when there is a lack of physicians, to quote the English version of *Sir John Harington* (1607): 'Use three Physicians still: first Doctor *Quiet,* Next Doctor *Merry-man,* and Doctor *Dyet.*'

So it is appropriate to end with diet!

References

1 American Health Foundation Committee on Food and Nutrition: Position statement on diet and coronary heart disease. Prev. Med. *1:* 225–286 (1972).
2 American Heart Association Nutrition Committee of the Steering Committee: Diet and coronary heart disease: a statement for physicians and other health professionals (Am. Heart Ass., Dallas 1978).
3 American Medical Association Council on Foods and Nutrition, joint statement with Food and Nutrition Board of the National Academy of Sciences – National Research Council: Diet and coronary heart disease. J. Am. med. Ass. *222:* 1647 (1972); Nutr. Rev. *30:* 223–225 (1972).
4 Anderson, J. T.: Dietary carbohydrates and serum triglycerides. Am. J. clin. Nutr. *20:* 168–175 (1967).
5 Aschoff, L.: Lectures in pathology (Hoeber, New York 1924).
6 Begg, T. B.; Preston, S. R.; Healy, M. J. R.: Dietary habits of patients with occlusive arterial disease. Atti V Convegno Internazionale sugli aspetti dietetici dell'infanzia e della senescenza, vol. 2, pp. 66–75 (Soc. Ed. Universo, Roma 1967).
7 Bronte-Stewart, B.; Keys A.; Brock, J. F.: Serum cholesterol, diet and coronary heart disease. Lancet *ii:* 1103–1108 (1955).
8 Cleave, T. L.: The saccharine disease – the master disease of our time (Keats, New Canaan 1975).
9 Committee on Diet and Cardiovascular Disease (J. F. Mustard, Chairman): Recommendations for prevention programs in relation to nutrition and cardiovascular disease. Report of the Committee as amended and adopted by the Department of National Health and Welfare (Ottawa 1977).
10 Conner, W. E.: The effects of sucrose in the diet on serum lipid levels and experimental atherosclerosis. Näringsforsk. *17:* suppl. 9, pp. 7–11 (1973).
11 Connolly, J. F. for the Health Advisory Committee of An Foras Taluntais: The prevention of coronary heart disease (An Foras Taluntais, Dublin 1977).
12 Council on Rehabilitation, International Society of Cardiology: Myocardial infarction: how to prevent, how to rehabilitate (Boehringer, Mannheim 1973).
13 De Langen, C. D.: Cholesterol metabolism in racial pathology. Geneesk. Tydschr. Nederl. Indie. *56:* 1–34 (1916).
14 De Langen, C. D.: Significance of geographic pathology in race problems in medicine (Dutch). Geneesk. Tydschr. Nederl. Indie. *73:* 1026–1044 (1933).
15 Den Hartog, C.; Buzina, R.; Fidanza, F.; Keys, A.; Roine, P. (eds): Dietary studies and epidemiology of heart diseases (Sticht. Wentensch. Voorl. Voedingsgebied, The Hague 1968). Also in separate parts in Voeding (1964–1967).
16 Dyer, A. R.; Stamler, J.; Berkson, D. M.; Lindberg, H. A.: Relationship of relative body weight and body mass index to 14-year mortality in the Chicago Peoples Gas Company Study. J. chron. Dis. *28:* 109–123 (1975).
17 Ebner, P.: La scuola di medicina di Velia, fasc. 1, p. 11 (Panorama Medico 1964).
18 Finegan, A.; Hickey, N.; Maurer, B.; Mulcahey, R.: Diet and coronary heart disease: dietary analysis on 100 male patients. Am. J. clin. Nutr. *21:* 143 (1968).
19 Food and Agriculture Organization, World Health Organization: Dietary fats and oils in human nutrition. Report of an expert consultation. Food and Nutrition paper No. 3 (Rome 1978).

20 Grande, F.: Sugar and cardiovascular disease. Wld Rev. Nutr. Diet., vol. 22, pp. 248–269 (Karger, Basel 1975).
21 Grande, F.; Anderson, J.T.; Keys, A.: Sucrose and various carbohydrate-containing foods and serum lipids in man. Am. J. clin. Nutr. 27: 1043–1051 (1974).
22 Joint Working Party of the Royal College of London and the British Cardiac Society: Report. Prevention of coronary heart disease. J. R. Coll. Physns 10: 214–275 (1976).
23 Keys, A.: The relation in man between cholesterol levels in the diet and in the blood. Science 112: 79–81 (1950).
24 Keys, A.: The cholesterol problem. Voeding 13: 539–555 (1952).
25 Keys, A.: Atherosclerosis – a problem in newer public health. J. Mt. Sinai Hosp. 20: 118–139 (1953).
26 Keys, A.: Obesity and degenerative heart disease. Am. J. publ. Hlth 44: 864–871 (1954).
27 Keys, A.: Obesity and heart disease. J. chron. Dis. 1: 456–461 (1955).
28 Keys, A.: Diet and the epidemiology of coronary heart disease. J. Am. med. Ass. 164: 1912–1919 (1957).
29 Keys, A.: Arteriosclerotic disease in a favored community. J. chron. Dis. 19: 245–254 (1966).
30 Keys, A.: Arteriosclerotic heart disease in Roseto, Pennsylvania. J. Am. med. Ass. 195: 93–95 (1966).
31 Keys, A.: Dietary factors in atherosclerosis; in Blumenthal, Arteriosclerosis; 2nd ed., pp. 576–619 (Macmillan, New York 1967).
32 Keys, A.: Official collective recommendation on diet in the Scandinavian countries. Nutr. Rev. 26: 259–263 (1968).
33 Keys, A.: Sucrose in the diet and coronary heart disease. Atherosclerosis 14: 193–202 (1971).
34 Keys, A.: Letter to the Editor. Atherosclerosis 18: 352 (1973).
35 Keys, A.: Coronary heart disease – the global picture. Atherosclerosis 22: 149–202 (1975).
36 Keys, A.: Is overweight a risk factor for coronary heart disease? Cardiovasc. Med. December: 1233–1241 (1979).
37 Keys, A.: Alpha lipoprotein (HDL) cholesterol in the serum and the risk of coronary heart disease and death. Lancet ii: 603–606 (1980).
38 Keys, A.: Coronary heart disease, serum cholesterol and the diet. Acta med. scand. 207: 153–160 (1980).
39 Keys, A.: Overweight, obesity, coronary heart disease and mortality. Nutr. Rev. 38: 297–307 (1980); Nutr. Today 15: 16–22 (1980).
40 Keys, A., et al.: Probability of middle-aged men developing coronary heart disease in five years. Circulation 45: 815–828 (1972).
41 Keys, A., et al.: Seven countries – a multivariate analysis of death and coronary heart disease (Harvard University Press, Cambridge 1980).
42 Keys, A.; Fidanza, F.; Scardi, V.; Bergami, G.; Keys, M.H.; Lorenzo, F. di: Studies in serum cholesterol and other characteristics on clinically healthy men in Naples. Archs intern. Med. 93: 328–336 (1954).
43 Keys, A.; Grande, F.; Anderson, J.T.: Bias and misrepresentation revisited: 'perspective' on saturated fat. Am. J. clin. Nutr. 27: 188–212 (1974).

44 Keys, A.; Karvonen, M.J.; Fidanza, F.: Serum cholesterol studies in Finland. Lancet *ii:* 175–178 (1958).
45 Keys, A.; Kimura, N.: Diets of middle-aged farmers in Japan. Am. J. clin. Nutr. *23:* 212–223 (1970).
46 Keys, A.; Kimura, N.; Kusukawa, A.; et al.: Serum cholesterol in Japanese coal miners: a dietary experiment. Am. J. clin. Nutr. *5:* 245–250 (1957).
47 Keys, A.; Kimura, N.; Kusukawa, A.; et al.: Lessons from serum cholesterol studies in Japan, Hawaii and Los Angeles. Ann. intern. Med. *48:* 83–94 (1958).
48 Keys, A.; Mickelsen, O.; Miller, E.v.O.; Hayes, E.R.; Todd, R.L.: The concentration of cholesterol in the blood and its relation to age. J. clin. Invest. *29:* 1347–1353 (1950).
49 Keys, A.; Taylor, H.L.; Blackburn, H.; Brozek, H.; Anderson, J.T.; Simonson, E.: Mortality and coronary heart disease among men studied for 23 years. Archs intern. Med. *128:* 201–214 (1971).
50 Keys, A.; Vivanco, F.; Miñon, J.L.R.; Keys, M.H.; Mendoza, H.C.: Studies on the diet, body fatness and serum cholesterol in Madrid, Spain. Metabolism *3:* 195–212 (1954).
51 Keys, A.; White, P.D. (eds): World trends in cardiology. I. Cardiovascular epidemiology (Hoeber-Harper, New York 1956).
52 Little, J.A.; Shanoff, H.M.; Csima, A.; et al.: Diet and male survivors of myocardial infarction. Lancet *i:* 933–935 (1967).
53 Malmros, H.: The relation of nutrition to health – a statistical study on the effect of war-time on atherosclerosis, cardiosclerosis, tuberculosis and diabetes. Acta med. scand. *246:* suppl., pp.137–153 (1950).
54 Malmros, H.: Debate: diet, lipids and atherosclerosis. Acta med. scand. *207:* 145–149 (1980).
55 Mann, J.I.; Truswell, A.S.: Effects of isocaloric exchange of dietary sucrose and starch on fasting serum lipids, postprandial insulin secretion and alimentary lipaemia in human subjects. Br. J. Nutr. *27:* 395–405 (1972).
56 National Heart Foundation of Australia, Committee on diet and heart disease: Dietary fat and coronary heart disease. A review. Med. J. Aust. *i:* 575–579, 616–620, 663–668 (1974).
57 National Heart Foundation of New Zealand: Coronary heart disease: a progress report (National Heart Foundation of New Zealand, Wellington 1977).
58 Netherlands Nutrition Council: Recommendation on amount and nature of dietary fats. Voeding *34:* 552–557 (1973).
59 Papp, O.A.; Padilla, L.; Johnson, A.L.: Dietary intake in patients with and without myocardial infarction. Lancet *ii:* 259–261 (1965).
60 Paul, O.; Macmillan, A.; McKeen, H.; Park, H.: Sucrose intake and coronary heart disease. Lancet *ii:* 259–261 (1968).
61 Reiser, R.: Perspective in nutrition: saturated fat in the diet and serum cholesterol concentration: a critical examination of the literature. Am. J. clin. Nutr. *26:* 524–555 (1973).
62 Roine, P.; Pekkarinen, M.; Karvonen, M.J.: Dietary studies in connection with epidemiology of heart diseases: results in Finland. Voeding *25:* 384–393 (1964).
63 Schornagel, H.E.: The connection between nutrition and mortality from coronary sclerosis during and after World War II. Docum. Med. Geogr. Trop. *5:* 173–183 (1953).

64 Scott, E. M.; Griffith, I. V.; Haskins, D. D.; Whaley, R. D.: Serum cholesterol levels and blood pressure of Alaska Eskimo men. Lancet *ii:* 667 (1958).
65 Seltzer, C. C.: Some re-evaluations of the build and blood pressure study. New Engl. J. Med. *274:* 254–259 (1966).
66 Shurtleff, D.: The Framingham study. An epidemiological investigation of cardiovascular disease. Section 30. Some characteristics related to the incidence of cardiovascular disease and death: Framingham study, 18-year follow-up (DHEW Publ. No. (NIH) 74–599, US Government Printing Office, Washington 1974).
67 Snapper, I.: Chinese lessons to western medicine (Interscience, New York 1941).
68 Stamler, J.: Debate. The established relationship among diet, serum cholesterol and coronary heart disease. Acta med. scand. *207:* 433–446 (1980).
69 Stamler, J.: Personal communication (1980).
70 Szanto, S.; Yudkin, J.: The effect of dietary sucrose on blood lipids, serum insulin and body weight in human volunteers. Post-grad. med. J. *45:* 602–607 (1969).
71 Truett, J.; Cornfield, J.; Kannel, W.: A. multivariate analysis of the risk of coronary heart disease in Framingham. J. chron. Dis. *20:* 511–524 (1967).
72 Truswell, A. S.: Diet in the pathogenesis of ischemic heart disease. Post-grad. med. J. *52:* 424–432 (1976).
73 Vartiainen, I.; Kanerva, K.: Arteriosclerosis and wartime. Annls Med. intern. Fenn. *36:* 748–758 (1947).
74 Walker, A. R. P.: Sugar intake and coronary heart disease. Atherosclerosis *14:* 137–152 (1971).
75 Werkö, L.: Diet, lipids and heart attacks. Acta med. scand. *206:* 435–439 (1979).
76 Working-party, Medical Research Council: Dietary sugar intake in men with myocardial infarction. Lancet *ii:* 1265–1271 (1970).
77 Yudkin, J.: Diet and coronary thrombosis: hypothesis and facts. Lancet *ii:* 152–162 (1957).
78 Yudkin, J.: Dietary carbohydrate and ischemic heart disease. Am. Heart J. *66:* 835–836 (1963).
79 Yudkin, J.: Dietary fat and dietary sugar in relation to ischaemic heart disease. Lancet *ii:* 4–5 (1964).
80 Yudkin, J.; Roddy, J.: Levels of dietary sucrose in patients with occlusive disease. Lancet *ii:* 6–8 (1964).

A. Keys, University of Minnesota, Minneapolis, MN 55455 (USA)

Diet and Hypertension: Anthropology, Epidemiology, and Public Health Implications

Henry Blackburn, Ronald Prineas

Laboratory of Physiological Hygiene, School of Public Health, Medical School, University of Minneapolis, Minneapolis, Minn., USA

Introduction

We attempt here to summarize and synthesize the evidence about diet and hypertension from the disciplines of anthropology, sociology, medicine, physiology, and epidemiology. These dietary factors include calorie balance and obesity, sodium, potassium and alcohol intake, and water supply.

The anthropological observations suggest that habitual physical activity and the composition of diet of affluent people today are very different from that of their hunter-gatherer forebears during the major stages of human evolution. Evolution is thought to have fine-tuned human adaptations to survival through eons of hunter-gatherer existence. That life-style is believed to have involved regular physical activity with alternating scarcity and abundance of foods predominantly of plant origin, high in potassium and low in sodium. Among hunter-gatherers there was homeostasis in a salt-poor environment, a distinctly smaller, leaner body and, it seems likely, low blood pressure and low blood lipid levels. The anthropological evidence leads further to the idea that human maladaptations in affluent society are due to a dissonance between the modern culture and the evolutionary legacy. These maladaptations make up many of our present day public health concerns: obesity, hypertension, hyperinsulinism, hyperglycemia, hyperuricemia, and hyperlipidemia. They are mass phenomena, thought to result in mass premature disease and death.

The sociologic and historic observations cited here about diet and blood pressure over-develop the issue of salt use deliberately because it is a subject less familiar to medical readers than the history of obesity. According to our synthesis, salt use is recent, both in historic and evolutionary terms. It is believed to have developed only with formal agriculture, the domestication and slaughter of animals, trade, and the need to preserve and store food. Its use as seasoning may have risen in part to counter the monotony of staple, single-grain diets. Later, automation and mass sedentation appeared. These sociocultural influences lead to obesity as well as to the acquisition, very early in life, of an elevated salt taste threshold and salt preference. This preference is maintained, in turn, by traditional eating styles which require salting and by the addition of unneeded sodium in the processing of foods. Most of the evidence suggests that salt use, individually and in populations, is culturally rather than physiologically determined.

Physiological and clinical observations suggest that body weight and salt intake contribute independently to blood pressure level and adult hypertension, though neither makes a very large contribution to the variance of blood pressure level within affluent populations. Experimental and therapeutic effects of substantial salt restriction and weight loss are impressive in lowering blood pressure among hypertensives. Severe hypertension is often 'cured' on salt-free diets. Milder hypertension is often controlled by low-salt diets and weight reduction. The dose of drugs required for control of moderate hypertension is frequently reduced and sometimes eliminated altogether. Other than sodium and potassium intake and calorie balance, excess alcohol intake now appears important among dietary influences in hypertension. No less interesting are observations on the mineral content of water supplies and perhaps even the effect of vegetable versus animal protein and fat.

Epidemiologic observations indicate a strong and largely consistent relationship between the population frequency of adult hypertension, or median population arterial pressure levels, and habitual salt intake. Population rates of hypertension are less consistently related to average relative body weight or the frequency of obesity. In contrast, *within* cultures, blood pressure level is more strongly associated with individual measures of body weight and obesity, and with their change over time, than with sodium intake. These then are the observations to be enlarged upon in this presentation. An attempt is also made to develop from this evidence a public health view and approach to hypertension control and prevention.

At the outset we would try to diminish unnecessary and unproductive controversy about the relative importance of nature versus nurture in the origin of hypertension. To reconcile these conceptual conflicts about causation, that is, inherent mechanisms versus sociocultural influences, we propose the following postulates: *All* disease results from an interaction between host and environment. *Mass* diseases, such as hypertension, result from an interaction between *widespread* susceptibility in the population and *powerful* sociocultural influences.

Corollaries of these postulates are that an extremely favorable environment should result in *minimal* exhibition of susceptibility in the population and a lower prevalence of the disease. In contrast, an extremely unfavorable environment should result in the *maximal* expression of clinical disease among susceptibles in the population.

We find that these concepts are widely confirmed by clinical, laboratory, and population observations as well as by direct experiment. We will discuss limitations in the specific epidemiological findings concerning obesity, salt and potassium consumption, and hypertension in populations. However, substantial congruence remains between these and the clinical and experimental findings. There are also logical mechanisms which strengthen the inference of causation. All these provide a theoretical base and a practical potential for prevention. The evidence suggests to us the presence of a large degree of susceptibility in the human genome as well as strong environmental influences which encourage the wide exhibition of the condition of elevated blood pressure.

We deal mainly here with our assignment about the population findings. But just as our expert clinical and laboratory colleagues in this volume touch on the epidemiology of hypertension, so we may touch on clinical-experimental evidence where together they are relevant to cause, to practice and to the public health. All these disciplines are essential to the broadest understanding of hypertension. A public health approach, we suspect, is necessary for its optimal control. Surely this is true for 'primal' prevention – that is, prevention of elevated blood pressure in the first place. From the congruent evidence, clear implications and guidelines emerge for a rational, feasible, and safe public health strategy in the control and prevention of hypertension.

Anthropological Observations

Diet, Salt and Evolution
Water, solutes, and osmolarity of plasma are intimately tied to the evolution of the mammalian kidney from the sea to the land and some-

times back to the sea. Many and engaging are the theories. They include the idea that the plasma of a species may reflect the actual composition of ancient sea water at the time the species 'emerged'. They include the idea that survival of our forebears on a salt-poor diet required exquisite renal conservation of salt and water. They include the theory that survival through famine alternating with abundance required avid body mechanisms to lay down fatty deposits during the good times. The theory is that natural selection during eons of hunter-gatherer subsistence resulted in survival characteristics for that manner of living. A modern life of perpetual abundance and sedentation, so the theory would go, overwhelms the evolutionarily fine-tuned survival characteristics of humans and leads to maladaptations such as obesity and hypertension.

There is little question that an affluent life-style is vastly different from the hunter-gatherer's. This begins with the recent use of calves' milk for infant feeding in contrast to the milk optimally adapted for human growth and development. It extends to affluent adult eating and activity patterns in which the quantity and quality of diet have vastly changed. This different mode of life is a very recent phenomenon dating merely from the onset of agriculture and civilization, 10,000 or 12,000 years, or 500–600 generations, ago. In contrast, upright posture and bipedal gait developed among our hominid ancestors sometime between 15 and 4 million years before the present. The human skeleton was modern as long ago as a million years. It is thought that the brain was evolved to its modern state 50,000 years before the present. Though neither soft tissues nor much of cultural artifacts is 'fossilized' for direct observation, it appears reasonable to consider from the existing evidence that most of the major metabolic, enzymatic, neurohumoral, and physiologic adaptations of humans had occurred by the time the modern brain developed, in other words, with *Homo sapiens sapiens.* They surely developed long before agriculture, civilization, and abundance. These are thought to be reasonable surmises from the archaeological and anthropological evidence and from the remarkable similarity of human nucleic acid configuration and metabolic processes to those of extant higher apes. The strong implication, then, is that modern man is still metabolically a hunter-gatherer. Many of the precursors of adult disease, and the diseases themselves, may result from the interaction of recent affluent cultural patterns with the genetic legacy of natural selection in the sparse hunter-gatherer existence.

The mammalian kidney, with its exquisite capacity for solute and water regulation, takes us much further back in time. A famous example is the Snowy Mountain Goat of Australia whose sodium intake in 100–200 days of grazing and browsing is roughly equivalent to the daily intake of Melbourne residents [22]. If humans, in early African and Middle East surroundings, were to survive the long spare periods and the low-sodium, high-potassium diets, would they not efficiently conserve sodium and water, with a similarly adapted neuroendocrine regulation of blood pressure? Could it be that calorie and salt loading, so recent and thus so stressful of this evolutionary legacy, result in the maladaptations which lead, in turn, to hypertension in man: that is, obesity, hypervolemia, dysregulation in cardiac output, and dysregulation of the resistance-compliance vessels?

All this makes a good 'story'. But just how good is the evidence that early humans ate little salt and much potassium and that they were, in fact, small and lean and had low blood pressure? And what, if any, logical mechanisms make this story plausible?

Cultural Variations in Salt Use

Based on archaeology, and paleontology, and on anthropology among extant hunter-gatherers, the possible diets of Early Man are believed to have ranged from almost entirely plant foods having very low sodium-potassium ratios, to subsistence over large parts of the year on the meat, marrow, blood and milk of animals. Generally, hunting and opportunistic meat-eating are thought to have forwarded the species. Hunting was rendered possible by the basic gathering culture with its predominance of staple plant food sources. In this manner of subsistence, men and women specialized and cooperated; basic human traits. But we calculate that even a largely meat, blood, and milk diet would provide a maximum of only 3 or 4 g of salt daily. This level barely overlaps the lower end of the distribution of the salt intake of affluent, urban humans which ranges from 5 to 30 g. If the hypothetical limit of salt intake in early humans were truly below 5 g daily, this would be quite compatible with levels now found in a number of unacculturated populations in the Pacific Islands, in Central and South America, and among hunter-gatherer tribes in Africa and Australia. Among them, adult hypertension is not found, nor is there a trend upward in blood pressure with age. In these places the people function well, often in tropical environments, usually on less than 3 g of salt a day [108].

We have conjured up the following 'luxurious' menu which might have occurred uncommonly in the orgy following a successful hunt by hunter-gatherer forebears: Suppose that 3,000 kcal were consumed at one sitting, and 40% of these calories were from fresh meat or animal products (for example, deer or antelope, rabbit, bird or quail, and eggs) and that 60% of the calories came from fresh plant foods (for example, a combination of yams, other root vegetables, beans, a leafy vegetable, nuts, seeds, berries, and fruits). This unusual celebration would amount to an intake of about 900 mg sodium and 7 g of potassium, a 1:7 ratio. Less than 2 g of salt would have been consumed, even under these feasting conditions!

The anthropological evidence about the diet and activity of early humans is both direct and inferential. Though fossils of our immediate predecessor, *Homo erectus,* are found widely over the earth, earliest forms of *Homo* have been found only in East and South Africa, in apparently salt-poor environments. Hunter-gatherer observations are merely windows on the past, but probably relevant to it. The northern half of the American Pacific Coast Indians used no salt, while the southern half did, along a sharp line of demarcation from the Columbia River to Nevada [73]. None of the theories posed about their type of agriculture, diet or climate satisfactorily explains this demarcation. Salt use also is very low in the Melville Islands and among the Aborigines in Australia and the Indians of Tierra Del Fuego in South America and appears quite unrelated to the type of agriculture there [62]. The Siriono Indians of Bolivia, predominantly hunters, apparently knew nothing of salt until it was introduced recently by an American cattle rancher [44]. From an initial reaction of distaste, they later developed craving. One of the more remarkable documented adaptations is the 'no-salt' culture of the Yanomamo Indians of Brazil and Venezuela [103]. They excrete sodium at the rate of 2 mEq or less daily and potassium at 200 mEq! And they have near-maximal stimulation of the renin-angiotensin-aldosterone system and of the sodium resorptive organs, the kidney, GI tract, and sweat glands. This stimulation is exaggerated in their pregnant and lactating women. All this occurs in the presence of average blood pressures of 107/68 mm Hg in men and 100/63 mm Hg in women, with no apparent disadvantage in growth, development, strength, or endurance [102].

Other cultures having documented very low salt intakes and low arterial pressures are the Polynesians, with sodium excretion around 60 mEq daily [114], the Eskimo with 80 mEq or less daily [89], and the Kalahari Desert Bushmen at 30 mEq, or less than 2 g of salt per day [58, 136].

The major lines of the evolutionary argument from these remarkable observations about salt are that human adaptations were toward sodium homeostasis in a salt-poor environment; that despite the well-developed sense of salt taste, early humans obtained sufficient sodium from natural foods without a mineral source [23]; that elevated levels of renin and aldosterone are the 'norm' for humans under primal conditions, to maintain blood pressure and conserve filtered sodium; and that 'depressed values' for these hormones in urban affluent man are the result of chronic sodium excess and potassium insufficiency [102]. Potassium fits reasonably well into all these arguments [88].

Ontogeny Recapitulates Phylogeny

Phylogenetically, when the sea was abandoned, salt remained a principal inorganic constituent of extracellular fluid. Ontogenetically, humans spend the first months of life in a saline sea and even pass through an embryonic stage showing gill-like ridges and a tail. The body is approximately 70% fluid and has a concentration of sodium in the extracellular fluid only slightly less than that of the present ocean [35]. It was speculated long ago that 'the sea within us' is the same salinity as sea water of the ecological time in which the species evolved [81]. *Macallum* [81] developed paleo-ratios for numerous inorganic elements in plasma over a wide range of animals, from numerous species, and supposed that the primary function of the kidney was to maintain this ratio, excretion being a secondary function. For example, it is demonstrated that the shark kidney maintains the plasma paleo-ratio of Cambrian times by retaining urea at 10 times the plasma concentration found in humans. Without this high concentration of urea, the shark would rapidly shrivel, dehydrated in the ocean. This teleological explanation is bolstered by the predominance and great physiological importance of sodium and chloride ions in the plasma and their crucial role in the electrical activity of cells and regulation of extracellular fluid volume. These observations suggest that renal mechanisms evolved to assure a constant plasma sodium level in a salt-poor environment [35].

What can be learned from the variability of salt intake between animals? Ruminants and ungulates supplement their feed with salt whenever available. These and other herbivores which subsist on high-potassium, low-sodium intake show deficiencies in growth and milk output and become ill without supplemental sodium. Carnivores, in contrast, need no added salt and appear to show no salt craving [35]. Why then should humans desire salt?

The early German physiologist *von Bunge* [139] postulated that the longing for salt by herbivores was an attempt to balance their high potassium diet. He extrapolated that idea to the desire for salt in man having risen from the consumption of predominantly plant foods high in potassium. He noted parallels in the salt intakes of nomadic societies and carnivores and between agricultural societies and herbivores. Finally, he linked these with the physiological observations that salt intake is accompanied by rapid potassium excretion. But his idea about human salt appetite is challenged by the finding of a vegetarian African tribe which uses only plant potash rather than salt as its main condiment. Moreover, people of the sub-Sahara show a lack of salt preference whether they live in sweet water or saline areas. The historical observations of desert caravans which carried salt may have led to misguided notions about the local salt consumption. The transport of salt does not imply heavy use [110].

Less information is available about the control of sodium intake than about the hemodynamic and hormonal control of sodium loss. In fact, it is not known in most species whether sodium intake is controlled at all. It is thought that some species control only output (man) and others, such as sheep and rabbits, intake. Others may control both output and intake while others control only that sodium intake which is in fluids (rats). *Fregly* [35] suggests that: 'perhaps the appetite for salt, which has served animals and man so well when salt is in scarce supply, has become a liability in a time when salt is so readily obtainable... once some people are conditioned to salt, they cling to its use stubbornly and may go to great lengths to fulfill an appetite beyond physical necessity.'

Denton [21], who has studied most the evolution of the mammalian kidney, summarizes: 'My own view of the evidence at present is that a pleasurably determined high salt intake, inappropriate to needs and metabolism, could act over a period of years as one condition in a causal mosaic determining hypertension in some individuals in the adult community. It may not be a simple relation but require other severally necessary and jointly sufficient factors to produce the effect. Such factors may be genetic predisposition and emotionally determined chronic inappropriate secretion of hormones influential on salt balance – particularly aldosterone.'

These then are the anthropological and evolutionary arguments on salt. On the other hand, some scientists question whether the control of salt intake is important at all for species in which sodium excretion is so effectively controlled. This idea may influence unfavorably the practice of many physicians today in a high-salting, pill-dispensing culture [134]. They may be inclined to say, 'Not to worry, the kidney will excrete any excess salt. If not, a diuretic will do the job.' Thus, they imply, 'You may have your salt and eat it, too.'

Obesity

The anthropological and social-historical observations about obesity make up a vast literature compared with the information on salt intake. We believe it is familiar to the medical reader, and because obesity and its causes have been so extensively reviewed, we will attempt here only a summary. The anthropological evidence indicates that hunter-gatherer man, during major evolutionary stages to *Homo sapiens* and since, has generally been smaller in stature, probably more muscular, stronger, and more physically active. Perhaps also hunter-gatherers were more efficient mechanically and, though leaner, had substantial seasonal weight variations which followed the alternating scarcity and abundance of calories. Hunter-gatherer adaptations, stressed by perpetual calorie abundance and a sedentary life-style, result in obesity and overweight. Large body size, overweight, and obesity are thereby mass phenomena of affluent cultures, related to overnutrition and sedentation of entire populations. The Seven Countries observation that habitual activity accounts for much of the difference in obesity between those populations [65] is directly relevant to a public health view on the causes and prevention of hypertension.

Social and Historical Observations

The Development of Salt Taste

General learning theory confirms the predominance of culture in determining salt intake. Learning early in development may be more efficient than later [79]. 'The developmental calendar on which an organism is exposed to environmental events may play an important role in determining how that exposure affects perpetual functioning subsequently' [143]. Early environmental influences are held to be effective when they occur at critical periods. 'Early introduction means not only that the stimuli act on a less mature and perhaps more malleable organism but that cumulative exposure at all subsequent periods is increased' [143]. Thus, regulation of sodium in the infant may be influenced by early exposure and the perception of saltiness.

There is little question but that infant exposure to saltiness is of very recent historical origin. Prior to 1900, almost all infalts were totally breast-fed for 6–9 months and thus experienced no salting. *Fomon* [32] reports that now only 20% of infants in the USA are breast-fed at all and by 6 months of age only 2%. The sodium intake of a nursing infant is only 5–10 mEq a day while cow's milk formula has about three times the sodium, potassium, and chloride. Bottle-fed infants also tend to be started on solid foods earlier than breast-fed infants and are more likely to be fed commercially prepared than home-made food. Sodium composition in the infant diet, therefore, may rise rapidly from about 13 mEq at 2 months to 29 mEq at 6 months, and 63 mEq at 12 months [32]. The concentration of salt in the mouth and saliva may determine the perception thresholds of salt and, thereby, participate in the early development of salt appetite or preference. The influence of cow's milk, formula, introduction of solid foods, and the age at which birth weight doubles, may be equally significant in setting the stage for large body mass and obesity, as the Anlage for mass hypertension and for other chronic diseases as well [130].

Subsequent to infancy, exposure to salt is increased by snack foods and in the chain eateries, in the vending machine products which compete with school lunches, by salty canned soups now specifically targeted to appeal to children, and by the greater frequency of meals taken outside the home. For example, a single 'quarter-pounder' cheeseburger at McDonald's may provide over 3 g of salt – this is without counting the usual accompaniment of a huge milk drink and salted french fried potatoes [50]. Finally, there is an increasingly wide exposure to salt which enters the water supply through water softening processes and from the tons of salt used to reduce accidents on wintry highways.

Social Currency

Darby [18] suggests that the importance of a particular food to man is reflected in the intensity of historical quests for the food and in allusions to the subject in language and literature. He indicates that salt is unique in this regard, 'ranking with sugar and spices', and points out, for example, that salt occupies three pages in the unabridged Oxford Dictionary while sugar occupies two and spices, one. The connotations of the word salt in the English language include: excellence, 'to be worth one's salt', worthiness, 'salt of the earth', and hospitality, 'above the salt', in which a large porcelain or silver salt container denoted the position of honor at a formal table in medieval times [35]. Language and folklore offer conflicting symbolic meanings of salt as being related to wit, wisdom, virility, fertility, barrenness, friendship, or food for good or evil spirits. Jews seal covenants by exchanging salt. Bedouins will not attack a man whose salt they have eaten. Bread and salt are gifts in Slavic countries. It was one of the first medicines used. Newborn babies in difficult times were washed with salt to ward off evil. Salt is the symbol of purity in the Roman Catholic Church today, and is used in baptism. Spilling salt is considered unlucky and it is counteracted by tossing a pinch of it over the left shoulder. Roman soldiers were given a salt allowance, salarius, which led to the word salary [35].

The importance of salt sources in history is documented in the names of communities such as Salzburg and Salisbury. Salt was a powerful factor in trade. 'As late as 1882, a British traveler was offered a girl for four loaves of salt' [35]. Kings maintained rule through a salt monopoly in contrast to the relative independence of countries where salt was plentiful. In the Nile region, in Babylon, in India, China, Mexico and Peru the populations were hopelessly dependent. The Ptolemies in Egypt, the ancient Romans, and their medieval successors had government monopolies on salt. Ghandi's revolution was against the British salt tax; Americans complained against the salt tax of George III; and the French revolutionists' outcry against the gabelle, all were reactions to salt tyranny. The first known agricultural settlements of man arose in the salty areas of Mt. Sodom, the Dead Sea, and Jericho. Earliest civilization grew up around salty springs and salt boiling places. 'This coveted object changed history and played a role in altering the balance of power among nations' [35]. Again, we are reminded that this coveting of salt was apparently culturally, not physiologically determined.

Salt Availability

Another dramatic example of the sociocultural element in salt consumption is its increased availability from commercial production, a recent industrial phenomenon [94]. Contrast the hunter-gatherer mode of obtaining salt with that of a later trading economy and that, in turn, with current industrial methods. Early Man, if he looked for salt at all, would have had to travel long distances to salt deposits, perhaps navigating precipitous trails, working waist-deep in powerful brine solutions or in dank, deep deposits, all this to fetch relatively small quantities of salt for wide sharing back at camp. Contrast this to the organized harvest by agricultural man of salt from the large surface beds of the sub-Sahara or from the rich mines of the Tyrol; then sculpting, wrapping, and tediously transporting a few such salt blocks over long trade routes. The effort entailed was so great that a single camel caravan even into this century might be sufficient for the purchase of several wives. Finally, contrast these early modes of harvest and transport with the commercial pursuit of salt in vast interconnecting subterranean chambers, carved from immense reserves of the ancient seas, harvested by massive bulldozers, and transported in long chains of railroad cars to modern refineries. Then consider the efficient mass distribution by wholesalers to markets, and the wide commercial use and encouragement of salting. The cultural effect of cheap, available, socially acceptable salt is indeed pervasive and overwhelming.

Dietary Sources of Salt

The dietary sources of salt are the naturally occurring sodium in foods, that added by food processing, that in the drinking water, and that added by the consumer. The dietary sources are often considered about one third from food and water, one third from addition in processing, and one third from that added just prior to eating. However, these traditional proportions for sources of salt intake must now be modified in the light of the 1979 Report on Sodium Chloride by an advisory panel of the Select Committee on GRAS (Generally Regarded As Safe) Substances of the US Food and Drug Administration. They found that, on the average in the USA, naturally occurring sodium now accounts for only one sixth of the sodium consumed, whereas salt in cooking and used at the table is one third and that in processed foods about one half! More precisely, average daily salt intake is 3.2 g from natural sources, 8.8 g from salt and other sodium compounds added in processing, and 5.2 g added by the consumer.

Sodium intake from sources other than sodium chloride constitutes less than 10% of the total sodium intake. The largest salt intake in food occurs in connection with grain and cereal products; the next largest in meat, fish, and poultry; the next in milk and cheese products; the next in oils and fats; and the least in fresh legumes, garden fruits, and vegetables. In the United States, commercially prepared foods, dinners, meats, snacks, and chain restaurant foods are increasing dietary sources of sodium, while they reduce the intake of potassium. For example, fresh vegetables contain little sodium and 10–100 times more potassium. Similarly, fish and meats, unprocessed, contain 2–10 times more potassium than sodium salts. The sodium/potassium ratio of the affluent diet is now 2:1 or greater, which is more than a reversal of that ratio over a very short time historically [135].

Based only on disappearance of salt in food processing, consumption per capita may have decreased in the USA since 1970. However, food producers continue to monitor consumer desires and respond in the marketplace on the basis of expectations of a 'salty taste'. Taste tests suggest to them that a salt concentration around 0.6–0.8% be considered 'standard saltiness', a level now found in commercial soups: chicken, rice, vegetable, and beef. This may be contrasted with a 'standard saltiness' of a 1.5% solution for the Japanese, based on the content of their staple 'miso' soup. Informal taste tests suggest that salting of low-salt foods up to 'reasonable saltiness' may add back only a fourth to a third of the salt used in standard brands of canned tuna, chicken soup, tomato juice, and vegetables [50]. Consumer surveys by industry report the presence of salt in foods as desirable both for preservation and flavor. In recent past history when food was less plentiful than now, flavoring with salt lent variety to cereal dishes and provided a whole group of foods that might otherwise have been wasted, such as sausage. In turn, these 'inferior' foods quickly became diet staples [33].

The functional role of salt in foods, over and above palatability, is less well-known. Salt strengthens gluten and permits more even rising of dough, while controlling yeast activity and improving crust color in bread baking. It serves to preserve meat by lowering water activity and limiting microbial growth. It adds to gel structure and color in the formation of emulsions. It destroys tapeworm. It plays an important role in the manufacture of butter and margarine and in their longer shelf-life as well as in the ripening, flavor, and texture of cheeses where it decreases the rate of protein decomposition. Cheeses must contain 2% salt to accomplish these purposes. Salt plays a central role in controlling the bacterial flora of sauerkraut, pickles, soy sauce, and other condiments. Processed meats are a substantial source of sodium in the form of phosphates, nitrates, caseinates, hydrolyzed vegetable protein, soy isolates, and monosodium glutamate [84]. If some of the current 'official recommendations' for lowered salt intake were to come about in the USA, most of these processed foods would not be compatible with the recommended dietary and thus might disappear from the market [33].

However, the variations in current commercial practices of food processors are enormous. For example, brands of tomato paste range in content from 731 mg of sodium to 25 mg; a typical American 'marshmallow' bread loaf contains 355 mg of sodium versus 81 mg for another. The companies making a concerted voluntary effort to reduce salt in their products, including the baby foods, apparently find acceptability and good economics for the change [50].

All these essentially recent eating patterns may be summed up in the hedonistic view of the French gastronome, *Brillat-Savarin:* 'The pleasures of the table are of all types and all ages; they go hand in hand with all other pleasures, outlast them, and in the end console us for their loss.' In this common view of whatever kind of traditional eating pattern, any officially recommended changes in the diet are considered by some as invasive pressures, meant to deprive people of a well-earned and desirable lifestyle. It is often not considered that they offer instead a positive educational process, with more varied and attractive, and presumably more healthy, alternatives.

Obesity

The social aspects of obesity and its history are documented in a rich literature. Until recent decades, the energy cost of occupational activity and the cost of food were major determinants of the population distribution of weight and obesity. Now, with cheap and plentiful food plus mass sedentation at work in Western affluent societies, the determinant of population obesity or leanness is rather how much activity can be encouraged or engineered into leisure. But this subject is extensively treated elsewhere [76].

Epidemiological Observations

Diet-Blood Pressure Relationships in Populations

Epidemiological observations on the relationship of diet to hypertension include measures of population salt intake versus mean population blood pressures on the one hand and the frequency of adult hypertension and stroke on the other; they include correlations within populations between individual sodium intake or sodium/potassium intake ratio and levels of blood pressure; they involve studies of the relationship of relative body weight and obesity, alcohol intake, of soft versus hard water to blood pressure levels and hypertension; finally, they include observations on migrants and on population changes in these variables over time, either spontaneously or experimentally.

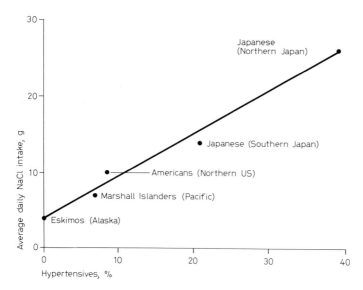

Fig. 1. Correlation of average daily salt intake with prevalence of hypertension. This incomplete data set, and the less than ideal evidence on which it is based, is given here because of its historic precedence. From *Dahl* [16].

Population Blood Pressure and Salt Intake

In populations selected for contrast in habitual salt intake, there is a strong general relationship between intake and the frequency of adult hypertension, though the design and comparability of the measurements in these comparisons leaves much to be desired [36, 106]. Here in figure 1, an old illustration from *Dahl* [16], the correlations are not adjusted for such obvious confounding variables as age and relative body weight. Nevertheless, the findings are consistent: Diets habitually low in salt, that is, under 5 g daily, are associated with little or no adult hypertension or age trends upward in blood pressure. The findings are compatible with the hypothesis that salt and hypertension are related and that the relationship explains some of the difference in population rates of hypertension.

Prior et al. [114] suggest the following features which distinguish populations having low pressure from those where hypertension is frequent and where blood pressure increases with age: (1) they are usually in isolated, traditional small communities; (2) they are largely in subsistence

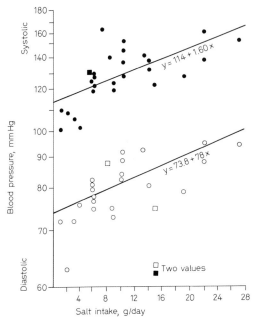

Fig. 2. Correlation of average daily salt intake with average population systolic and diastolic blood pressures. From *Gleiberman* [41].

economies having very little trade; (3) their physical activity levels are relatively high and obesity is uncommon; (4) they have strong communal support systems in which individuals have clear responsibilities toward the community; (5) they have a habitually low salt intake, usually around 40–50 mEq of sodium a day, or less; (6) their calorie intake overall is relatively low, with periods when food is in short supply.

Similarly, average population blood pressure values are consistently and positively related to population salt intake as summarized in figure 2 from *Gleiberman* [41]. Some of these studies paid great attention to the standardization of methods while some did not. Some measured urinary sodium and some did not. The population sodium-blood pressure correlations were of a higher order for men ($r=0.69$ systolic blood pressure, SBP; 0.66 diastolic blood pressure, DBP) than for women ($r=0.54$ SBP and 0.63 DBP). The crude results here are unadjusted for age or body mass. Nevertheless, they are compatible with the hypothesis that different population levels of salt intake may account for a substantial proportion of the differences in the average blood pressure of populations.

Generally similar, high sodium intakes are found among Western cultures of continental Europe, the British Isles, Scandinavia, North America, Australia, and New Zealand where from 8 to 15 g are consumed on average daily as sodium chloride. The age-related rise in average blood pressure in those cultures has long been observed and often assumed to be a normal aging phenomenon. Hypertension, variously defined, occurs in 10–15% of adult populations there. Detailed study of age trends in blood pressure among cohorts of these populations indicates a subpopulation which is responsible for the cross-sectional age trends. For example, about two thirds of healthy adults followed over the years show either no change or a reduction in their blood pressure with increasing age; the trends are apparently due to the one third which show a distinct pressure rise with age.

Joossens [54] and *Sasaki* [120] have consistently found high order correlations between population salt intake, blood pressure, and death rates from cerebrovascular disease. Japanese studies in particular point out the extremes in salt intake and adult rates of hypertension and stroke. In parts of Japan, for example, the average salt intake is over 20 g daily, with individual intakes of up to 50 g. The staple soy paste, and 'miso soup', soy sauce, and salt fish, are complemented by the use of pickled fruits and vegetables. The whole is a culture literally bathed in salt by tradition. Similarly, some of these areas report up to 60% adult hypertension and among the highest rates of stroke deaths found anywhere [70, 119].

Population Trends in Salt Intake and Hypertension

Here the epidemiological data are sparse and poor. The 'distance' between national salt consumption data and vital statistics on hypertensive deaths is simply too great to be useful. The data from direct examinations of Finnish populations show a greater prevalence of hypertension in eastern Finland in the 1950s and 1960s than in the 1970s. But they also suggest important sources of systematic error and pronounced effects of selection by deaths among the hypertensives [61]. Thus, relating population trends of salt intake to trends of blood pressure and disease rates requires the difficult method of parallel surveillance [40]. So far, no serious attempt has been made to relate the falling rates of hypertensive disease in Western countries to change in salt intake, or obesity, or in any other cultural patterns. However, *Kimura* et al. [67] have commented anecdotally on the parallel and significant decrease in deaths from stroke in northern Japan and the apparent reduction there of average salt intake from 35 to 25 g daily, between 1959 and 1971.

Other systematic attempts are lacking to document population blood pressure trends and their possible causes. In the few studies reported which take some care toward comparable methodology, many of the known correlates of blood pressure have not been measured. For example, an apparent secular increase in mean blood pressure levels in Yugoslavia was unaccounted for by change in overweight, but sodium intake was unmeasured [72]. Similarly, in Belgium where downward trends in sodium intake may have been documented, trends in blood pressure measurements were not simultaneously recorded [64]. The trend downward in hypertension-related deaths in US whites antedated effective drugs and is not explained.

Migrant studies are important because they are thought to stand somewhere in importance between epidemiologic observation and experiment for the reasonable inference of causation. However, dietary and physical factors have rarely been studied systematically in them. The Tokelau Islanders Study is surely the most complete in its establishment of major pedigrees for the populations of migrants and non-migrants and its systematic examination of diet calories, protein, sodium, and sodium/potassium ratios. The Na/K ratios are significantly higher in the acculturated Islanders having migrated to New Zealand, as are their average values for arterial pressure. However, weight gain, life-style, and social interactions are calculated to contribute more to the blood pressure differences [115].

In the Cook Islands, the more acculturated populations have a 28% rate of adult hypertension compared to 3% in the more traditional groups. This is not explained by differences in height, weight, or prevalence of infectious diseases. Nevertheless, salt intake in the acculturated group is double that in the traditionalists [114]. Similarly, the one of five New Guinea populations in which adult hypertension and upward age trends of blood pressure were prevalent was a coastal tribe. It consumed canned goods and had potassium/sodium excretion ratios significantly lower than the interior groups [82]. But in the Solomon Islands, salt consumption was found to vary independently of acculturation. Under these conditions, blood pressure was more closely correlated with sodium intake than with the other cultural factors measured [108].

Other migrant studies of particular interest are those of the Easter

Islanders to South America [15], Atlas Mountain Jews to Israel [24], and the Japanese to Honolulu and to the United States [56, 83]. In the first two cases the pressure was higher in migrants and in the latter case there was apparently no difference in pressure; if anything it was lower among the acculturated than traditional Japanese. Most migrant studies are less complete, particularly in respect to sodium intake measurement. However, psychosocial stresses were suggested by *Shaper* et al. [126] in studies on Kenyan warriors at different times after their entry into the Army. The greatest rise occurred some time after the period of maximal weight gain or diet change [126].

Cassel [10] reviewed many of the major migrant studies, including Irish Brothers, Easter Islanders, Chicago Blacks, Atlas Mountain Jews, Palauans, and Cape Verde Islanders. He found design and data limitations in all, but surmised that eventually, 'We will find diet, salt intake and the degree of adaptation to one's niche in society *all* important in the determination of blood pressure levels.' Unfortunately, information on sodium and potassium intakes, crucial to his and others' interpretations of the effect of migration on blood pressure, is generally poor or unavailable.

All these observations are consistent in turn with the view that many susceptibles exist in human populations, and that powerful sociocultural influences act on them. Indeed, the Japanese data suggest that up to 60% of people may bear the genetic susceptibility to hypertension which is exhibited in those regions having a maximally unfavorable environment [120].

Page [107] summarizes as follows: 'When all individuals in a population are habitually using very small amounts of sodium, blood pressure does not rise with age, and hypertension is virtually absent. When all members of the population are ingesting very large amounts of sodium, a high percentage, reflecting the maximum number of susceptible individuals, develop hypertension. Between these two extremes the relationship between blood pressure and sodium intake is difficult to perceive because of wide variation in genetic susceptibility and other types of "noise" introduced by other variables' (fig. 3).

Clearly, details of individual cellular sodium transport or other markers may be of great interest in practice, to define intrinsically high individual risk. However, the Japanese data speak for the likelihood that human susceptibility to hypertension is so widespread that sociocultural influences should be the predominant focus and concern of the public health.

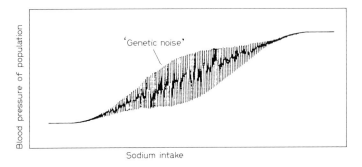

Fig. 3. The relationship between population salt intake and average blood pressure is sharp at the extremes and obscured, according to *Page* [107], by genetic variability.

Discrepant Population Data for Race

Inconsistent epidemiological data may reveal important leads for future research and better understanding of diet-hypertension relationships. For example, what characterizes populations that lie furthest from the regression line of salt intake against average blood pressure of hypertension rates? Attention has been directed particularly to Blacks on St. Kitt's Island in the Caribbean [121], in the Bahamas [52], and in Evans County, Georgia [137]. In these Black groups, the rates for hypertension may be disproportionately higher than predicted from estimated salt intake and little related to obesity. Racial differences in hypertension have not been sufficiently studied systematically to indicate the relative contribution of obesity and salt and potassium intake. However, the Evans County Study [42] suggests a relatively stronger effect of weight in Whites and of potassium intake in Blacks. Of note is the finding in HANES, the US Nutrition Survey, that twice as many Black as White school children consumed a salty snack food one or more times a day.

There are also hypothetical effects of genetic selection which could lead to an unusual population susceptibility to salt. Could an exaggerated hypertensive response to salt be related, for example, to selective effects in Blacks of slavery? First came the forced recruitment of Africans from central, low salt-use areas; subsequently came a selective wastage from heat stress and salt and water deprivation during the brutal voyage across the sea. Among survivors, likely those most fit to withstand the acute stress, there followed an abrupt exposure to a poor quality diet with heavy salting on which southeastern American slaves subsisted. Salt-saving, renal-adrenal adaptations to a low salt environment, and selection, would be, by this idea, overwhelmed in the new salt-rich environment. Excess pressor responses to sodium might result not only from these sorts of selection process, but from the stress of social dissonance among the slaves. But all these ideas too are broadly speculative.

Regional Differences

Regional differences in diet and blood pressure within cultures are not as widely documented as differences between populations, and where found are often not well-explained. Community differences in Yugoslavia were possibly associated with ethnic origin, alcohol intake, and socio-economic class, but were unexplained by obesity; unfortunately, sodium intake was not measured [72]. However, regional differences in blood pressure in Newfoundland were significantly related to measured salt consumption [31].

Among nine populations in Japan, random samples were taken of men aged 40–59, and sodium excretion was measured. High correlations ($r = 0.69$) were found between excretion and the frequency of adult hypertension [71]. *Komachi* et al. [70] make the interesting suggestion that in populations having little atherosclerosis, such as the Japanese, high sodium intake is strongly associated with elevated blood pressure and the incidence of cerebrovascular disease. In contrast, Japanese regional rates for myocardial infarction and angina pectoris are found to be parallel to dietary fat intake. The range of incidence of each entity in selected populations in Japan, under conditions in which salt and fat intake vary independently, are on the order of 10-fold for apoplexy and 3-fold for coronary heart disease.

Diet-Blood Pressure Relationships within Populations

Individual Salt Intake

Comparisons of salt intake among hypertensives and non-hypertensives in the same culture give inconsistent results. For example, studies in Framingham [60], Wales [90], and Scotland [2] find no excess sodium excretion in hypertensives, while in Goteburg [4] and Belgium [63] the contrary is found. However, in all these cultures salt intake is high, far above the 5-gram intake of societies having little adult hypertension. Moreover, in most of these studies, salt intake was characterized by only one urine collection.

Individual sodium-blood pressure relationships *within* populations are similarly weak or inconsistent. However, these 'zero correlation' findings are beset by multiple problems of measurement and other sources of variation for sodium intake and for blood pressure, as well as by the homogeneously high salt intake characteristic of Western populations [49, 80].

The Framingham data on individual salt intake and blood pressure illustrate some of the limitations of most earlier studies *within* affluent populations [20]. In the first place, it was difficult and expensive to get sodium excretion studies, and they were obtained only once on 185 participants on the fifth biennial examination. Thus, nothing is known about intraindividual variability; it is assumed to be large. Only about 22% of the population had intakes of less than 8 g of sodium chloride daily, so the population was indeed homogeneously high in intake. Despite the absence of significant pressure differences by category of salt intake, the authors did not comment on their finding of a trend upward from 20% hypertensive in those eating less than 8 g of sodium chloride daily to 36% hypertensives among those eating more than 13 g daily. Detailed plots of individual 24-hour urine sodium, potassium, and sodium/potassium ratios against individual blood pressures revealed no association. However, it is now estimated that nine or more 24-hour urine sample determinations are required to reduce to 10% the error of estimate of sodium output. Five or more determinations are needed to 'adequately' separate individuals into tertiles, high, medium, or low, of sodium intake [80]. Recent evidence suggests that questionnaire ascertainment of salt intake can be used to characterize populations, but it is clearly not valid for the individual even with the use of detailed food diaries about salty food use and salting behavior [38]. Beyond the technical sources of variation, there is biological variability in salt excretion which is sizable, some being diurnal and others related to temperature and sweating, as well as salt intake. The sum of all these sources of variation results in the situation in which intraindividual variability approaches the level of interindividual variability. This particularly obtains for salt intake in Western cultures. This phenomenon assures weak correlations between individually measured variables and disease or risk. The other major factor producing weak correlations in such populations is the homogeneity of diet, i.e. the limited range of salt consumption and the fact that the whole range of intake is set high, well above 5 g of salt a day. It has, therefore, become increasingly apparent that the proper epidemiologic study of diet-blood pressure-disease relationships requires the greatest attention to design and methodology, the reduction of measurement variability, and the study of relationships over the widest range of variables. Such undertakings become progressively more sophisticated, but more fruitful as well.

Langford and Watson [75] were among the early groups to find, due to an improved methodology, significant, if low-order correlations, between urinary constituents and blood pressure. For Black females these correlations were on the order of 0.37 for sodium/potassium ratio and diastolic pressure, based, in turn, on 8 days pressure values and 6 days urine measures!

Similarly, *Kesteloot* et al. [63], with the earlier work by *Joossens* et al. [55], made another advance in methodology with the training of subjects in repeat home blood pressure measurement, repeat urine-sodium determinations, multiple-variable prediction, and study of the relationships over a wide range of intake and blood pressures in Koreans and Belgians. Under these improved conditions, from 10 to 45% of home blood pressure differences were explained by urinary electrolytes, heart rate, age, and weight. The relationship to arterial pressure was as great for potassium as for sodium in the urine.

Kesteloot et al. [64] have recounted the following factors which lead to the absence of significant correlations between sodium and potassium intake and blood pressure *within* populations: poor reliability of measures, large within-individual variability in sodium intake, small between-individual variation in sodium intake, counter-effect of potassium intake, bias from health campaigns (conceivably decreased intake in known hypertensives and increased intake in those hypertension-free).

Age Trends

Epstein and Eckhoff [26] have proposed useful models for different levels and slopes of population age trends in systolic blood pressure. For example, nomadic herdsmen in Iran have levels and age slopes for blood pressure comparable to industrialized populations, despite their leanness and absence of an increase in body weight with age [109]. Generally, however, much of the trend upward in blood pressure can be accounted for by an increase in relative weight [12].

Experimental Change in Sodium Intake

The rich literature on the independent experimental effects of sodium and potassium intake and weight change is reviewed in this symposium proceedings and elsewhere [68, 92, 95, 132]. One unique community-based intervention in California suggests that populations exposed to a 2-year campaign on sodium reduction achieved a downward shift in the means and distributions of blood pressures in reference to comparison towns [28].

Overweight-Obesity-Blood Pressure Relationships in Populations

Cross-cultural comparisons of measures of relative weight and population blood pressure levels are characterized by low order relationships. The Seven Countries and other population comparisons suggest that relative weight contributes little to population differences in blood pressure [65, 66]. The lean Japanese have among the highest rates of adult hypertension. However, in other populations where there is little upward age trend in relative weight, there is often no significant age trend in blood pressure [12].

The weight-blood pressure relationship is stronger, however, in migrant population studies. Much of the increase in average blood pressures in those migrating from unacculturated traditional conditions to affluent or Western communities can be accounted for by an increase in relative weight, among both children and adults [1, 30, 114, 115, 144]. However, only rarely have the independent weight effects been examined while simultaneously accounting for calorie excess, salt, potassium, and protein intake or the psychosocial stresses of acculturation [115].

The relationship of blood pressure and body mass is greater still in cross-sectional studies *within* populations, with correlations between relative weight or skinfold thickness and blood pressure on the order of 0.2 to 0.3 [12, 57, 65, 129]. These are highly significant correlations for large populations, but the proportion of the total variance in blood pressures accounted for by relative weight within these populations is only 4–9%. A much greater proportion of the variance in children's blood pressure is accounted for by body mass [113, 140]. Obese children have higher average pressures and more 'hypertension' than the non-obese [14]. Even the blood pressure of children within the highest quintile of body mass is greater with increasing thickness of the triceps skinfold [113].

Stronger yet is the relationship of weight and weight gain, studied longitudinally in cohorts, to future risk of hypertension [46, 51, 59, 105, 127]. Figure 4 displays this relationship most effectively and shows an increased relative risk of new hypertension (>95 mm Hg DBP) for Whites and Blacks by class of relative weight and/or weight gain [137]. The weight

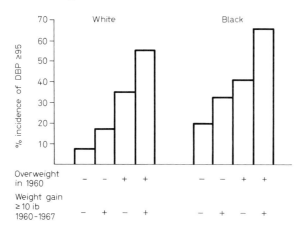

Fig. 4. The 7-year incidence of diastolic blood pressure ≧ 95 in individuals with diastolic blood pressure < 90 in 1960 in relation to baseline weight and weight gain by race, Evans County residents, age 35–59 in 1960. From *Tyroler* et al. [137].

effect is more striking in Whites. This is despite the fact that the average change in blood pressure accounted for by change in weight is not great (△ 20 lb. = △ 4 mm Hg DBP). Attributable risk computations from these Evans County data suggest that the potential for primary prevention of hypertension by population weight control is on the order of 28% in Blacks and 58% in Whites. They suggest that the potential for reduction in incidence of hypertension by weight reduction only among the 12–14% of the population already obese is 17% in Blacks and 28% in Whites. Similarly, observational, in contrast to experimental, weight loss is related to decreasing blood pressure levels among adults followed longitudinally [60, 99].

But experimental weight loss gives the most powerful evidence linking body weight to blood pressure. A vast literature documents the facts, if not the mechanisms, of a sustained fall in blood pressure with weight reduction, in 'normals and hypertensives' [12, 29, 117, 128, 133, 137]. *Dahl* et al. [17] showed in a small but tightly controlled study that the pressure fall with weight reduction was largely attributable to decreased sodium intake during caloric restriction. *Reisen* et al. [117] attempted to maintain a high salt intake in a free-living population undergoing weight loss. Indeed, 24-hour urine sodium excretion was similar between the weight loss and control group on the last visit. But their conclusion of an effect of weight loss on pressure, independent of salt, is uncertain because sodium intake was not measured at the outset, nor was compliance to anti-hypertensive drugs compared. *Gillum* et al. [39] from this laboratory find an independent and distinct additive effect of weight loss to sodium restriction in blood pressure lowering among mild hypertensives. In contrast, the effects of physical conditioning on blood pressure, independent of weight loss, are not yet firmly established [5, 77].

The global effect of caloric excess and the associated absolute sodium intake may have been insufficiently considered and tested. It is on the order of 2 g sodium per 1,000 kcal in the usual North American diet. Further, it is not demonstrated whether sodium intake at 'ideal or desirable' levels should be relative to body mass, of individuals or populations. Finally, the general health consequences of the hypervolemia, cardiac, venous and arterial, of obesity and excess sodium consumption are inadequately understood in their contribution to cardiovascular workload. Clearly again, wider public health implications emerge for cultural obesity and salting.

Drinking Water-Blood Pressure Relationships

The association between drinking water and cardiovascular diseases was found more than 20 years ago by *Kobayashi* [69], who observed in Japan a strong positive association between acidity of water (sulfate/bicarbonate ratio) and mortality from cerebral hemorrhage. Since that time, similar studies in other countries have focused on the hardness and softness of water versus the risk for cardiovascular diseases [43, 53, 85, 93, 97, 122]. Drinking water represents a very small fraction of the total dietary intake of minerals. However, the importance of the water source may be in the lack of chelation and better absorption (bioavailability) than from food [97].

Schroeder and Kramer [123] found no single element of water quality that would explain differing morbidity or death rates from cerebral hemorrhage or hypertensive diseases in a study of 94 US cities and 35 constituents of the water supply. *Punsar* et al. [116], in areas of Finland with greatly contrasting cardiovascular disease mortality, found that no mineral clearly differentiated the drinking water between regions.

Blood pressure levels reported from hard and soft water areas give widely conflicting results. 'Natural waters', hardness or softness have no bearing on sodium content. On the other hand, in artificially softened water with base-exchange treatments in home water softening, calcium and magnesium are replaced by sodium in a ratio of 1:2; thus softening can add substantially to the sodium content of drinking water.

In a US study by *Calabrese and Tuthill* [8], a significant difference in SBP and DBP was found between groups of adolescents consuming home drinking water having differences in sodium concentration of 10–100 mg/l; the higher sodium content was associated with higher levels of blood pressure. Nevertheless, it is very difficult to explain how such a small difference in sodium intake in the water supply, representing less than 10% of the estimated daily sodium intake, could explain such substantial differences in blood pressure level. Observer bias and other sources of error may have existed.

Larger studies of drinking water and mortality in cardiovascular diseases are now under way in the UK and interesting Italian studies are reported in this volume. In the UK, prospective studies are planned for 7,500 middle-aged men in 24 towns to examine long-term relationships of hard and soft waters, the individual water constituents, and their change over time to be related to change in levels of blood pressure. The differences of blood pressure between these communities are now largely accounted for by differences in water hardness, rainfall, temperature, and social factors. The water hardness will also be experimentally manipulated [112].

Alcohol

Clinical, experimental, and population studies suggest a strong association between alcohol use, blood pressure, and hypertension. *Wallace* et al. [141] recently reviewed five population studies as summarized in table I. A consistently higher prevalence of hypertension was found in heavy drinkers versus light users of alcohol with ratios on the order of two. Similarly, table II shows the consistently higher average blood pressure levels in heavy versus light users in eight population studies, with the suggestion of a continuous relationship – a dose response – in six of the eight. In most of these studies, adjustment for age, weight, and body build had little influence on hypertension differences by alcohol use. The association also does not seem to depend on a common hereditary mechanism tending toward alcoholism and hypertension. In a twin study of 30 male pairs discordant for alcohol consumption, the heavy drinkers had significantly higher blood pressures [96].

Dyer et al. [25], in the only systematic population study of incidence in respect to alcohol use, found a greater than double 4-year rate of new diastolic hypertension in heavy versus non-heavy alcohol users (90/1,000 versus 40) in 871 normotensive Western Electric male employees.

Table I. Alcohol use and hypertension prevalence ratios in population studies[1]

Study		Ratio of heavy to light use
Dupont		2.3
Peoples Gas		1.6
W. Electric		1.8
Framingham		2.0
Kaiser[2]	WM	2.4
	WF	1.8
	BF	1.6
	BM	1.5

[1] BP ≥ 160/95 mm Hg.
[2] Ratio of heavy use to *no* use (cited in *Wallace* et al. [141]).

Table II. Alcohol use and blood pressure in population studies. Heavy versus other use: BP difference (mm Hg)

Study	SBP	DBP	Dose response
LA Civil Service	1.9	1.6	No
UK executives	4.3	3.2	Yes
Peoples Gas	8.5	4.7	No
W. Electric	9.7	5.9	Yes
Swede twins	7.1	4.4	Yes
Dane workers[1]	8.1	4.8	Yes
Kaiser[1]	10.9	4.5	Yes
LRC	8.0	–	Yes

[1] Heavy use versus *no* use (cited in *Wallace* et al. [141]).

Clinical studies in a heavy-drinking population of brewery workers [91] found a significantly independent association of alcohol intake and mean blood pressure, with some suggestion of a 'plateau'. Above 180 g of alcohol a week, there was little further increase in blood pressure. Experimental studies of the acute effects of alcohol on blood pressure are difficult to carry out and their results inconsistent. One study found an increase, another a decrease in measured pressure [141]. Experiments in people and animals provide little confirmation of the epidemiologic findings.

Despite the fact that all these studies are severely limited by the absence of an objective, quantitative measure of alcohol intake, the association of alcohol use and hypertension is substantially demonstrated in populations. The consistency and strength of the association suggest that the association may be causal. But, the issue is potentially confounded by the acute effects of alcohol, i.e. the interval since last consumption, as well as by weight, diet composition, salt intake, coffee habits, and social class influences. Nevertheless, there are plausible physiologic mechanisms to explain the alcohol-hypertension relationship found in populations. Alcohol affects blood cortisol levels [87], catecholamines [101], the renin-angiotensin system [138], and vasopression and aldosterone [78].

Some have speculated that the relationship is a function of alcohol withdrawal because (a) the acute alcohol experiments do not suggest a consistent pressure-elevating effect, (b) plasma catecholamines rise substantially between 12 and 24 h after alcohol withdrawal, and (c) because the systolic blood pressure effect is greater than diastolic – all consistent with adrenergic mechanisms [141]. Finally, 'detoxification' results in a return toward normal pressure levels.

The apparent increase in alcohol consumption in affluent countries, its contribution to caloric excess and obesity (and perhaps also to salt intake), and the possibility of a direct effect of regular heavy drinking on blood pressure, all suggest the public health importance of alcohol use to the control and prevention of hypertension.

Blood Pressure Relationships to Other Dietary Items

The Framingham nutrient analysis, based on a single 7-day food record, showed no difference in cross-sectional analyses of mean blood pressure by class of average daily intake of total calories, protein, fat, cholesterol, or alcohol [60]. However, a recent study of 'macrobiotic' vegetarians in a Boston commune suggests that a habitually high intake of animal protein may be related to elevated blood pressure, independent of weight and salt intake [118].

Essential Fatty Acids

Iacono et al. [47] have developed an hypothesis that dietary fatty acids influence arterial pressure through prostaglandin effects on renal handling of water and sodium. The mechanisms and connections are not well-established in the chain between dietary linoleic acid and the arachidonic acid cycle to glycerolphospholipids back to arachidonic acid, then to prostaglandins and a blood pressure effect. However, in several cross-over feeding experiments in man, varying the P/S fatty acid ratio among mild hypertensives (while controlling fat, calories, weight, and sodium intake) was associated with a consistent fall in diastolic pressure. Nevertheless, several aspects of measurement technique and necessary blinding of the observations are not described. This idea developed out of observations on blood pressure lowering effects of infusions of arachidonic acid and of prostaglandins in animals. It is clearly an idea requiring more research. In other experiments, blood pressure rose among rabbits with increases in dietary lipids, regardless of the type of fat [7].

Protein

Our Japanese colleagues, *Kimura* et al. [67], have long suspected the importance for hypertension in Japan of other aspects than salt in the traditional diet. They have explored the nutrients and serum proteins in population-based samples of men aged 40–59, farmers and fishermen of Kyushu, and followed them 15 years. They find significant associations within these populations of individual serum A/G protein ratio, and alcohol intake, with blood pressure. The data also suggest the predictive import of low serum proteins, i.e. below 3.6 g/dl albumen, for 15-year stroke incidence. The findings merit more study in themselves, and also because of the analogous picture of salt, traditional diets, and hypertension in many underdeveloped rural areas worldwide. Our own evaluation, discussed with these colleagues, is that all elements are not yet present to support strongly the thesis of 'low' protein, traditional diets leading to hypertension independently of salt. In the first place, these Southern Island Japanese are not 'very hypertensive', 8% with diastolic pressures greater than 95 mm Hg in farmers versus 13.5% in fishermen; the lower rate was in the population with the *lower* protein diets and lower serum albumen. In the second place, the serum values are not actually very low, 6.5 g/dl of total protein in farmers versus 7.7 in fishermen. Moreover, the diet protein is 'adequate' both in quality (60% from fish) and quantity (12 versus 17% of daily calories). Finally, sodium-potassium intakes were not measured in these comparisons.

Host-Environmental Factors and Interactions

The genetic component of hypertension is more appropriately considered elsewhere. It is only mentioned here to provide background to a preventive and public health approach to hypertension. The genetic defects related to individual susceptibility are not yet demonstrated in humans. *Tobian* [132] suggests a renal defect in ability to excrete sodium, requiring a higher perfusion pressure, based on his experiments in rats. A series of studies on possible genetic differences in the sodium, lithium, and potassium countertransport were initiated by the work of *Canessa* et al. [9].

However, *Schull* et al.'s [124] family aggregation studies suggest that hypertension aggregates only weakly in family sets, and that prior studies within populations may have overestimated a genetic contribution to blood pressure, possibly by confounding age and time changes. Their findings are not consonant with any simple genetic model distinguishing normotensives and hypertensives, and they conclude environmental origin.

The low heritability of hypertension is confirmed in twin studies where about 20% of variation in hypertension is accounted for. This is lower than the figure for height or hair color, and does not support the idea that inheritance is a primary factor in population rates of hypertension 'or even a particularly strong one' [111]. Neither can genetic findings account for the large differences in frequency of adult hypertension between populations, or the marked and rapid changes of blood pressure in migrants. Change in population rates of hypertension occurs in less than 20 years, whereas the rates of genetic change in human populations, even with strong selection pressures, might require 1,000 years or more to achieve a 5-fold increase in population gene frequency [11]. Modern sociocultural change also is thought to override any possible shift in the expression of susceptibility to hypertension brought out by the improving public health. Even the apparent susceptibility of Blacks to hypertension may disappear when they are matched with Whites in comparisons according to socioeconomic circumstances. They may then have similar mean blood pressures [27]. However, this was not confirmed in the larger experience of the Hypertension Detection and Follow-up Study: Black-White differences remained after adjustment for social status [45].

'Thus, the polygenic or multifactorial model for blood pressure seems to be the result of the interaction between strong environmental factors and the weak genetic component' [15].

Findings in Youth Populations
Essential to the broadest understanding of the causes, as well as the eventual prevention of elevated blood pressure, are the observations in

youthful populations. They include the following [3, 13, 14, 34, 113, 142]: Children and youth can be ranked over a considerable blood pressure range. Blood pressure in youth is significantly related to weight and body size and their change over time. Blood pressure in youth is correlated with sodium/potassium excretion ratios in a few careful studies. Blood pressure in very young children is not different between populations or races. Blood pressure values tend to 'track' through adolescence – having before-after correlations on the order of 0.7–0.8. Since body size, weight, and eating habits, which contribute so directly to obesity and salt intake in adults, are importantly determined by cultural factors during youth, the implications are clear for 'primal' prevention, that is, prevention of elevated blood pressure in the first place.

Probabilities: Strengths of the Evidence

The salient points of the diet-blood pressure relationship lie in these observations and conclusions: (1) Population rates of adult hypertension, mean population blood pressures, and pressure trends with age, all correlate positively and significantly with population values of habitual salt intake. In cultures consuming less than 3 g of salt daily there are few adult hypertensives and no blood pressure trends upward with age. (2) Individual blood pressure values within populations correlate significantly with personal salt and potassium intakes, and with overweight and weight gain, where these relationships are examined across an appropriate range of the variables and where care is taken to reduce major sources of variability. (3) Experimental salt restriction and weight reduction independently reduce blood pressure in normals and hypertensives, youth and adults, where care is taken to reduce sources of bias and variability. (4) Different genetic composition of populations is unlikely to be a large factor in the sizable population differences in adult hypertension rates, in the blood pressure changes among acculturated migrant groups, and in the significant change in hypertensive death rates in recent decades in many societies, which antedated medical control of hypertension. (5) The importance of individual differences in susceptibility is probably exaggerated in family studies within cultures in which the environmental factors are ubiquitous and powerful.

The observational and experimental data suggest the potentially important influence in prevention of hypertension in the first place of multiple strategies achieving a diminution of salt use and frequency of overweight.

Discrepancies: Limitations in the Evidence

The discrepant, uncertain, or partially resolved issues about the relationship of diet and hypertension include these findings and issues: Population differences of overweight and salt intake are insufficient to explain population differences in adult hypertension or mean blood pressure. Departures from *Dahl's* [16] tentative 'regression line' (fig. 1) are, however, largely anecdotal and poorly documented. A major discrepancy in the population relationship of salt and hypertension may be the Saharan Morocco nomads, living in regions of little water of high salinity. *Paque and Veylon* [110] suggest that selection pressure there contrasted with that in most parts of the world where there was little salt. His observations need confirmation by systematic field studies throughout desert Africa, perhaps with automated blood pressure measurement. Black-White differences are found in hypertensive disease in Evans County, Georgia, residents consuming the same (relatively high) amount of salt, but different amounts of potassium [86].

These population discrepancies in salt, obesity, hypertension relationships may be in part real, in part an absence of relevant data, and in part a matter of the large variability of measurement. Within populations, the correlations between individual levels of body weight, salt intake, and blood pressure are relatively weak. However, this can be explained by intrinsic and measurement variability and the homogeneously high salt intake of the Western diets measured. The correlations consistently improve when care is taken to reduce variability. However, major questions remain about extrapolating the epidemiological and anthropological evidence to preventive strategy in affluent populations. The findings of hypertension prevalence are clear only for the extremes of salt intake, both between and within populations. For intermediate values of intake, the data provide little support for much effect of a modest reduction in sodium intake now considered feasible for affluent cultures. However, it still may be that a relatively small shift downward in the population mean and distribution of sodium intake could have more profound effects than estimated. Such a shift lowers the 'lower limit' and involves more people in truly low sodium intake patterns. It also lowers the upper end of the distribution, thus potentially reducing the population frequency of salt-sensitive hypertension.

Experiments control inadequately for confounding and interaction among the major known risk characteristics for hypertension. However, experimental design is improving. Human experiments with salt-loading are not overwhelmingly impressive in the extent of blood pressure effect achieved compared to the high salt dose required. However, as they expand to larger size and better design, the results appear compatible with a salt effect related to susceptibility defined by family history of hypertension. So far, the effects of hygienic interventions are relatively small in the reduction of moderate hypertension. However, compliance has not been good and behavioral strategies have rarely been applied. The crucial experiment about the primary preventability of elevated blood pressure has only begun to be made in putatively susceptible populations. The prevention of hypertension risk factors, and raised blood pressure in the first place, begs a proper test. Integrated approaches to the primary prevention of hypertension in the whole community are insufficiently explored. They probably require multiple, concurrent educational strategies, professional and public. Real concern about any substantial risk of reducing salt intake at the population level is expressed by few knowledgeable investigators [131]. This is principally manifest in concern for inadequate individual responses to a rapid depletion of sodium and water such as in hot climes, or with diarrhea, vomiting, or diuretic use.

The absence of well-established mechanisms by which diet, nutrients, and minerals including sodium, influence neuroendocrine control of renal function and blood pressure is the greatest deficiency in the otherwise increasingly congruent evidence. Nor are they now plausible enough to provide strong confirmation to the epidemiological and clinical evidence. Many newer ideas about mechanisms, some of which are touched on in this volume (renin-prostaglandin, insulin receptors, Na+/K+ cellular cotransport, central nervous system catecholamines, and volume – third factor – ADP-ase relationships) remain to be tied to excess dietary sodium or Na+/K+ intake, to arterial wall sodium content and contractility, and to renal sodium handling. The approach to hypertension prevention remains, therefore, pragmatic and simplistic, thus, intellectually unsatisfying.

Preventive Practice and the Public Health

Introduction
We consider that the potential for primary prevention and improved control of hypertension is real, based on shifts occurring in habitual diet

and activity of populations which affect body weight, sodium, potassium, and alcohol intake and their effects, in turn, on heart rate, blood volume, and vascular resistance. These changes are demonstrated to be feasible and safe. However, salting behaviors, as well as heavy 'non-discretionary' exposure to sodium, are highly ingrained in many cultures. Plans, policies, and educational programs are required for evolutionary change in the culture. But the educational strategies needed to bring about change are not yet sufficiently developed or tested on a community scale. The question may reasonably be asked whether changes in health behavior, of the order which might be anticipated by a successful community strategy, would be sufficient to reduce population levels of blood pressure or its risk characteristics. Would the changes be sufficient to modify the prevalence or the new occurrence of adult hypertension and its complications? We do not know. However, we suggest that the hygienic preventive measures discussed here are appropriate, promising, and safe alternatives. They are available now for preventive practice and public health program. They meet criteria for action, i.e. congruent evidence from all medical disciplines, a real potential for benefit, feasibility, and safety, in the presence of a major and costly burden: hypertension.

Moreover, the long-term prospect of a solely medical approach to the mass detection and drug treatment of hypertension is not an optimistic one, even if it were possible to identify most cases, and those at future risk, and to target effective efforts to them. But, for now, there is little evidence to suggest that this is feasible. Rather, there is a huge number of poorly distinguished genotypes susceptible to hypertension in the population. If environmental-behavioral factors are important to the induction and maintenance of elevated blood pressure in these susceptible persons, as we believe, then a broader approach is needed than hypertension detection and control. If there are so many people who need to change behavior and exposures which lead eventually to hypertension, if these people are not supported in their needed change due to an unfavorable environment, then efforts need to be applied by the health care system not only to the individual but to the culture and environment itself. Such efforts involve education, persuasion and, where appropriate and agreed upon democratically, legislation.

What are the component steps to an improved preventive practice and public health action in hypertension? They include professional education toward improved assessment of risk of high blood pressure with systematic diagnosis and advice for individual patients. A systematic approach is also needed to the diagnosis and treatment of personal behaviors, and to effective self-help procedures. Finally, efforts in the larger system are required to support the needed individual change in the surroundings of family, worksite, school, and community.

Eating Behavior Change

The ability to achieve and maintain weight loss among the obese is insufficiently studied, as are large-scale efforts to reduce salt intake. It is likely that weight reduction and its maintenance among adults through calorie restriction will be more successful if accompanied by increased physical activity [74]. One of the problems is that little is known about the basic mechanisms which control appetite [104]. But the need for weight and appetite control surely starts early in life. The avoidance of obesity and of salt preference in infancy and youth, as well as throughout life, is a primary goal of prevention.

Neumann [98] has set out these guidelines to primary prevention strategies for mothers and young children: (a) encourage breast-feeding for at least 3 months; (b) add no solids prior to 4 months; (c) never force a child to eat; (d) plot carefully the weight gain in infancy and childhood and observe whether it exceeds the average; (e) encourage free movement and physical activity by the infant; (f) limit weight gain during pregnancy to 25–30 pounds; (g) control obesity in young women.

Professional Attitude Change

The effectiveness of detection and control of hypertension by the medical profession is increasing, and favorable professional attitudes toward screening, treatment, and follow-up are in rapid evolution. With respect to salt intake, the conservative view of the leading American medical nutritionist is compared with that of a leading contemporary clinical researcher in hypertension:

'General, moderate reduction in dietary salt intake cannot predictably be anticipated to decrease the development of hypertension. The role of the diet in the cause or prevention of hypertension has not yet been defined' [19].

'Although the mechanism of essential hypertension is still obscure, the evidence is very good if not conclusive that reduction of salt in the diet to below 2 g a day would result in the prevention of essential hypertension and its disappearance as a major public health problem' [36].

Both statements are accurate. But one distinguished expert has an academic, conservative, laissez-faire outlook; the other a broad, anthropological, positive view of the public health.

Primary Prevention Trials in Hypertension

A group of North American investigators has initiated a multi-center trial of the primary prevention of hypertension in healthy individuals, about half of which would have a family history of hypertension and half would not. They would be assigned randomly to control and to a multiple strategy education program involving modifications of body weight, physical activity, and salt and potassium intake in a factorial design [100]. Other important studies are under way in youth and adults on the combined effects of dose reduction of medications, plus hygienic approaches to reduce weight and salt intake [128].

A 20-year community comparison in Japan has been directed toward reduction in miso soup consumption in the northern villages. In the sizable sample of the Nishime village, sodium consumption and the Na/K ratio in urine were reduced, associated with significant mean blood pressure lowering. However, details of the sample selection and response rates are insufficient [120].

Several well-designed, quasi-experimental, community research and demonstration programs, funded as competitive NIH grants, include in their programs a strong component of hypertension education, detection, control, and primary prevention. Hygienic changes are attempted community-wide on weight, physical activity, and salt intake. Multiple educational and motivational strategies are used: screening and direct education, mass media, school programs, and community organization; all are focused on health behavior to reduce the risk characteristics for hypertension and atherosclerosis [6].

National Education Programs

The National High Blood Pressure Education Program, initiated by Secretary Richardson, and in effect under the NHLBI since 1973, appears to have had a significant effect on identification of hypertensives and their follow-up and control in the United States. This effort has involved the participation of the drug industry and voluntary agencies with active professional and public education programs.

We believe that the major current challenge of hypertension research and prevention is to go beyond drug control to the earlier identification and the primary prevention of elevated blood pressure. Many official bodies, health agencies, and consumer groups have proposed strategies having the following elements:

Professional and public education about the link between salt, overweight, alcohol intake, and hypertension; about the relatively low salt and high potassium content of natural foods, particularly plant foods; about the extremely high sodium content of many processed foods; about salt added in food preparation and at the table; and about the role of exercise in the regulation of appetite and control of body weight.

Legislation

Legislation has been proposed for food labelling in precise terms (mg and mEq of sodium) and in terms readily comprehensible to all consumers such as 'low, medium, high and very high salt', and use of warnings of high content such as a salt shaker emblem. In addition, consumer agencies have proposed to limit the salt content in processed foods and to declare salt an additive rather than a GRAS substance [48]. The latter proposal is based on the observation that the labelling of sugar content in breakfast cereals has not significantly influenced voluntary change by industry, nor significantly affected buying decisions of the public. Thus, it is considered unlikely that salt labelling itself would influence the majority shopping habits or food choices, increasingly dependent on convenience, processed, or restaurant foods. Because of the 'non-discretionary' sodium content of so many commercial products, such foods being a major and increasing source for many consumers, a public health policy is recommended based on ethical and democratic pressures to reduce the sodium unnecessarily added to foods in processing.

The Select Committee on GRAS Substances, of the Federation of American Societies for Experimental Biology, under contract to the FDA, no longer considers salt a GRAS substance. It also accepts the potential that reduction of sodium chloride consumption by the population may reduce the frequency of hypertension. It proposes that salt should be treated like any other food additive, by prescribing the conditions under which it may be safely used [125]. This principle has also been endorsed by the Committee on Nutrition of the American Academy of Pediatrics.

Under the definition of the Food and Drug Act, substances which are not classified GRAS are automatically considered food additives. Similarly, if there is a 'genuine difference of expert opinion' on whether a substance is GRAS, the FDA Commissioner must, nonetheless, conclude that the substance is *not* GRAS. Although some individuals may ingest large quantities of sodium without serious effects, excess sodium has adverse consequences to the public health at large. Labelling regulation may be justified on this basis [48].

Consumer agencies also propose restrictions on salt as an additive because of the concept that preloading food with high levels of salt deprives people of the opportunity for choice. The choice is now being made by the manufacturer; they propose that it be returned to the individual [50]. These agencies further suggest that a 50% reduction of salt in products in which it is not possible for the consumer to add salt would probably not be noticed by most people. In contrast, for the large number of prepared foods to which salt may readily be added, the commercially available content should be reduced more drastically so that salt could be added to taste by the individual. The latter foods include soups, vegetables, gravies, stews, and casseroles.

Industry Action

The American food industry tends to defend its current salt-adding customs. In the process, it takes vigorous stands at public hearings on diet and health, maintaining that the American people will not tolerate low salt or unsalted foods and that it is irrational to think of any alternative. They insist that the public health recommendations on salt intake, and regulatory restrictions to adding salt in processed foods, are neither needed nor appropriate for the whole population. This, in our view, reflects a significant lack of comprehension of the excess salting issue and mass susceptibility to hypertension. It shows the serious lag between a demonstrated public health concern and response by industry. There are notable exceptions to these views among producers of baby foods and a few others who have already acted. It is anticipated that these producers will reap the eventual benefits, both economically and in good will.

These are some of the recommendations which have been made to industry by consumer groups and their consultants [50]: Use less salt and sodium containing additives. Use herbs and spices, vinegar and lemon juice in place of some of the salt. Use potassium chloride as a replacement for half of the sodium and most or all of the sodium nitrate. Add more real food to products.

Technology

There is a school, comparable perhaps to the 'safer-cigarette' school, that sincerely considers the most practical approach to the public health issue is 'to develop a safe and taste satisfying salt substitute that will make a truly "no-salt" diet acceptable...' [37].

Government Action

Recommendations are made for government action to include specifications for sodium content in government feeding programs in its office buildings, military and other institutional messes, elderly and child nutritional programs, all to be consistent with the 1990 National Goals of the US Department of Health and Human Services.

Voluntary Agencies

Official public recommendations have been made by the following agencies for reduction in personal salting habit, for food labelling, and for the restriction of salt as a food additive by industry:

White House Conference on Food, Nutrition and Health, 1969
American Academy of Pediatrics Committee on Nutrition
    (Pediatrics, Springfield 53, 1974)
Senate Select Committee on Nutrition and Human Needs, 1974
Grocery Manufacturer's of America Task Force with the National
    Nutritional Policy Conference, 1974
The Select Committee on GRAS Substances, 1979
The Surgeon General of the United States: Healthy People, 1979
The American Heart Association, 1978 (now under revision)
The Food and Nutritional Board of the National Academy of Sciences:
    Recommended Dietary Allowances, 1980
US Departments of Agriculture and Health and Human Services:
    Dietary Guidelines for Americans, 1980
US Department of Health and Human Services:
    Objectives for the Nation, 1990

Public Health Recommendations

In our view, an 'ideal' population intake of salt would be a median value around 3 g daily with a range of 2–5 g. This distribution is compatible with normal growth and development, with physiological function, and with the absence of adult hypertension. A Na/K ratio of $1/2$ or smaller is proposed for its possible enhanced effect on the ideal sodium intake. This ratio would occur automatically from the general dietary recommendations.

A 'desirable' median population salt intake would be 5 g a day, with a range of 2–8 g, and this is thought to be compatible with a very significant reduction in the burden of hypertensive disease in the community.

A 'feasible' value for the median population intake of salt within the next decade or so is considered to be 8 g a day with a range of 2–10 g. The potential effect of this population change is likely less than some other speculations. But this 'feasible' goal is thought likely to be a worthwhile and supportive development if combined with specific education targetted for those at high risk of hypertension.

These population values would be achieved by a greatly diminished use of sodium in commercially prepared meats and snacks, public education, and labelling of foods toward decreased use of salty and processed foods, less salting in food preparation and at table, the use of salt substitutes, and the increased use of fresh and natural foods, particularly those of plant origin.

These recommendations include the control and prevention of overweight and of excess alcohol intake, and the protection of water supplies from increasing salination. All these changes require enhanced facilities for preventive care of adults, direct educational programs for youth, and a community-wide strategy to support individual change in eating and activity patterns.

References

1 Beaglehole, R.: Social factors and blood pressure in children; in Lauer, Shekelle, Childhood prevention of atherosclerosis and hypertension, pp. 313–321 (Raven Press, New York 1980).
2 Beevers, D. G.; Hawthorne, V. M.; Padfield, P. L.: Salt and blood pressure in Scotland. Br. med. J. 281: 641–642 (1980).
3 Berenson, G. S.; Voors, A. W.; Webber, L. S.: Importance of blood pressure in children: distribution and measurable determinants; in Kesteloot, Joossens, Epidemiology of arterial blood pressure (Nijhoff, The Hague 1980).
4 Berglund, G.; Wickstand, J.; Wallentin, I.; Wilhelmsen, L.: Sodium excretion and sympathetic activity in relation to severity of hypertensive disease. Lancet i: 324–328 (1976).
5 Blackburn, H.: Non-pharmacologic treatment of hypertension. Ann. N. Y. Acad. Sci. 304: 236–242 (1978).
6 Blackburn, H.; Carleton, R.; Farquhar, J.; Stunkard, A.: Community models of cardiovascular disease prevention. International Communications Ass., Minneapolis 1981.

7 Burstyn, P.G.; Firth, W.R.: Effects of three fat-enriched diets on the arterial blood pressure of rabbits. Cardiovasc. Res. 9: 807–810 (1975).
8 Calabrese, E.J.; Tuthill, R.W.: Elevated blood pressure levels and community drinking water characteristics. J. Environ. Sci. Hlth A13: 781–802 (1978).
9 Canessa, M.; Adragna, N.; Solomon, H.S.; Connolly, T.M.; Tosteson, D.C.: Increased sodium-lithium countertransport in red cells of patients with essential hypertension. New Engl. J. Med. 302: 772–776 (1980).
10 Cassel, J.: Studies of hypertension in migrants; in Paul, Epidemiology and control of hypertension, pp. 41–58 (Symposia Specialists, Miami 1975).
11 Cavalli-Sforza, L.; Bodmer, W.: Genetics of human populations. (Freeman, San Francisco 1971).
12 Chiang, B.M.; Perlman, L.V.; Epstein, F.H.: Overweight and hypertension. A review. Circulation 39: 403–421 (1969).
13 Cooper, R.; Soltero, I.; Liu, K.; Berkson, D.; Levinson, R.; Stamler, J.: The association between urinary sodium excretion and blood pressure in children. Circulation 62: 97–104 (1980).
14 Court, J.; Hill, G.J.; Dunlop, M.; Boulton, T.J.C.: Hypertension in childhood obesity. Aust. Paediat. J. 10: 296–300 (1974).
15 Cruz-Coke, R.: Genetics of arterial hypertension in man; in Gross, Robertson, Arterial hypertension. WHO Expert Committee Report (Pitman, Bath 1979).
16 Dahl, L.K.: Possible role of salt intake in the development of essential hypertension; in Reubi, Bock, Cottier, Essential hypertension, pp. 53–65 (Springer, Berlin 1960).
17 Dahl, L.; Silver, L.; Christie, R.: Role of salt in the fall of blood pressure accompanying reduction of obesity. New Engl. J. Med. 258: 1186–1192 (1958).
18 Darby, W.J.: Foreword; in Kare, Fregly, Bernard, Biological and behavioral aspects of salt intake (Academic Press, New York 1980).
19 Darby, W.J.: Why salt? How much? Contemp. Nutr. 5: 1–2 (1980).
20 Dawber, T.R.; Kannel, W.B.; Kagan, A.; Donabedian, R.K.; McNamara, P.M.; Pearson, G.: Environmental factors in hypertension; in Stamler, Stamler, Pullman, The epidemiology of hypertension, pp. 255–288 (Grune & Stratton, New York 1967).
21 Denton, D.A.: Salt appetite; in Handbook of physiology, vol. 1, sect. 6, pp. 433–459 (Williams & Williams, Baltimore 1967).
22 Denton, D.: Instincts, appetites and medicine. Aust. N.Z.J. Med. 2: 203–212 (1972).
23 Denton, D.A.: Sodium and hypertension; in Sambhi, Mechanisms of hypertension, pp. 46–54 (Excerpta Medica, Amsterdam 1973).
24 Dreyfuss, W.; Hamosh, B.; Adam, Y.; Kallner, B.: Coronary heart disease and hypertension among Jews emigrated to Israel from the Atlas Mountains Region in North Africa. Am. Heart J. 62: 470–477 (1961).
25 Dyer, A.; Stamler, J.; Paul, O.; Berkson, D.M.; Lepper, M.H.; McKean, H.; Shekelle, R.B.; Lindberg, H.A.; Garside, D.: Alcohol consumption, cardiovascular disease risk factors, and mortality in two Chicago epidemiologic studies. Circulation 56: 1067–1074 (1977).
26 Epstein, F.H.; Eckhoff, R.D.: The epidemiology of high blood pressure. Geographic distributions and etiological factors; in Stamler, Stamler, Pullman, The epidemiology of hypertension, pp. 155–166 (Grune & Stratton, New York 1967).

27 Eyer, J.: Hypertension as a disease of modern society. Int. J. Hlth Serv. 5: 539–558 (1975).
28 Farquhar, J.W.; Wood, P.D.; Haskell, W.L.: The relationship of urinary sodium/potassium ratios to systolic blood pressure change: Stanford Three Community Study. CVD Epidemiology Newsletter (Am. Heart Ass., New York, 1978).
29 Fletcher, A.P.: The effect of weight reduction on the blood pressure of obese hypertensive women. Q. J. Med. 23: 331–345 (1954).
30 Florey, C. du V.; Cuadrado, R.R.: Blood pressure in native Cape Verdeans and in Cape Verdean immigrants and their descendants living in New England. Hum. Biol. 40: 189–211 (1968).
31 Fodor, J.G.; Abbotts, E.C.; Rustad, J.E.: An epidemiological study of hypertension in Newfoundland. Can. med. Ass. J. 108: 1365–1368 (1973).
32 Fomon, S.J.: What are infants fed in the United States? Pediatrics, Springfield 56: 350–354 (1975).
33 Forsythe, R.H.; Miller, R.A.: Salt in processed foods; in Kare, Fregly, Bernard, Biological and behavioral aspects of salt intake, pp. 221–228 (Academic Press, New York 1980).
34 Frank, G.C.; Berensen, G.S.; Webber, L.S.: Dietary studies and the relationship of diet to cardiovascular disease risk factor variables in ten-year-old children – Bogalusa Heart Study. Am. J. clin. Nutr. 31: 328–340 (1978).
35 Fregly, M.S.: Salt and social behavior; in Kare, Fregly, Bernard, Biological and behavioral aspects of salt intake, pp. 3–11 (Academic Press, New York 1980).
36 Freis, E.D.: Salt, volume and the prevention of hypertension. Circulation 53: 589–595 (1976).
37 Freis, E.D.: Salt and hypertension; in Yamori, Lovenberg, Freis, Prophylactic approach to hypertensive diseases, pp. 539–543 (Raven Press, New York 1979).
38 Gillum, R.F.; Elmer, P.J.; Prineas, R.J.: Changing sodium intake in children: The Minneapolis Children's Blood Pressure Study. Hypertension 3: 698–703 (1981).
39 Gillum, R.F.; Prineas, R.J.; Jacobs, D.R.; Jeffery, R.W.; Elmer, P.J.; Gomez, O.; Blackburn, H.: Non-pharmacologic therapy of hypertension: The effects of weight reduction and sodium restriction in overweight borderline hypertensive patients. Am. Heart J. (in press 1982).
40 Gillum, R.F.; Prineas, R.J.; Luepker, R.V.; Jacobs, D.R.; Taylor, H.L.; Blackburn, H.: The decline of coronary deaths: A search for explanations. Minn. Med. 65: 235–238 (1982).
41 Gleiberman, L.: Blood pressure and dietary salt in human populations. Ecol. Food Nutr. 2: 143–156 (1973).
42 Grim, C.E.; Luft, F.C.; Miller, J.Z.; Meneely, G.R.; Battarbee, H.D.; Hames, C.G.; Dahl, L.K.: Racial differences in blood pressure in Evans County, Georgia: relationships to sodium and potassium intake and plasma renin activity. J. chron. Dis. 33: 87–94 (1980).
43 Heyden, S.: The hard facts behind the hard-water theory and ischemic heart disease. J. chron. Dis. 29: 149–157 (1976).
44 Holmberg, A.R.: Nomads of the Long Bow. Smithson. Inst. Soc. Anthropol. 10: 35 (1950).
45 Hypertension Detection and Follow-Up Cooperative Group: Race, education and prevalence of hypertension. Am. J. Epidem. 106: 351–361 (1977).

46 Hsu, P.; Matthewson, F.A.L.; Rabkin, S.W.: Blood pressure and body mass index patterns - a longitudinal study. J. chron. Dis. *309:* 93–113 (1977).
47 Iacono, J.M.; Judd, J.T.; Marshall, M.W.; Canary, J.J.; Dougherty, R.M.; Mackin, J.F.; Weinland, B.T.: The role of dietary essential fatty acids and prostaglandins in reducing blood pressure. Prog. Lipid Res. *20:* 349–364 (1981).
48 Institute for Public Interest and Center for Science in the Public Interest. Petitions to the FDA for: A rule to label sodium content of foods. A rule to regulate sodium content of processed foods. Washington (1978).
49 Jacobs, D.R.; Anderson, J.T.; Blackburn, H.: Diet and serum cholesterol: Do zero correlations negate the relationship? J. Epidemiol. *110:* 77–87 (1979).
50 Jacobson, M.: Shaking out the truth about salt. Nutr. Act. *5:* 3–7 (1978).
51 Johnson, B.C.; Karunas, T.M.; Epstein, F.H.: Longitudinal changes in blood pressure in individuals, families and social groups. Clin. Sci. mol. Med. *45:* suppl. 1, pp. 35–45 (1973).
52 Johnson, B.C.; Remington, R.D.: A sampling study of blood pressure in White and Negro residents of Nassau, Bahamas. J. chron. Dis. *13:* 39–51 (1961).
53 Joossens, J.V.: Salt and hypertension, water hardness and cardiovascular death rates. Triangle *13:* 9–16 (1973).
54 Joossens, J.V.: Stroke, stomach cancer and salt; in Kesteloot, Joossens, Epidemiology of arterial blood pressure (Nijhoff, The Hague 1980).
55 Joossens, J.V.; Willems, J.; Claessens, J.; Claes, J.; Lissens, W.: Sodium and hypertension; in Fidanza, Keys, Ricci, Somogyi, Nutrition and cardiovascular diseases, pp. 91–110 (Morgagani Edizioni Scientifiche, Rome 1971).
56 Kagan, A.; Marmot, M.G.; Kato, H.: The Ni-Hon-San study of cardiovascular disease epidemiology; in Kesteloot, Joossens, Epidemiology of arterial blood pressure (Nijhoff, The Hague 1980).
57 Kahn, H.A.; Medalie, M.H.; Neufeld, H.N.; Riss, E.; Goldbourt, V.: The incidence of hypertension and associated factors. The Israel Ischemic Heart Disease Study. Am. Heart J. *84:* 171–182 (1972).
58 Kaminer, B.; Lutz, W.: Blood pressure in Bushmen of the Kalahari Desert. Circulation *22:* 289–295 (1960).
59 Kannel, W.B.; Brand, N.; Skinner, J.J.; Dawber, T.R.; McNamara, P.M.: The relation of adiposity to blood pressure and development of hypertension. The Framingham Study. Ann. intern. Med. *67:* 48–49 (1967).
60 Kannel W.B.; Sorlie, P.: Hypertension in Framingham; in Paul, Epidemiology and control of hypertension, pp. 553–592 (Symposia Specialists, Miami 1975).
61 Karvonen, M.; Punsar, S.: Sodium excretion and blood pressure of West and East Finns. Acta med. scand. *202:* 501–507 (1977).
62 Kaunitz, H.: Causes and consequences of salt consumption. Nature, Lond. *178:* 1141–1144 (1956).
63 Kesteloot, H.; Park, B.C.; Lee, C.S.; Brems-Heyns, E.; Joossens, J.V.: A comparative study of blood pressure and sodium intake in Belgium and Korea; in Kesteloot, Joossens, Epidemiology of arterial blood pressure (Nijhoff, The Hague 1980).
64 Kesteloot, H.; Vuylsteke, M.; Costenoble, A.: Relationship between blood pressure and sodium and potassium intake in a Belgian male population group; in Kesteloot, Joossens, Epidemiology of arterial blood pressure (Nijhoff, The Hague 1980).

65 Keys, A. (ed.): Coronary heart disease in seven countries. Circulation 41/42 suppl. I (1970).
66 Keys, A.: Seven countries: a multivariate analysis of death and coronary heart disease (Harvard University Press, Cambridge 1980).
67 Kimura, N.; Toshima, H.; Nakagama, Y.; Takayama, K.; Tashiro, H.; Takagi, M.: Fifteen-year follow-up population survey on stroke: a multivariate analysis of the risk of stroke in farmers of Tamishimaru and fishermen of Ushibuku; in Yamori, Lovenberg, Freis, Prophylactic approach to hypertensive diseases, pp. 505–510 (Raven Press, New York 1979).
68 Kirkendall, W. M.; Connor, W. E.; Abboud, F.; Rastogi, S. P.; Anderson, T. A.; Frey, M.: The effect of dietary sodium chloride on blood pressure, body fluids, electrolytes, renal function and serum lipids of normotensive man. J. Lab. clin. Med. 87: 418–434 (1976).
69 Kobayashi, K.: Geographical relationships between the chemical nature of river water and death rate from apoplexy. Ber Ohara Inst. Launch. Biol. 11: 12–21 (1957).
70 Komachi, Y.; Iida, M.; Ozawa, H.; Shimamoto, T.; Chikayama, Y.; Takahashi, H.; Konishi, M.; Ueshima, H.: Comparisons of risk factor of CHD and CVA in several groups in Japan with special reference to dietary intake; in Asahina, Shigiya, Physiological adaptability and nutritional status of Japanese (Tokyo Press, Tokyo 1975).
71 Komachi, Y.; Shimamoto, T.: Salt intake and its relationship to blood pressure in Japan; in Kesteloot, Joossens, Epidemiology of arterial blood pressure (Nijhoff, The Hague 1980).
72 Kozarevic, D.; McKee, D.: Epidemiology of essential hypertension in Yugoslavia; in Kesteloot, Joossens, Epidemiology of arterial blood pressure, pp. 207–216 (Nijhoff, The Hague 1980).
73 Kroeber, A. L.: Cultural element distribution. Salt, dogs, tobacco. Anthropol. Rev. 6: 1–20 (1942).
74 Krotiewski, M.; Mandroukas, K.; Sjostrom, L.; Sullivan, L.; Wetterqvist, H.; Bjorntorp, P.: Effects of long-term physical training on body fat, metabolism and blood pressure in obesity. Metabolism 28: 650–658 (1979).
75 Langford, H. G.; Watson, R. L.: Electrolytes and hypertension; in Paul, Epidemiology and control of hypertension, pp. 119–130 (Symposia Specialists, Miami 1975).
76 Leon, A.; Blackburn, H.: Physical inactivity and coronary heart disease; in Kaplan, Stamler, Preventive cardiology (Saunders, Philadelphia 1981).
77 Leon, A.; Gillum, R. F.; Blackburn, H.: The effect of physical conditioning on blood pressure. Findings, concepts and preventive practice; in Freis, E. (ed.), Hypertension (1982).
78 Linkola, J.: Alcohol and hypertension. New Engl. J. Med. 300: 680 (1979).
79 Lipsitt, L. P.; Mustaine, M. G.; Ziegler, B.: Effects of experience on the behavior of the young infant. Neuropädiatrie 8: 107–133 (1977).
80 Liu, K.; Cooper, R.; McKeever, J.; McKeever, P.; Byington, R.; Soltero, I.; Stamler, R.; Gosch, F.; Stevens, E.; Stamler, J.: Assessment of the association between habitual salt intake and high blood pressure. Am. J. Epidem. 110: 219–226 (1979).
81 Macallum, A. B.: The paleopathology of the body fluids and tissues. Physiol. Rev. 6: 316–367 (1926).
82 Maddocks, I.: Blood pressure in Melanesians. Med. J. Aust. i: 1123–1126 (1967).
83 Marmot, M. G.; Kagan, A.; Kato, H.: Hypertension and heart disease in the Nihonsan

Study; in Kesteloot, Jossens, Epidemiology of arterial blood pressure (Nijhoff, The Hague 1980).
84 Marsden, J. L.: Sodium containing additives in processed meats. Am. Med. Ass. Symp., Washington 1978.
85 Masironi, R.; Pisa, Z.; Clayton, D.: Myocardial infarction and water hardness in the WHO myocardial infarction registry network. Bull. WHO *57:* 291–299 (1979).
86 McDonough, J. R.; Garrison, G. E.; Hames, C. G.: Blood pressure and hypertensive disease among negroes and whites. Ann. intern. Med. *61:* 208–228 (1964).
87 Mendelson, J. H.; Ogata, M.; Mello, N. K.: Adrenal function and alcoholism. I. Serum cortisol. Psychosom. Med. *33:* 145 (1971).
88 Meneely, G. R.; Battarbee, H. D.: High sodium-low potassium environment and hypertension. Am. J. Cardiol. *38:* 768–785 (1976).
89 Meneely, G. R.; Dahl, L. K.: Electrolytes in hypertension: the effects of sodium chloride. Med. Clins. N. Am. *45:* 271–283 (1961).
90 Miall, W. E.: Follow-up of arterial pressure in the populations of a Welsh mining valley. Br. med. J. *2:* 1204–1210 (1959).
91 Mitchell, P. I.; Morgan, M. J.; Boadle, D. J.; Batt, J. E.; Marstand, J. L.; McNeil, H. P.; Middleton, C.; Rayner, K.; Lichiss, J. N.: Role of alcohol in the etiology of hypertension. Med. J. Aust. *ii:* 198 (1980).
92 Morgan, T.; Gilles, A.; Morgan, G.; Adam, W.; Wilson, M.; Garvey, S.: Hypertension treated by salt restriction. Lancet *i:* 227–230 (1978).
93 Morris, J. N.; Crawford, M. D.; Heady, J. A.: Hardness of local water supplies and mortality from cardiovascular disease in the country boroughs of England and Wales. Lancet *i:* 860–862 (1961).
94 Multhauf, R. P.: Neptune's gift (Johns Hopkins Press, Baltimore 1978).
95 Murray, R. H.; Luft, F. C.; Bloch, R.; Weyman, A. E.: Blood pressure responses to excesses of sodium intake in normal man. Proc. Soc. exp. Biol. Med. *159:* 432–436 (1978).
96 Myrhed, M.: Blood pressure: alcohol consumption in relation to factors associated with ischemic heart disease. Acta med. scand. suppl. *567:* 40 (1974).
97 National Academy of Sciences. Drinking water and health (National Academy of Sciences, Washington 1977).
98 Neumann, C. G.: Prevention of obesity in infancy and childhood; in Lauer, Shekelle, Childhood prevention of atherosclerosis and hypertension, pp. 373–376 (Raven Press, New York 1980).
99 Oberman, A.; Lane, N. E.; Harlan, W. R.; Graybiel, A.; Mitchell, R. E.: Trends in systolic blood pressure in the thousand aviator cohort over a twenty-four-year period. Circulation *36:* 812–822 (1967).
100 Oberman, A.; Prineas, R.: The primary prevention of hypertension. A proposal to National Institutes of Health (1981).
101 Ogata, M.; Mendelson, J. H.; Mello, N. K.; Majchrowicz, E.: Adrenal function and alcoholism. II. Catecholamines. Psychosom. Med. *33:* 159 (1971).
102 Oliver, W. J.: Sodium homeostasis and low blood pressure populations; in Kesteloot, Joossens, Epidemiology of arterial blood pressure (Nijhoff, The Hague 1980).
103 Oliver, W. J.; Cohen, E. L.; Neel, J. V.: Blood pressure, sodium intake, and sodium-related hormones in the Yanomamo Indians, a 'no-salt' culture. Circulation *52:* 146–151 (1975).

104 Oscai, L.: Recent progress in the possible prevention of obesity; in Lauer, Shekelle, Childhood prevention of atherosclerosis and hypertension, pp. 205–211 (Raven Press, New York 1980).

105 Paffenbarger, R. S.; Thorne, M. C.; Wing, A. L.: Chronic disease in former college students. VIII. Characteristics in youth predisposing to hypertension in later years. Am. J. Epidem. *88:* 25–52 (1965).

106 Page, L. B.: Epidemiological evidence on the etiology of human hypertension and its possible prevention. Am. Heart J. *91:* 527–534 (1976).

107 Page, L. B.: Dietary sodium and blood pressure: evidence from human studies; in Shekelle, Lauer, Childhood prevention of atherosclerosis and hypertension, pp. 291–303 (Raven Press, New York 1980).

108 Page, L. B.; Damon, A.; Moellering, R. C.: Antecedents of cardiovascular disease in six Solomon Island societies. Circulation *29:* 1132–1146 (1974).

109 Page, L. B.; Vandevert, D.; Nader, K.; Lubin, N.; Page, J. R.: Blood pressure of Quash'qui pastoral nomads in Iran in relation to culture, diet, and body form; in Kesteloot, Joossens, Epidemiology of arterial blood pressure, pp. 291–303 (Nijhoff, The Hague 1980).

110 Paque, C.; Veylon, R.: L'homme, l'eau et le sel. Nouv. Presse méd. *3:* 1617–1627 (1974).

111 Pickering, G.: The inheritance of arterial pressure; in Stamler, Stamler, Pullman, The epidemiology of hypertension, pp. 18–27 (Grune & Stratton, New York 1967).

112 Pocock, S. J.; Shaper, A. G.; Cook, D. G.; Packham, R. F.; Lacey, R. F.; Powell, P.; Russel, P. F.: British regional heart study: geographic variations in cardiovascular mortality and the role of water quality. Br. med. J. *280:* 1243–1249 (1980).

113 Prineas, R. J.; Gillum, R. F.; Horibe, H.; Hannan, P.: The Minneapolis children's blood pressure study. 2. Multiple determinants of children's blood pressure. Hypertension *2:* suppl. 1, pp. 124–128 (1980).

114 Prior, I. A. M.; Evans, J. G.; Harvey, H. P. B.; Davidson, F.; Lindsey, M.: Sodium intake and blood pressure in two Polynesian populations. New Engl. J. Med. *279:* 515–520 (1968).

115 Prior, I. A. M.; Stanhope, J. M.: Blood pressure patterns, salt use and migration in the Pacific; in Kesteloot, Joossens, Epidemiology of arterial blood pressure (Nijhoff, The Hague 1980).

116 Punsar, S.; Erametsu, O.; Karvonen, M. J.; Ryhanen, A.; Hilska, P.; Vonamo, H.: Coronary heart disease and drinking water. A search in two Finnish male cohorts for epidemiological evidence of a water factor. J. chron. Dis. *28:* 259–287 (1975).

117 Reisen, E.; Abel, R.; Modan, M.: Effect of weight loss without salt restriction on the reduction of blood pressure in overweight hypertensive patients. New Engl. J. Med. *298:* 1–6 (1978).

118 Sachs, F. M.; Rosner, V.; Kass, E. H.: Blood pressure in vegetarians. Am. J. Epidem. *100:* 390–398 (1974).

119 Sasaki, N.: Epidemiological studies on hypertension in the northeastern part of Japan. Jap. Circulation *4:* 1139–1142 (1977).

120 Sasaki, N.: Epidemiological studies on hypertension in northeast Japan; in Kesteloot, Joossens, Epidemiology of arterial blood pressure (Nijhoff, The Hague 1980).

121 Schneckloth, R. E.; Corcoran, A. C.; Stuart, K. L.; Moore, F. E.: Arterial pressure and hypertensive disease in the West Indian population. Report of a survey in St. Kitts, West Indies. Am. Heart J. *63:* 607–628 (1962).

122 Schroeder, H. A.: Relation between mortality from cardiovascular disease and treated water supplies. Variations in states and 163 largest municipalities of the United States. J. Am. med. Ass. *172:* 1902–1908 (1960).
123 Schroeder, H. A.; Kramer, L. A.: Cardiovascular mortality, municipal water and corrosion. Archs Envir. Hlth *28:* 303–311 (1974).
124 Schull, W. J.; Harburg, E.; Schork, M. A.; Weener, J.; Chape, C.: Hereditary stress and blood pressure, a family set method. III. J. chron. Dis. *30:* 659–669 (1977).
125 Select Committee on GRAS Substances. Substances evaluation of the health aspects of sodium chloride as a food ingredient (Life Sciences Research Office, Fed. Am. Society of Experimental Biology, Bethesda 1979).
126 Shaper, A. G.; Wright, D. H.; Kyobe, J.: Blood pressure and body build in three nomadic tribes of Northern Kenya. E. Afr. med. J. *46:* 273–281 (1969).
127 Stamler, J.; Berkson, D. M.; Dyer, A.: Relationship of multiple variables to blood pressure – findings from four Chicago epidemiological studies; in Paul, Epidemiology and control of hypertension, pp. 307–356 (Symposia Specialists, Miami 1975).
128 Stamler, J.; Farinaro, E.; Mojonnier, L. M.; Hall, Y.; Moss, D.; Stamler, R.: Prevention and control of hypertension by nutritional-hygienic means. J. Am. med. Ass. *243:* 1819–1823 (1980).
129 Stamler, R.; Stamler, J.; Riedlinger, W. F.; Algera, G.; Roberts, R. H.: Weight and blood pressure. Findings in hypertension screening of 1 million Americans. J. Am. med. Ass. *240:* 1607–1610 (1978).
130 Stini, W.: Early nutrition, growth, disease and human longevity. Nutr. Cancer *1:* 31–39 (1978).
131 Swales, J. D.: Dietary salt and hypertension. Lancet *i:* 1177–1179 (1980).
132 Tobian, L.: Salt and hypertension. Ann. N.Y. Acad. Sci. *304:* 178–197 (1978).
133 Tobian, L.: Hypertension and obesity. New Engl. J. Med. *298:* 46–48 (1978).
134 Trocmé, C.: Le régime et les médicaments dans l'hypertension artérielle. Nouv. Presse méd. *3:* 91 (1974).
135 Truswell, H. C.: Hypertension and salt. Lancet *ii:* 204 (1978).
136 Truswell, A. S.; Keneely, B. M.; Hanse, J. D. L.; Lee, R. B.: Blood pressure of Kung bushmen in Northern Botswana. Am. Heart J. *84:* 5–12 (1971).
137 Tyroler, H. A.; Heyden, S.; Hames, C. G.: Weight and hypertension: Evans County studies of blacks and whites; in Paul, Epidemiology and control of hypertension, pp. 197–204 (Symposia Specialists, Miami 1975).
138 Vannucchi, H.; Campana, A. O.; Oliveira, J. E. D. de: Alcohol and hypertension. Lancet *ii:* 1365 (1977).
139 von Bunge, G.: Lehrbuch der physiologischen und pathologischen Chemie (Verlag von FCW Vogen, Leipzig 1894).
140 Voors, A. W.; Webber, L. S.; Berenson, G. S.: Time course studies of blood pressure in children – the Bogalusa Heart Study. Am. J. Epidem. *109:* 320–334 (1979).
141 Wallace, R. B.; Lynch, C. F.; Pomrehn, P. R.; Criqui, M. H.; Heiss, G.: Alcohol and hypertension: epidemiologic and experimental considerations. The Lipid Research Clinics Program. Circulation *64:* suppl. III, p. 41 (1981).
142 Watson, R. L.; Langford, H. G.; Abernathy, J.; Barnes, T. Y.; Watson, M. J.: Urinary electrolytes, body weight, and blood pressure. Pooled cross-section results among four groups of adolescent females. Hypertension *2:* 193–198 (1980).

143 Weiffenbach, J.M.; Daniel, P.A.; Conwart, B.J.: Saltiness in developmental perspective; in Kare, Fregly, Bernard, Biological and behavioral aspects of salt intake (Academic Press, New York 1980).
144 Winkelstein, W.; Kagan, A.; Kato, H.; Sacks, S.T.: Epidemiologic studies of coronary heart disease and stroke in Japanese men living in Japan, Hawaii and California: blood pressure distributions. Am. J. Epidem. *102:* 502–513 (1975).

H. Blackburn, MD, Professor and Director, Laboratory of Physiological Hygiene, School of Public Health, Professor of Medicine, Medical School, University of Minneapolis, Minneapolis, MN 55455 (USA)

Relation between Coronary Heart Disease and Certain Elements in Water and Diet[1]

Ermanno Lanzola, Giovanna Turconi, Massimo Allegrini, Roberto de Marco, Alessandra Marinoni, Primo Miracca

Istituto di Scienze Sanitarie Applicate, University of Pavia, Pavia, Italy

Introduction

Interest in research concerning a correlation between coronary heart disease (CHD) and the mineral characteristics of local water supplies dates back about 20 years to the first studies by *Kobayashi* [6] in Japan and by *Schroeder* [14, 15] in the United States and has long been supported by the observation that the incidence of CHD has a wide geographic variation. Since 1968, numerous research studies and ample reviews [3, 7, 9–11, 13, 17] have been published on this subject. However, even today, the responsibility of any of the numerous elements found in water has not been clearly proven except for an inverse relationship between water hardness and mortality rates from cardiovascular diseases in various countries [8]. Numerous hypotheses to explain this relationship have been formulated and, recently, the importance of the Zn/Cu ratio [4, 5,], of Mg [16], and of Cd [1] have been considered; this last element was considered in particular because of its importance in the etiology of hypertension but, now, its correlation with CHD is controversial.

On the other hand, considering the total diet, elevated fat consumption seems to be a more important risk factor than water softness. This correlation, however, is not always confirmed as, for instance, in the United Kingdom where mortality rates across the country may differ twice as much, and the communities with high mortality rates have diets with a lower percentage of calories from fats [2]. Moreover, according to *Masi-*

---

[1] Supported by the Commission of the European Communities (Contract 201-77-1 ENV1). The authors wish to acknowledge the contribution of Mr. *Sergio Comizzoli* in carrying out the analyses.

roni et al. [8], it is similarly unlikely that marked dietary differences exist within the United States or Canada to explain the marked geographical differences in cardiovascular mortality rates that occur across these countries in association with water hardness. So, while the dietary fat hypothesis may hold true on an international basis to explain in part national differences in cardiovascular mortality rates, it does not seem to apply within countries, whereas the water hardness hypothesis is valid both internationally and intranationally in different studies [8]. The intent of the present work is to contribute to the epidemiological studies in this field, taking into account the intake of various elements both with water and diet, since it is well-known that the dietary contribution to total mineral intake is far superior to that of water.

Materials and Methods

The area included in the survey is in the province of Pavia (Lombardia, Northern Italy) which has the following characteristics: one third of the province is situated in a mountainous area and the other two thirds in the Po plain; water supplies have different origins and different chemical characteristics. As for the mortality rates for CHD they are widely varied in the different communities.

The research was carried out in two phases. First, the association between CHD and the mineral content of tap water was investigated; second, a dietary survey of case-control families was undertaken, followed by a measurement of the mineral content of their total diets.

CHD Mortality and Water Survey

The area was subdivided into communities presenting homogeneous mortality rates for CHD in order to establish meaningful guidelines for the water samples to be taken from individual water sources throughout the province of Pavia. Information concerning CHD mortality rates was collected from the local city halls and the analysis of water was restricted to those adjacent areas exhibiting significant differences in these rates. The mortality rates were standardized according to age and sex using the population of Pavia as a reference point; a 6-year period (1970–1975) was considered in order to have a larger number of cases and to avoid the annual fluctuations possible when dealing with small numbers. The characteristics of the municipalities involved in this survey are set forth in table I. The following parameters and elements in water were determined: hardness, Ca, Mg, Zn, Cu, Cr, Cd, and Pb.

Medical and Dietary Survey

The approach used was the family survey. The family was chosen as the epidemiological unit because it can be considered a homogeneous cluster regarding the food consumed. Those families in which a death resulting from CHD had occurred in the period 1970–1979, or in which one member suffered from that disease, were defined as case families.

Table I. Characteristics of the municipalities

Municipalities	Population n	Families n	Mortality[1] (per 100,000)
Bagnaria	647	199	576
Val di Nizza	921	323	144
Varzi	4,394	1,399	146
Casatisma	891	273	592
Casteggio	7,813	2,334	132
Montescano	366	124	561
Montù Beccaria	2,282	786	150
Canneto Pavese	1,683	608	180
Filighera	802	271	517
Vistarino	1,009	276	164

[1] Standardized rate for age and sex.

For the nosographic classification, the VIII revision of the WHO (410–414; 400–404) was used. All individuals over 70 years of age were screened out. The cases to be studied were chosen at random from previously compiled lists of families with heart patients. Every case was paired with two control families.

The survey was carried out in three rural communities with a high mortality rate, namely Casatisma, Montescano, and Filighera. As a control, a number of families without incidents of death from CHD was chosen. The total sample consisted of 99 families: 33 cases and 66 controls. The number of controls was purposely high since the possibility of finding CHD in some of these families, and thus ultimately reducing the number of controls, could not be excluded.

There were several reasons for this type of an approach: first, to increase the probability of having cases in the sample; second, to collect as many families as possible in a few communities; and third, to gather information about eating habits in rural areas, where one is more likely to find foods produced locally and therefore more closely related to the mineral content of the local water.

Members of each family unit were questioned about their length of residence in the community in order to screen out those families who had not resided there for at least 15 years. Each family was asked to reply to a questionnaire to obtain the following information: (1) Family medical history. (2) Possible incidence of CHD among members of the family unit. (3) Eating habits, with particular attention paid to where the foods came from. In order to establish the weight of the food consumed by each family member, a food atlas presenting various-sized food portions was employed and the per capita average daily food consumption weighted for family member body weight was then calculated.

The information thus gathered was used to reclassify the family units into cases and controls. The matching was done according to size of the family unit and occupation of the head of household.

Sample Collection and Preparation

Water. In each community samples were taken from three different points of the same aqueduct in spring and winter. The water samples were taken directly from the taps after running the water at a constant flow for 5 min. At the time of sampling, 1 ml HNO_3 65% was added for each liter of water.

Food. Market baskets representing a 1-week diet for families in the communities under consideration were collected in local stores and supermarkets. Preference was given to the sampling of locally produced food items. When the market baskets arrived in the laboratory, the food items were processed as for home consumption, homogenized, dried, and powdered.

Analytical Methods

Water. Pb, Cu, Zn, and Cd were concentrated by solvent extraction using the diethyl-ammoniumdiethyldithiocarbamate (DDDC)/xylene system. Ca, Mg, and Cr were determined directly on the water solution after dilution or concentration by evaporation. The water hardness was determined using ethylenediamine tetraacetic acid (EDTA).

Food. The wet-ashing procedure using $HNO_3/HClO_4$ (1:3) mixture was used for sample dissolution. Lead and cadmium were extracted using the DDDC/xylene system and measured by atomic absorption spectrophotometry. Calcium, magnesium, zinc, and copper were measured in the acid solution after diluting. The method was validated performing recovery and precision studies. The accuracy was checked measuring the aforesaid elements in the National Bureau of Standards Standard Reference Materials (SRM): Oyster (1566), Spinach (1570), and Rice (1568).

Results

Water

The results are summarized in table II, the single values are spread over a wide range. The communities were divided into two categories: A, with a high mortality rate (500 per 100,000), and B with a low mortality rate (200 per 100,000). The comparison between the values of each element in the two areas shows a significant difference for Cu, with higher values in area B, and for the Zn/Cu ratios, with higher values in area A. The linear correlation coefficients calculated between the values of the elements and the mortality rates do not show a significant correlation; only in the case of Cu does the coefficient arrive at a value lower than 20%. In addition, a nonparametric correlation was carried out between the mean values for the element concentrations and the mortality rates in the different communities; the negative correlations with Ca ($T=-0.60$, $S=-27$) and with water hardness ($T=-0.55$, $S=-25$) proved to be significant. As for the values of Cd, Cr, and Pb, no inferential analysis was carried out because in most of the water samples the levels were below the

Table II. Mineral characteristics of tap water in communities with different mortality rates

		Category A	Category B	Significance of comparison*	Correlation coefficient
Ca, mg/l	$\bar{x}$	83.66	93.50	$0.20 < p < 0.30$	$r = -0.069$
	SD	39.25	35.53		ns
Mg, mg/l	$\bar{x}$	29.85	29.15	ns	$r = +0.105$
	SD	15.71	14.80		ns
Hardness, mg/l	$\bar{x}$	372.54	399.166	ns	$r = -0.029$
	SD	177.95	155.90		ns
Zn, µg/l	$\bar{x}$	313.87	307.30	ns	$r = +0.027$
	SD	689.75	563.61		ns
Cu, µg/l	$\bar{x}$	7.50	11.27	< 0.05	$r = -0.418$
	SD	5.19	7.87		$p < 0.20$
Zn/Cu	$\bar{x}$	61.82	54.30	$\simeq 0.05$	$r = +0.036$
	SD	67.31	60.00		ns

* Analysis on the data transformed in log.
Category A (mortality for CHD $\geq$ 500 per 100,000); category B (mortality for CHD $\leq$ 200 per 100,000); ns = not significant.

Table III. Water Cd, Cr, and Pb content in the communities of categories A and B

	Category A			Category B		
	mean[1] µg/l	min µg/l	max µg/l	mean[1] µg/l	min µg/l	max µg/l
Cadmium	–	< 1.0	< 1.0	1.5(2)	< 1.0	1.5
Chromium	6.25(8)	< 1.0	9.0	5.5(13)	< 1.0	12.0
Lead	3.75(12)	< 1.0	9.0	8.0(12)	< 1.0	28.0

[1] The mean was calculated only in the case of detectable levels. The number of samples is indicated in parentheses.

detection limit. Therefore, only the mean, maximum, and minimum values have been reported for these elements in table III.

Food

Of the original 99 families selected, 86 complied with the survey (30 cases and 56 controls). The percentage of those refusing to comply was not significantly different for cases or controls.

The head of household in 51.2% of the sample (44 families) was more than 60 years old; in 40.7% of the sample (35 families), he was between 45 and 60 years of age, and in 8.1% (7 families) he was between 30 and 45 years old. As far as the occupational categories are concerned, in 51 families (59.3%) the head was either retired or a housewife, in 20 families (23.2%) a laborer, in 9 families (10.5%) a farmer, and in 6 families (7%) a clerk or otherwise occupied. The family units were composed for the most part of couples with children (37.2%) and of childless couples living with one or more other adults of the same age group (36%). 39 families consisted of 2 members, 23 of 3 members, and 12 of 4 members. The total number of individuals in the sample was 232.

The personal family medical history confirmed that diabetes is a probable risk factor. In fact, even though the difference between the cases and the controls is not completely significant ($p = 0.08$), among nine families in the sample presenting instances of diabetes, six were found among the cases and only three among the controls. As far as other diseases are concerned, no significant differences were found because these were distributed uniformly among both the cases and the controls.

Contrary to preliminary expectations about food habits, the majority of the foodstuffs is purchased in stores and does not come from local production. The foods of local production are fowl, rabbit (consumed in very small amounts), and vegetables. There is no significant difference between the amounts consumed either by the cases or controls. Locally produced fruit and wines are consumed in very few families. Only 23 families, without significant differences between cases and controls, drink tap water; all the others drink mineral water coming from nonlocal sources. The dietary intake findings for the six elements are shown in tables IV and V.

Conclusions and Comments

Water

The significant difference found between the values for Cu and for the Zn/Cu ratio of the two areas shows a tendency towards a negative correlation with mortality from CHD. Zinc has not been shown in association with CHD. This finding agrees with the hypothesis of *Klevay* who states that, in the Zn/Cu ratio, the important role belongs to copper [4, 5]. The negative association of CHD with water hardness and Ca can be shown only by using nonparametric correlation according to a graded scale.

Table IV. Per capita average daily dietary intake of Ca, Mg, Zn, Cu, Cd, and Pb (case and control groups)[1]

Element	Casatisma			Filighera			Montescano		
	cases (11)[2]	controls (21)	total (32)	cases (13)	controls (24)	total (37)	cases (6)	controls (11)	total (17)
Calcium, mg/day*	577 ± 142[3]	573 ± 174	575 ± 161	591 ± 198	551 ± 164	565 ± 175	497 ± 210	585 ± 201	554 ± 202
Magnesium, mg/day**	177 ± 26	194 ± 36	188 ± 34	160 ± 37	155 ± 38	156 ± 37	167 ± 27	199 ± 40	187 ± 38
Zinc, mg/day***	8.9 ± 1.0	9.8 ± 1.4	9.5 ± 1.3	8.2 ± 1.9	8.3 ± 1.6	8.3 ± 1.7	10.3 ± 1.8	10.9 ± 2.0	10.6 ± 1.9
Copper, mg/day**	1.3 ± 0.3	1.5 ± 0.3	1.5 ± 0.3	1.3 ± 0.3	1.3 ± 0.3	1.3 ± 0.3	1.4 ± 0.3	1.5 ± 0.3	1.5 ± 0.3
Cadmium, μg/day***	46 ± 14	48 ± 12	47 ± 13	32 ± 19	26 ± 13	28 ± 15	54 ± 5	60 ± 14	58 ± 12
Lead, μg/day***	149 ± 41	164 ± 33	159 ± 36	62 ± 19	58 ± 21	59 ± 20	82 ± 20	102 ± 34	95 ± 31

[1] Based on average consumption in the three communities over a 7-day period.
[2] Numbers in parentheses refer to the number of families.
[3] Each value represents mean ± standard deviation.
Probability of F (among communities): * (not significant); ** ($p < 0.05$); *** ($p < 0.001$).

Table V. Per capita average daily content of several elements in the total diet for case and control groups

	Cases (30)	Controls (56)	Total (86)
Calcium, mg/day	567 ± 179	566 ± 172	566 ± 175
Magnesium, mg/day	167 ± 31	178 ± 42	174 ± 38
Zinc, mg/day	8.9 ± 1.7	9.4 ± 1.9	9.2 ± 1.8
Copper, mg/day	1.3 ± 0.3	1.4 ± 0.3	1.4 ± 0.3
Cadmium, μg/day	42 ± 18	41 ± 19	41 ± 18
Lead, μg/day	98 ± 49	107 ± 56	104 ± 53

Diet

The food consumption profile of the case families and controls shows the following: (1) There are no differences between cases and controls in the food supply sources. (2) The diets of the two groups do not differ qualitatively. (3) There are no statistically significant differences in the quantity of foods consumed by the two groups.

Two-way unbalanced analysis of variance [12] was used to examine any significant differences in the intake of the six elements between the case and control groups within each community and among the three communities. There was no significant difference between the case and control groups. There was, however, a significant difference in the per capita intake of Mg, Zn, Cu, Cd, and Pb among the three communities. Filighera presented the lowest intake of these five elements.

The foods which are the largest contributors to total intake of the six elements are the following: milk and dairy products (Ca); milk, dairy, and cereal products (Mg); meat, dairy, and cereal products (Zn); dairy and cereal products, fruit (Cu); cereal products (Cd) cereal products, fruit and vegetables, wine (Pb).

Drinking water was not included in the calculation owing to the great variability in the concentrations of the elements present as has already been pointed out. An approximation of the contribution drinking water makes to the total intake of elements can be made supposing an average per capita consumption of 1 liter per day. In this case, water would supply the following percentages of the total daily mineral intake: Ca, 6–19%; Mg, 5–18%; Cu, <0.1–1.7%; Zn, <0.1–24%; Pb, <1–18%; Cd, <1–3%.

In conclusion, there is no statistical difference between the case and control groups both for food consumption and mineral intake. The significant difference in the mineral intake among the three communities cannot be compared with the mortality rates for CHD since they are very similar in the three communities. Tap water is not usually an important dietary source of the six elements. Very hard water is responsible for the higher percentage contribution levels of Ca and Mg, while those of the other elements are probably the result of pipeline leakages.

References

1 Bierenbaum, M. L.; et al.: Possible toxic water factor in coronary heart disease. Lancet *i:* 1008–1010 (1975).
2 Clayton, D. C.: Water hardness and cardiovascular mortality in England and Wales; in

Amavis et al., Proc. Eur. Scientific Coll. on Hardness of Drinking Water and Public Health, Luxembourg 1975, pp. 323–340 (Pergamon Press, Oxford 1976).
3   Crawford, M.D.: Hardness of drinking water and cardiovascular disease. Proc. Nutr. Soc. *31:* 347–353 (1972).
4   Klevay, L.M.: Coronary heart disease: the zinc/copper hypothesis. Am. J. clin. Nutr. *28:* 764–777 (1975).
5   Klevay, L.M.; Forbush, J.: Copper metabolism and the epidemiology of coronary heart disease. Nutr. Rep. int. *14:* 221–228 (1976).
6   Kobayashi, J.: in Neri, L.C.; Johanson, H.L.: Ann. N.Y. Acad. Sci. *304:* 203–219 (1978).
7   Neri, L.C.; Johansen, H.L.: Water hardness and cardiovascular mortality. Ann. N.Y. Acad. Sci. *304:* 203–219 (1978).
8   Masironi, R.: Water quality, trace elements, and cardiovascular disease. WHO Chron. *27:* 534–538 (1973).
9   Masironi, R.; Piša, Z.; Clayton, D.: Myocardial infarction and water hardness in the WHO myocardial infarction registry network. Bull. WHO *57:* 291–299 (1979).
10 Müller, G.: Probleme der epidemiologischen Beurteilung von Wasserinhaltsstoffen. Schriftenr. Ver. Wasser. Boden. Lufthyg. *40:* 39–52 (1973).
11 Neri, L.C.; Hewitt, D.: Review and implications of ongoing and projected research outside the European communities; in Amavis et al., Proc. Eur. Scientific Coll., Luxembourg, pp. 443–466 (Pergamon Press, New York 1975).
12 Neri, L.C.; Hewitt, D.; Schreiber, G.B.: Can epidemiology elucidate the water story? Am. J. Epidem. *99:* 75–88 (1974).
13 Nie, N.H.; Hull, C.H.; Jenkins, J.G.; Steinbrenner, K.; Bent, D.H.: Statistical package for the social sciences; 2nd ed. (McGraw-Hill, New York 1975).
14 Punsar, S.: Cardiovascular mortality and quality of drinking water. An evaluation of the literature from an epidemiological point of view. Work Environ. Hlth *10:* 107–125 (1973).
15 Schroeder, H.A.: Degenerative cardiovascular disease in the orient. II. Hypertension. J. chron. Dis. *8:* 312–333 (1958).
16 Schroeder, H.A.: Relation between mortality from cardiovascular disease and treated water supplies. Variations in states and 163 largest municipalities of the United States. J. Am. med. Ass. *172:* 1902–1908 (1960).
17 Seelig, M.S.; Heggtveit, H.A.: Magnesium interrelationships in ischemic heart disease: a review. Am. J. clin. Nutr. *27:* 59–79 (1974).
18 Sharrett, A.R.; Feinleib, M.: Water constituents and trace elements in relation to cardiovascular diseases. Prev. Med. *4:* 20–36 (1975).

E. Lanzola, MD, Istituto di Scienze Sanitarie Applicate, University of Pavia, Via Taramelli 1, I-27100 Pavia (Italy)

Nutrition and Lipoproteins

Nutrient – High-Density Lipoprotein Relationships: An Overview

Basil M. Rifkind

Lipid Metabolism and Atherogenesis Branch, Division of Heart and Vascular Diseases, National Heart, Lung, and Blood Institute, National Institutes of Health, Bethesda, Md., USA

Background and Objectives –
The Lipid Research Clinics Program

This paper reviews the background and objectives of the Lipid Research Clinics (LRC) Program and describes the LRC Prevalence Study, a two-stage population survey in which dietary and lipid and lipoprotein data were collected by 12 North American LRCs. Prevalence Study results are highlighted, especially the relationships of high-density lipoprotein (HDL) levels and selected nutrient (e. g. dietary cholesterol, fat, alcohol, total carbohydrate) and anthropometric (e. g. weight, Quetelet index) variables.

In June 1970 an expert panel on hyperlipidemia and premature atherosclerosis met to recommend to the then National Heart and Lung Institute ways to achieve the goal of preventing premature atherosclerosis through the diagnosis and treatment of hyperlipidemia. One of their recommendations led to the creation of the LRC Program in 1971 [1].

This recommendation stemmed from research indicating that lipoprotein patterns could provide important information not provided by blood cholesterol and triglyceride levels alone. The specific identification of distinct lipid transport disorders (hyperlipoproteinemias) that were formerly lumped under the general heading of 'familial hyperlipidemia' or 'hypercholesterolemia' offered a more systematic approach to the study, understanding, and treatment of these conditions.

Hyperlipoproteinemias are an important public health concern because they are frequently associated with coronary heart disease (CHD), especially in young persons. Many researchers have attempted to measure the prevalence and the extent of the relationship of hyperlipoproteinemias with CHD. However, few such studies have been population-based or have involved sufficient numbers; therefore, a better understanding of the overall contribution of hyperlipoproteinemias to CHD is still needed.

Moreover, an adequate interpretation of the significance of hyperlipoproteinemia requires a description of the distribution of lipids and lipoproteins within populations, and an examination of those factors that appear to affect lipid and lipoprotein levels (e.g. diet, life-style, genetic factors).

In light of these research needs, a network of LRCs, and their requisite support facilities, were established with the following major objectives:

(1) Evaluation of current techniques for diagnosis of hyperlipoproteinemia and development of better ones. (2) Acquisition of data on the prevalence of different types of hyperlipoproteinemia, especially among young age groups, with special emphasis on the nature and frequency of genetic forms. (3) Collection of better data on the prevalence and incidence of atherosclerosis in different types of hyperlipoproteinemia. (4) Improvement of detection, diagnosis, and medical care for hyperlipidemic patients by providing guidance and assistance to physicians on the management of these patients. Such 'service' should be in the context of study and evaluation. (5) Testing and development of improved therapy (both *dietary* and *drug*) for specific disorders. (6) Acquisition of information bearing on the ultimate question of how dietary and drug intervention designed to lower plasma lipid concentrations may be applicable to both the 'high-risk' and general population in regard to the prevention or control of premature atherosclerosis.

To achieve these goals, two major sets of collaborative studies were undertaken by the LRC Program:

Population Studies [15]

The general objectives of this group of studies were to determine the prevalence of hyperlipidemias and hyperlipoproteinemias, and to describe the distribution of lipids and lipoproteins in major population groups. Additional major objectives of this group of studies were to explore the relationship between nutrient intake and blood lipid and lipoprotein levels, and between the latter and CHD.

The major components of the Population Studies are (1) the Prevalence Study [14], (2) the Family Study, and (3) the Follow-up Study. Field work has been completed for the Prevalence Study and the Family Study; the Follow-up Study is still active in the field.

Coronary Primary Prevention Trial [16]

The objective of this trial was to test the hypothesis that, if blood lipids are reduced, CHD morbidity and mortality would also be reduced. This long-term clinical trial is scheduled for completion in 1983.

The LRC Program Prevalence Study [7, 14, 15]

The LRC Prevalence Study has aimed to collect adequate lipid and lipoprotein data on a population basis. It was designed to provide for acquisition and analysis of data on the prevalence of different types of hyperlipoproteinemia, to provide reliable information on the distribution of lipids and lipoproteins in defined populations, to relate diet to lipid and lipoprotein levels, and to establish correlates with CHD. Care has been taken in the selection of study populations to include different age, race, and socioeconomic groups.

The Prevalence Study involved two sequential examinations. The first (visit 1) was a brief screen to collect information on sociodemographic variables and on usage of five types of lipid-altering medication and to measure plasma cholesterol and triglyceride levels in fasting participants. Participants who were part of a 15% randomly selected sample or who had elevated lipid levels or were taking lipid-altering medication were asked to return for a second, more extensive examination (visit 2). Those selected for visit 2 included approximately 25% of the visit 1 participants. Personal and family histories relevant to atherosclerosis, a detailed drug history, and nutrient intake by means of a 24-hour dietary recall were recorded at visit 2, as were measurements of lipids and lipoproteins, blood pressure, clinical chemistries, resting and exercise ECGs, and anthropometric variables.

HDL – Nutrient Relationships

When the LRC studies were in progress, it became apparent that HDL is also an important risk factor for CHD but acting in a direction opposite to the other known risk factors, namely the higher the HDL levels, the lower the CHD risk. *Miller and Miller* [11] in 1975 noted that there was much evidence, including prospective epidemiological studies, pointing to such a role for HDL. Subsequent prospective studies have generally confirmed these observations [6]. Laboratory studies have suggested several mechanisms through which HDL might exert a protective effect [9].

That HDL might be under some degree of dietary control is suggested by a number of observations. Several of the apolipoproteins of HDL appear to originate in the intestinal wall. Chylomicron and very low-density lipoprotein (VLDL) remnants are a source of components of HDL particles. Alcohol intake has been reported to be directly correlated with HDL levels. HDL levels are inversely correlated to obesity. It was therefore considered appropriate that the relationship between diet and HDL levels be explored in the LRC Prevalence Study.

Highlights of Findings from the LRC Prevalence Study

The purpose in this report is to describe some of the findings of the LRC Prevalence Study in relation to HDL, especially the relationship of selected nutrient and anthropometric variables. The data reported here are from white participants of both sexes, who were recalled to visit 2 as part of the 15% random sample of visit-1 participants in the North American LRCs. This report is based primarily on data analyses done by three LRC working groups: the Alcohol-Nutrition, Anthropometry, and Lipoprotein Working Groups, and is based on their findings, which have been reported in considerable detail elsewhere [3].

Distributions of HDL Cholesterol in North American Populations
[3, 15]

Figure 1 presents some distributional characteristics of HDL cholesterol (HDL-C) by sex and age for white participants in the North American LRC. Below age 12 years, median and mean HDL-C values are higher in males than females. Thereafter, the female levels are higher, owing mainly to an abrupt fall in HDL-C in the males. HDL-C shows a small linear increment in females with age up to 60 years. Thereafter, no differences in mean values by age are discernible. In males, fairly stable HDL-C values below age 12 are followed by a decline during puberty and adolescence, relatively stable values in adulthood, high values in the age group 55–60 years, and a plateau thereafter.

Percentile Distributions for the Quetelet Index of Body Mass
[5, 7, 15]

Tables I and II provide the percentile distributions of the Quetelet index (calculated as weight (kg)/height^2(cm)×1,000) for white males and females, aged 4–70 years. For white males, the cross-sectional data reveal a gradual increase in Quetelet index with age; the highest levels were in the 40- to 49-year-old group, with modest declines seen thereafter. For white

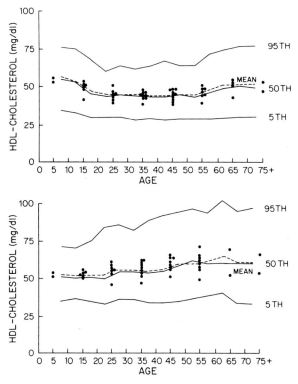

Fig. 1. Overall mean (interrupted line), median, and 5th and 95th percentile values of HDL-C levels by age (years) for males (above) and females (below). Filled circles represent mean HDL-C values for each population sampled by the participating LRCs [from ref. 3 with permission].

females, however, the Quetelet index gradually increased with age throughout the distribution.

Correlations of Age, Weight, Height, and Ponderosity Indexes with HDL-C and Triglyceride [5]

Simple correlations between various anthropometric measures and HDL-C and log triglyceride with age groups for each sex are summarized in tables III and IV. In the age groups 4–11 years, there were no significant relationships between measures of height and weight or various ponderosity indexes and HDL-C. Plasma triglyceride levels were, however, positively and significantly related to weight and weight/height^2 in all 4- to 11-year-old children.

Table I. Percentile distribution of the Quetelet index for white males ages 4–79 years [from ref. 3 with permission]

Age years	n	Quetelet index (kg/cm^2) × 1,000, percentiles						
		5th	10th	25th	50th	75th	90th	95th
4–11	246	1.42	1.45	1.52	1.62	1.73	1.96	2.10
12–16	368	1.60	1.66	1.78	1.97	2.18	2.48	2.69
17–19	135	1.78	1.84	2.00	2.16	2.46	2.78	3.07
20–29	369	2.00	2.06	2.24	2.43	2.67	2.89	3.07
30–39	773	2.11	2.19	2.39	2.59	2.81	3.04	3.25
40–49	708	2.15	2.27	2.46	2.63	2.84	3.10	3.26
50–59	589	2.10	2.23	2.42	2.61	2.82	3.05	3.20
60–69	230	2.16	2.24	2.38	2.55	2.76	3.01	3.20
70–79	99	2.06	2.16	2.29	2.50	2.73	2.92	2.99

Table II. Percentile distribution of the Quetelet index for white females ages 4–79 years [from ref. 3 with permission]

Age years	n	Quetelet index (kg/cm^2) × 1,000, percentiles						
		5th	10th	25th	50th	75th	90th	95th
4–11	231	1.38	1.44	1.51	1.63	1.79	2.02	2.15
12–16	320	1.56	1.63	1.81	1.97	2.15	2.44	2.65
17–19	124	1.76	1.80	1.99	2.18	2.41	2.55	2.75
20–29	510	1.83	1.90	2.02	2.17	2.38	2.66	3.14
30–39	635	1.89	1.96	2.08	2.24	2.51	2.96	3.30
40–49	645	1.95	2.01	2.14	2.37	2.67	3.01	3.21
50–59	500	1.91	2.00	2.19	2.41	2.74	3.14	3.41
60–69	269	1.98	2.07	2.21	2.41	2.74	3.13	3.40
70–79	144	1.96	2.00	2.19	2.50	2.73	3.02	3.23

For 12- to 16-year-old children there were significant inverse associations between HDL-C, measures of weight, and ponderosity for both males and females. Moreover, for all children aged 12–16 years, significant positive correlations were observed between weight and triglyceride, and for the most part between the weight indexes and triglyceride levels. The same overall patterns were also seen for children 17–19 years old; measures of weight and ponderosity were inversely associated with HDL-C and positively associated with triglyceride. However, most of the inverse associa-

Table III. Pearson product-moment correlations between high-density lipoprotein, cholesterol, triglycerides and selected variables in white males [from ref. 3 with permission]

Age years		n	Age	W	H	W/H^2	W/H^3	W/H
4–11	HDL-C	236	0.09	0.02	0.07	–0.01	0.06	0.02
	ln TG	236	0.22***	0.20**	0.19**	0.18**	0.04	0.20**
12–16	HDL-C	363	–0.33***	–0.48***	–0.40***	–0.42***	–0.29***	–0.47***
	ln TG	362	0.12*	0.39***	0.22***	0.41***	0.35***	0.40***
17–19	HDL-C	134	–0.06	–0.14	–0.06	–0.12	–0.10	0.13
	ln TG	134	0.11	0.27**	–0.19**	0.36***	0.39***	0.32***
20–29	HDL-C	369	–0.02	–0.20***	0.04	–0.25***	–0.24***	–0.23***
	ln TG	369	0.03	0.29***	–0.04	0.35***	0.34***	0.33***
30–39	HDL-C	772	–0.11**	–0.22***	–0.06	–0.21***	–0.18***	–0.23***
	ln TG	772	0.13***	0.29***	–0.05	0.35***	0.34***	0.33***
40–49	HDL-C	704	0.02	–0.16***	–0.04	–0.15***	–0.13***	–0.16***
	ln TG	704	0.02	0.23***	–0.07	0.35***	0.34***	0.32***
50–59	HDL-C	589	0.17***	–0.26***	–0.15***	–0.21***	–0.16***	–0.25***
	ln TG	589	0.09	0.26***	0.05	0.28***	0.25***	0.28***
60–69	HDL-C	230	0.01	–0.26***	0.02	–0.31***	–0.30***	–0.29***
	ln TG	230	0.02	0.23***	0.06	0.29***	0.26***	0.30***
70–79	HDL-C	99	0.11	–0.04	–0.06	–0.01	0.01	–0.02
	ln TG	99	–0.22*	0.20	–0.01	0.22*	0.20*	0.21*

* p < 0.05; **p < 0.01; *** p < 0.001.
W = Weight; H = height; HDL-C = high-density lipoprotein cholesterol; ln TG = log triglyceride.

tions between HDL-C and the weight/height indexes were not statistically significant in children 17–19 years old. For each 10-year age group in the adult population, weight correlated inversely with HDL-C, and weight and measures of ponderosity correlated positively with triglyceride.

Relations of HDL-C to Quetelet Index after Adjustment for Covariables [5]

The relationships of HDL-C to Quetelet index were further explored by covariance adjustment for age, cigarette smoking, alcohol intake, and gonadal hormone use. Adjusted mean HDL-C levels for white males and females at approximately the 10th, 50th, and 90th percentiles of the Quetelet index for their age, race, and sex are shown in figure 2. The highest HDL-C levels were observed at the lowest Quetelet percentile and vice versa; overall, the mean HDL-C differences were about 3 mg/dl between

Table IV. Pearson product-moment correlations between high-density lipoprotein, cholesterol, triglycerides and selected variables in white females [from ref. 3 with permission]

Age years		n	Age	W	H	W/H^2	W/H^3	W/H
4–11	HDL-C	226	–0.06	0.10	–0.10	–0.05	0.01	–0.09
	ln TG	225	0.20**	0.24***	0.20**	0.22**	0.08	0.25***
12–16	HDL-C	320	–0.03	–0.15**	–0.09	–0.12*	–0.10	–0.14**
	ln TG	320	–0.06	0.14*	0.01	0.18**	0.18***	0.16**
17–19	HDL-C	122	–0.02	–0.20*	–0.13	–0.15	–0.12	–0.18*
	ln TG	122	0.19*	0.18*	–0.15	0.26**	0.29**	0.22*
20–29	HDL-C	510	0.12**	–0.24***	–0.04	–0.25***	–0.24***	–0.25***
	ln TG	510	–0.02	0.10*	–0.03	0.13**	0.13**	0.12**
30–39	HDL-C	630	0.06	–0.23***	0.001	–0.26***	–0.25***	–0.25***
	ln TG	630	0.14***	0.21***	–0.06	0.25***	0.26***	0.24***
40–49	HDL-C	642	0.07	–0.30***	–0.03	–0.32***	–0.38***	–0.31***
	ln TG	642	0.14***	0.22***	–0.06	0.26***	0.26***	0.24***
50–59	HDL-C	497	–0.03	–0.28***	–0.002	–0.28***	–0.26***	–0.28***
	ln TG	497	0.14**	0.18***	–0.07	0.23***	0.21***	0.20***
60–69	HDL-C	268	–0.08	0.39***	0.03	–0.41***	–0.39***	–0.41***
	ln TG	268	0.10	0.30***	–0.01	0.31***	0.30***	0.31***
70–79	HDL-C	133	–0.04	–0.27**	0	–0.27**	–0.26**	–0.28**
	ln TG	133	–0.03	0.16	–0.06	0.19*	0.19*	0.18

* $p<0.05$; ** $p<0.01$; *** $p<0.001$.
W = Weight; H = height; HDL-C = high-density lipoprotein cholesterol; ln TG = log triglyceride.

the 10th and 50th percentiles and about 3–4 mg/dl between the 50th and 90th percentiles.

Nutrient Intake Data [2, 4]

The nutrient intake data reported here are based on 24-hour dietary recalls obtained from the visit – 2 participants who were in the 15% random sample in the North American LRC.

The 24-Hour Dietary Recall

The strengths and weaknesses of different methods of assessing dietary intake in free-living populations have been well-documented. It is generally accepted that no method is best in all respects. The ultimate selection of the method depends on the design and objectives of the particular study and the deficiencies the investigator is willing to tolerate in order to obtain certain advantages. The dietary recall method is used frequently in large-

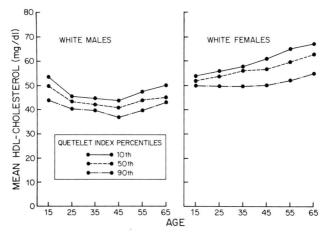

Fig. 2. Variance-adjusted mean HDL-C values by Quetelet index and age [from ref. 3 with permission].

scale studies. Its advantages are that it is inexpensive and imposes little burden on the respondent. (The 24-hour recall was usually completed in less than 30 min in the LRC study.) The method is less likely to distort customary intake and so may be more representative of habitual diet than methods based on diet diaries. A major drawback of the 24-hour dietary recall is that a single recall cannot be used to assess the usual diet of an individual. It does, however, yield useful data where the objective is to compare or classify groups of people. The LRC Program addressed the problems encountered with a method that relies on memory by strengthening the interview process through intensive training of interviewers and use of standardized food models and other aids.

The dietary recall method used by the LRC Program was part of an overall NHLBI nutrition data system developed for the LRC and Multiple Risk Factor Intervention Trial programs. The components of this system included formal training and certification procedures for the interviewers, the interview itself, coding of the recalls at a central facility (The Nutrition Coding Center), and use of a coding procedure manual and the NHLBI table of food composition, the data base that allows the translation of the coded food items into the nutrient intakes. For each item in the food table the following nutrients are listed: protein, total fat, total carbohydrate, saturated fatty acids, monounsaturated fatty acids, polyunsaturated fatty acids, cholesterol, starch, sucrose, crude fiber, alcohol, and selected vitamins and minerals. In the NHLBI table of food composition, carbohydrate is categorized as sucrose, starch, or other carbohydrate (lactose, fructose, glucose). Sweeteners, except when known to be sucrose, are classified as other carbohydrate. Accordingly, the sucrose intake may have been underestimated, but the impact of this underestimation on the analyses could not be determined.

Alcohol Intake Data

In addition to the 24-hour recall, alcohol intake data were obtained through specific interview questions regarding alcohol use during the previous week (7-day recall). In the 7-day recall, each participant was asked how many bottles of beer and glasses of wine, mixed drinks, and liqueurs had been consumed during the past week. We converted this information to ounces of alcohol by assuming each bottle of beer was 12 oz (0.432 oz alcohol), each glass of wine was 3 oz (0.366 oz alcohol), each mixed drink had 1.5 oz whiskey (0.645 oz alcohol), and each glass of liqueur was 1 oz (0.300 oz alcohol). A conversion factor of 29.574 ml/oz was used to express the alcohol consumption of either of these two sources as ml/day. Except where noted, the 7-day recall was used because the 24-hour dietary recall provided a less extensive record and because no 24-hour dietary recalls had been taken for Saturdays.

Statistical Analyses

The data were explored to identify variables that showed evidence of possible relationship with HDL-C levels. These variables are shown in table V, along with nutrient variables that were reportedly associated with HDL levels (even if our initial explorations did not appear to confirm such a relationship). Such variables included fatty acids (saturated, monounsaturated, polyunsaturated), total energy intake, and total fat. This approach was taken to screen nutrient variables because of the large intraindividual variation in the measurement of nutrient intake, which can result in an underestimation of the 'true' correlation coefficient.

Multiple linear regression analysis was used to explore the relationship between HDL and nutrient intake. This technique takes account of possible correlations between various nutrients and between each nutrient and age and weight. Steps were also taken to assess the possible significance of, and to adjust for, any population (clinic) differences in HDL-C and nutrient intake, because the LRC data represent the aggregate of a series of identically designed population studies.

Table V. Pertinent nutrient variables [from ref. 3 with permission]

Basic nutrients	Derived variables
Calories	cal/kg body weight
Protein, g	% cal from protein; protein/kg body weight
Fat, g	% cal from fat;
Polyunsaturated fatty acids (PFA)	% cal from PFA; % fat from PFA
Saturated fatty acids (SFA)	% cal from SFA; % fat from SFA
Monounsaturated fatty acids (MFA)	% cal from MFA; % fat from MFA
PFA/SFA ratio	
Cholesterol, mg	cholesterol/kg body weight; cholesterol/1,000 cal
Carbohydrate, g	% cal from carbohydrates
Sucrose, g	% cal from sucrose
Starch, g	% cal from starch; C/S ratio[1]
Other carbohydrates, g	% cal from other carbohydrates
Alcohol, ml	% cal from alcohol; alcohol/kg body weight; alcohol/Quetelet index[2]

[1] Ratio of the sum of sucrose and other carbohydrates to starch.
[2] Quetelet index = (weight/height²) × 1,000.

In the various analyses, women have been divided into those taking gonadal sex hormones and those not taking these drugs. This step was taken because gonadal sex hormones are known to have distinct effects on blood lipids and lipoproteins and because prevalence of such hormone consumption is high for our female population.

Correlations of HDL-C Levels and Selected Nutrient and Anthropometric Variables

Table VI shows, for each age- and sex- (hormone use) specific group, the correlation coefficients for HDL-C and alcohol, total carbohydrate, sucrose, starch, and dietary cholesterol, as well as weight and Quetelet index.

Table VI. Pearson correlation coefficients between high-density lipoprotein cholesterol level and the intake of selected nutrient variables.[1] Body weight and Quetelet index [from ref. 3 with permission]

	Men – age, years						Women not taking hormones – age, years						Women taking hormones – age, years					
	20–29	30–39	40–49	50–59	≥60	mean[2]	20–29	30–39	40–49	50–59	≥60	mean[2]	20–29	30–39	40–49	50–59	≥60	mean[2]
Number of participants	356	734	647	501	235		249	419	416	252	194		237	183	182	160	90	
Alcohol[3], ml	0.14**	0.13***	0.27***	0.23***	0.31***	0.21***	0.08	0.30***	0.28***	0.19***	0.38***	0.25***	0.10	0.10	0.20**	0.25**	0.31**	0.17***
Alcohol, ml	0.06	0.11**	0.20***	0.27***	0.33***	0.18***	0.07	0.22***	0.25***	0.19***	0.34***	0.21***	0.18**	0.15*	0.16**	0.18**	0.30**	0.18***
Alcohol, % cal	0.11*	0.13***	0.18***	0.28***	0.34***	0.19***	0.08	0.26***	0.26***	0.16**	0.33***	0.22***	0.20**	0.17*	0.20**	0.25**	0.31**	0.21***
Total carbohydrate, g	−0.05	−0.02	−0.07	−0.10*	−0.07	−0.06***	−0.15*	−0.09	−0.04	0.06	−0.11	−0.06*	−0.07	−0.08	−0.11	−0.13	−0.07	−0.09**
Total carbohydrate, % cal	−0.12***	−0.12***	−0.15***	−0.09	−0.25***	−0.13***	−0.18***	−0.18***	−0.10*	−0.11	−0.26***	−0.16***	−0.18**	−0.08	−0.14	−0.20*	−0.29***	−0.17***
Sucrose, g	−0.16**	−0.10**	−0.09	−0.15***	−0.04	−0.11***	−0.17**	−0.19***	−0.08	0.04	−0.24***	−0.13***	−0.10	−0.07	−0.09	−0.17*	−0.12	−0.11**
Sucrose, % cal	−0.18***	−0.14***	−0.13***	−0.16***	−0.12	−0.15***	−0.16**	−0.22***	−0.07	−0.01	−0.26***	−0.14***	−0.12	−0.05	−0.07	−0.18*	−0.14	−0.11**
Starch, g	0.02	−0.00	−0.07	−0.08	−0.19**	−0.06***	−0.16**	−0.07	−0.07	0.06	−0.12	−0.09***	−0.05	−0.08	−0.18*	−0.11	−0.18	−0.11**
Starch, % cal	−0.02	−0.05	−0.12**	−0.05	−0.29***	−0.09***	−0.14**	−0.13**	−0.13**	−0.07	−0.17*	−0.13***	−0.11	−0.07	−0.24***	−0.13	−0.33**	−0.16***
Cholesterol, mg	−0.02	−0.02	0.05	−0.00	−0.05	−0.00	0.13*	0.04	0.11*	0.04	0.07	0.08**	−0.13*	0.01	0.04	−0.06	0.10	0.05

Table VI. Continued

	Men – age, years						Women not taking hormones – age, years						Women taking hormones – age, years					
	20–29	30–39	40–49	50–59	≥60	mean[2]	20–29	30–39	40–49	50–59	≥60	mean[2]	20–29	30–39	40–49	50–59	≥60	mean[2]
Cholesterol, mg per 1,000 cal																		
	–0.01	–0.06	0.03	0.06	–0.09	–0.01	0.21**	0.05	0.10*	–0.01	0.02	0.08**	–0.14*	0.01	0.06	–0.08	0.12	0.05
Weight, kg																		
	–0.23***	–0.24***	–0.18***	–0.28***	–0.18**	–0.23***	–0.29***	–0.22***	–0.33***	–0.24***	–0.22**	–0.26***	–0.22***	–0.24***	0.19*	–0.21**	–0.31**	–0.23***
Quetelet index																		
	–0.27***	–0.22***	–0.16***	–0.22***	–0.20**	–0.21***	–0.31***	–0.25***	–0.33***	–0.28***	–0.27***	–0.29***	–0.23***	–0.25***	–0.19*	–0.19**	–0.33**	–0.23***

* $0.05 > p \geq 0.01$; ** $0.01 > p \geq 0.001$; *** $p \leq 0.001$.
[1] Nutrient data collected from 24-hour dietary recall unless otherwise indicated.
[2] Average of correlation coefficients weighted by their sample size.
[3] As reported on visit 2 interview questionnaire.

Table VII. Proportion of variance of high-density lipoprotein cholesterol explained by linear models [from ref. 3 with permission]

Variables[1] in model	Men	Women not taking hormones	Women taking hormones	
	(n=2,473)	(n=1,530)	(n=852)	
			20–39 (n=418)	≥ 50 (n=249)
Clinic, weight, age	0.139	0.172	0.179	0.186
Clinic, weight, age, alcohol	0.169	0.198	0.182[3]	0.208
Clinic, weight, age, alcohol, sucrose[2], % cal	0.178	0.206	0.184[3]	0.222
Clinic, weight, age, alcohol, sucrose[2], % cal, cholesterol[2]/1,000 cal	0.179[3]	0.212	0.189[3]	0.224[3]
Clinic, weight, age, alcohol, sucrose[2], % cal, cholesterol[2]/1,000 cal, starch[2] % cal	0.182	0.215	0.190[3]	0.237[3]

[1] Data reported on visit 2 interview questionnaire unless otherwise indicated.
[2] As reported on 24-hour dietary recall.
[3] Variable added at this level does not reach statistical significance.

No significant relationships were found between HDL-C levels and saturated, monounsaturated, or polyunsaturated fatty acids, total energy intake, total fat intake, or total protein intake. This was true for HDL-C level by nutrient level, nutrient level by HDL-C level, or when correlation coefficients were examined.

We calculated mean correlation coefficients for each of the three groups by averaging the age-specific correlation coefficients. All the nutrients in table VI, except dietary cholesterol, were significantly associated ($p < 0.05$) with HDL-C levels. Dietary cholesterol and HDL-C levels were significantly associated only for women not taking gonadal hormones.

Alcohol intake, whether measured by questionnaire for the previous 7 days or by the 24-hour dietary recall, was strongly and positively associated with HDL-C. Weaker inverse correlations were noted between HDL-C and total carbohydrate, sucrose, and starch, especially when the nutrients were shown as a percentage of total calories rather than in absolute amounts. Generally, the magnitudes of the correlation coefficients for each nutrient were similar for each of the three groups. The relationship of alcohol to HDL-C appeared to get stronger with age but clear age-related patterns were not discernible for the other nutrients. On the basis of these observations alcohol intake (reported for the 7-day interview) and sucrose (% calories), starch (% calories), and dietary cholesterol (per 1,000 calories), were selected for the multivariate analysis.

The multiple linear regression model for each sex group (appropriately pooled by age) controlling for clinic, weight, and age (table VII) was applied in an attempt to elucidate the proportion of interindividual variation in HDL-C levels that could be attributed to alcohol and the nutrient variables. For men, alcohol, sucrose, and starch made a statistically significant ($p<0.05$) addition to the proportion of the variance explained. For women not taking hormones, alcohol and all the nutrient variables made statistically significant ($p<0.05$) contributions. For women taking exogenous gonadal hormones, ages 20–39 years, none of the variables contributed significantly. For such women aged 50 years and older, alcohol and all the nutrient variables except cholesterol made statistically significant ($p<0.05$) contributions to the proportion of variance explained.

Alcohol Intake by Source

The data were also evaluated as to whether the impact of alcohol varied with the form of beverage used. We were particularly interested in exploring whether any particular source of alcohol had a specific effect on HDL-C in view of the suggestion that wine consumption might specifically protect against CHD death.

This was examined by calculating the regression of HDL-C on the amount of alcohol consumed for each age group (fig. 3). Indicated on the graphs are points (indexed by letters) for each alcohol consumption pattern. These points were plotted at the average HDL-C and alcohol intake for that pattern. Although the data suggest a higher than average gradient of HDL-C responses to alcohol when it is derived from wine, the difference between HDL-C levels (wine only drinkers) and other drinkers was trivial and not statistically significant ($p>0.05$). For example, for men the mean HDL-C level was only 0.7 ± 5.2 mg/dl higher in 'wine only' than in other drinkers. For women not taking gonadal hormones, the 'wine only' category was 1.9 ± 6.0 mg/dl higher and, for women on hormones, the level was

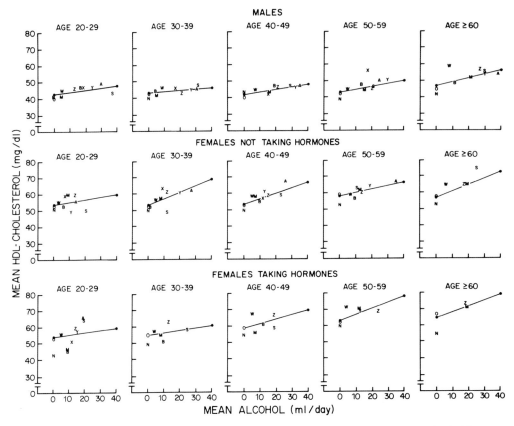

Fig. 3. Mean HDL-C by mean alcohol intake and type of alcohol consumed. N = No alcohol during past year; 0 = no alcohol during past week; B = only beer; W = only wine; M = only mixed drink; X = beer and wine; Y = beer and mixed drink; Z = wine and mixed drink; A = beer, wine and mixed drink; S = liqueurs with and without other alcohol [from ref. 3 with permission].

4.2 ± 8.6 mg/dl higher than that of the other categories of drinkers. Thus, though slightly higher HDL-C levels were indeed found in the wine drinkers, the differences were of low magnitude and did not achieve significance.

An interesting feature of these graphs is a comparison of subjects who never drink, that is, who have not drunk in the past year (N), and occasional drinkers, that is, those who did not drink in the past week but had drunk in the past year (0). Both sets of points tend to fall below the regression line, that is, such persons had lower than expected levels of

Table VIII. Mean high-density lipoprotein cholesterol (HDL-C) in relation to alcohol consumption by study for men ages 50–69 years [from ref. 3 with permission]

Alcohol ml/day	LRC Prevalence[1]		Albany[2]		Framingham[2]		Honolulu[2]		San Francisco[2]	
	n	mean	n	mean	n	mean	n	mean	n	mean
0	200	41.87	201	46.28	111	41.40	849	42.18	133	44.38
0.01–16.89	265	47.60	372	47.35	112	44.78	320	44.82	90	45.75
16.90–42.24	144	50.65	260	53.25	111	47.43	354	48.28	40	51.72
42.25–84.50	55	55.25	80	54.57	44	50.14	166	52.20	12	57.75
≧ 84.50	3	48.33	10	60.60	15	58.40	24	56.71	0	–
Correlation of HDL-C with alcohol consumption		0.28		0.28		0.30		0.28		0.25

[1] White only.
[2] As reported in Lancet ii: 153 (1977).

HDL-C. On an average, however, the HDL-C level for those who never drink was lower than the occasional drinkers, although the difference was statistically significant only for women ($p < 0.05$). This suggests that occasional alcohol drinking raises the HDL-C level and that this persists at least for some time in the absence of further drinking.

In view of several previous reports linking alcohol intake and HDL-C levels, we have compared in table VIII for males aged 50–69 years the mean HDL-C level in relation to alcohol consumption for the current and several other studies. The similarity of the correlations is impressive.

It may be asked how much variation in HDL-C levels occurs between persons in the general population on account of alcohol. Expressed as a percentage of the total population variance of HDL-C, alcohol consumption accounts for 4–6% of the variance. It may account for a greater proportion of the variance when allowance is made for unreliability of the history of alcohol consumption. A factor of comparable importance in the general population is weight and, in women, gonadal hormone use.

Alcohol appears to be one of several factors making an important contribution to the variance of HDL-C levels between individuals in populations. However, it should be emphasized that alcohol should not be regarded as a potential public health measure to reduce CHD risk through elevating HDL levels. Apart from the lack of direct evidence indicating that intervening in populations to raise HDL levels actually reduces CHD risk, there are many hazards in advocating alcohol use for such a purpose and any possible benefits would be certainly offset by many deleterious consequences of widespread alcohol use.

Carbohydrate Intake

There have been several reports of lower levels of HDL-C associated with higher intakes of carbohydrates. However, few investigators have reported results specific to the kind of carbohydrate consumed. The scant evidence available leads one to expect higher sucrose consumption to correlate with lower HDL-C levels as indicated by intake to HDL-C levels, as was observed in the present study. Since only a small proportion of the variance of HDL-C was explained by the addition of sucrose or starch to the multiple regression equation after age, weight, and alcohol had been entered (table VII), it could be asked whether these variables have any biological significance in determining HDL-C levels. On account of the limitations of the nutrient measures and other methodological considerations, the relationship is probably being understated. It is likely that the importance of sucrose or starch intake in explaining HDL-C levels is probably not great. Furthermore, certain possible confounding factors such as physical activity and cigarette smoking were not taken into account and might influence the conclusions. Although this means that the observed correlations with these nutrients may be artifacts, it does not demonstrate that they indeed are. Hence, these findings are presented as a spur for additional investigation rather than as strong evidence for a positive relation.

Dietary Cholesterol and Fat

The relationship of dietary cholesterol and fat to HDL-C is of particular interest in view of the frequent claims that dietary cholesterol and fat intake should be altered to lower total and LDL cholesterol levels, and thereby possibly to reduce CHD incidence. The question has been raised as to whether such dietary changes might influence HDL-C levels, with possible consequences on CHD incidence. The present study observed no significant association between polyunsaturated or saturated fatty acid intake and HDL-C. These results differ from findings in two clinical studies in which an extremely high intake of polyunsaturated fatty acids resulted in decreases in HDL-C. More moderate intake of polyunsaturated fatty acids was not found to change HDL-C levels in other investigations.

The effect of dietary cholesterol intake on HDL-C levels is also unclear. Two studies [12, 13] have reported increases in HDL-C as a result of high dietary cholesterol intakes. However, *Mahley* et al. [10] found that cholesterol intake led to little change in the level of HDL-C and little or no change was reported in other clinical studies. Our cross-sectional data showed negative or weakly positive associations only.

In summary, we conclude that alcohol consumption and certain nutrients may determine HDL-C levels. The most consistent and strongest association involved alcohol consumption. Sucrose and starch intake, especially when represented as a percentage of total calories, also seemed to have a modest influence on HDL-C. We observed no consistent relationships of HDL-C with dietary cholesterol and polyunsaturated or saturated fatty acid intake. Again it should be explained that these findings were based on the general validity and known pitfalls of cross-sectional studies, as well as the limitations of nutrient intake measurement. The findings did not derive from the observation of changes within participants who altered their nutrient intake or energy balance. The results would not necessarily parallel such observations, which would have to come from longitudinal studies or experimental manipulation. It is important that these findings be supplemented by those of other investigations in view of the potential public health implications of alcohol and nutrition effects on HDL cholesterol levels and the relationship of the latter to coronary heart disease.

Acknowledgments

The author thanks Janet Bungay for editorial assistance and Karen Wishnow for manuscript preparation.

Alcohol-Nutrition Working Group
Nancy Ernst, MS, RD; Marian Fisher, PhD; Tavia Gordon; Basil Rifkind, MD.

Anthropometry Working Group
Charles J. Glueck, MD; Henry L. Taylor, PhD; David Jacobs, PhD; John A. Morrison, PhD; Robert Beaglehole, MD; O. Dale Williams, PhD.

Lipoprotein Working Group
Gerardo Heiss, MD; Israel Tamir, MD; Clarence E. Davis, PhD; Herman A. Tyroler, MD; Basil Rifkind, MD; Gustav Schönfeld, MD; David Jacobs, PhD; Ivan D. Frantz, Jr., MD.

Nutrition Analysis Executive Committee
Fred Mattson, PhD, Chairman; George Beaton, MD; Elizabeth Brewer, RD; Tavia Gordon; William Insull, MD; J. Alick Little, MD; Robert E. Shank, MD; H. A. Tyroler, MD; O. Dale Williams, PhD; Nancy Ernst, MS, RD; Marian Fisher, PhD; Basil Rifkind, MD.

LRC Nutrition Committee
Fred Mattson, PhD, Chairman; Janice Henske, RD; Rhea Larsen, RD; Agnes Gordon Fry, RD; Linda Snetselaar, MS, RD; Katherine Salz, MS, RD; Eileen Taylor, RD; Elizabeth Brewer, RD; Katherine Moore, RD; Elizabeth Burrows, RD; Phyllis Ullman, RD; Susan Grimes, RD; Valerie McGuire; Nancy Ernst, MS, RD; Virginia Keating, MS, RD.

LRC Epidemiology Analysis Executive
William Insull, MD; John LaRosa, MD; Robert Wallace, MD; Henry L. Taylor, PhD; Jack Medalie, MD; Carl Rubenstein, MD; Thomas Sheffield, MD; Fred Mattson, PhD; Gerardo Heiss, MD, PhD; Richard Mowery; H.A. Tyroler, MD, Chairman; O. Dale Williams, PhD; Manning Feinleib, MD; Basil M. Rifkind, MD; Kathe Kelly.

LRC Epidemiology Committee
H. A. Tyroler, MD, Chairman; Paul Anderson, PhD; Elizabeth Barrett-Connor, MD; Mary Brockway, PhD; Gary Chase, PhD; Bobbe Christensen, PhD; Linda Cowan, PhD; Michael Criqui, MD; Michael Davies, MD; Alexander Deev, PhD; Ido deGroot, MPH; Leslie Ellis-Kirkland, MPH; Manning Feinleib, MD; Marian Fisher, PhD; Igor Glasunov, MD; Gaetan Godin, MS; S. T. Halfon, MD; Susan Harlap, MD; Robin Harris, MPH; William Haskell, PhD; Gerardo Heiss, MD, PhD; David Hewitt, MA; Judith Hill, MS; Joanne Hoover, MD; Donald Hunninghake, MD; David Jacobs, PhD; Kathe Kelly, MS; J. Alick Little, MD; Arden Mackenthun, PhD; Fred Mattson, PhD; Irma Mebane, MS; Jack Medalie, MD; Richard Mowery, MSPH; John Morrison, PhD; John B. O'Sullivan, MD; Basil Rifkind, MD; Carl Rubenstein, MD; William J. Schull, PhD; William Schwarz, MPH; Dimitri Shestov, MD; Israel Tamir, MD; Henry Taylor, PhD; Pearl Van Natta, MS; Gwen Waldman, MPH; Robert Wallace, MD; O. Dale Williams, PhD, Andrei Zadoja, MD.

LRC Directors Committee
François Abboud, MD; Edwin Bierman, MD; Reagan Bradford MD, PhD; Virgil Brown, MD; William Connor, MD; Gerald Cooper, MD,PhD; John Farquhar, MD; Ivan Frantz, MD; Charles Glueck, MD; Elena Gerasimova, MD; Antonio Gotto, MD, PhD; Victor Grambsch; James Grizzle, PhD; William Hazzard, MD; Donald Hunninghake, MD; Frank Ibbott, PhD; William Insull, MD; Anatoli Klimov, MD; Robert Knopp, MD; Peter Kwiterovich, MD; John LaRosa, MD; Basil Rifkind, MD; J. Alick Little, MD; Fred Mattson, PhD; Maurice Mishkel, MD; Gustav Schönfeld, MD; Helmut Schrott, MD; Robert Shank, MD; Thomas Sheffield, MD; Yechezkiel Stein, MD; Daniel Steinberg, MD; George Steiner, MD; O. Dale Williams, PhD.

References

1 Arteriosclerosis: A report by the National Heart and Lung Institute Task Force on Arteriosclerosis. DHEW Publication No. (NIH) 72–137, vol. 1 (June 1971).

2   Dennis, B.; Ernst, N.; Hjortland, M.; Tillotson, J.; Grambsch, V.: The NHLBI nutrition data system. J. Am. dietet. Ass. 77: 641–647 (1980).
3   Tyroler, H. A. (ed.): Epidemiology of plasma high-density lipoprotein cholesterol levels. The Lipid Research Clinics Program Prevalence Study. Circulation 62: suppl. IV, pp. 1–136 (1980).
4   Ernst, N.; Fisher, M.; Smith, W.; Gordon, T.; Rifkind, B. M.; Little, J. A.; Mishkel, M. A.; Williams, O. D.: The association of plasma high-density lipoprotein cholesterol with dietary intake and alcohol consumption. The Lipid Research Clinics Program Prevalence Study. Circulation 62: suppl. IV, pp. 41–52 (1980).
5   Glueck, C. J.; Taylor, H. L.; Jacobs, D.; Morrison, J. A.; Beaglehole, R.; Williams, O. D.: Plasma high-density lipoprotein cholesterol: association with measurements of body mass. The Lipid Research Clinics Program Prevalence Study. Circulation 62: suppl. IV, pp. 62–69 (1980).
6   Gordon, T.; Castelli, W. P.; Hjortland, M. C.; Kannel, W. B.; Dawber, T. R.: High-density lipoprotein as a protective factor against coronary heart disease. Am. J. Med. 62: 704–714 (1977).
7   Heiss, G.; Johnson, N. J.; Reiland, S.; Davis, C. E.; Tyroler, H. A.: The epidemiology of plasma high-density lipoprotein cholesterol levels. The Lipid Research Clinics Program Prevalence Study. Summary. Circulation 62: suppl. IV, pp. 116–136 (1980).
8   Heiss, G.; Tamir, I.; Davis, C. E.; Tyroler, H. A.; Rifkind, B. M.; Schonfeld, G.; Jacobs, D.; Frantz, I. D.: Lipoprotein-cholesterol distributions in selected North American populations. Circulation 61: 302–315 (1980).
9   Levy, R. I.; Rifkind, B. M.: The structure, function and metabolism of high-density lipoproteins. A status report. Circulation 62: suppl. IV, pp. 4–8 (1980).
10 Mahley, R. W.; Innerarity, T. L.; Bersot, T. P.; Lipson, A.; Margolis, S.: Alterations in human high density lipoproteins with or without increased plasma cholesterol, induced by diets high in cholesterol. Lancet ii: 807–809 (1978).
11 Miller, G. J.; Miller, N. E.: Plasma high density lipoprotein concentrations and development of ischaemic heart disease. Lancet i: 16–19 (1975).
12 Mistry, P.; Nicoll, A.; Niehaus, C.; Christie, I.; Janus, E.; Lewis, B.: Cholesterol feeding revisited. Circulation 54: suppl. II, p. 178A (1976).
13 Tan, M. H.; Dickinson, M. A.: High cholesterol diet raises HDL cholesterol in man. Clin. Res. 25: 703A (1977).
14 The Lipid Research Clinics Population Studies Data Book. I. The Prevalence Study. NIH Publication No. 80-1527 (July 1980).
15 The Lipid Research Clinics Program: Plasma lipid distributions in selected North American populations. The Lipid Research Clinics Program Prevalence Study. Circulation 60: 427–439 (1979).
16 The Lipid Research Clinics Program: The Coronary Primary Prevention Trial. Design and implementation. J. chron. Dis. 32: 609–631 (1979).

B. M. Rifkind, MD, FRCP, Deputy Associate Director for Etiology of Arteriosclerosis and Hypertension, Chief, Lipid Metabolism and Atherogenesis Branch, Division of Heart and Vascular Diseases, National Heart, Lung, and Blood Institute, National Institutes of Health, Room 4A14A – Federal Building, Bethesda, MD 20205 (USA)

Dietary Regulation of Plasma Lipoprotein Metabolism in Humans[1]

Richard J. Havel

Cardiovascular Research Institute and Department of Medicine, University of California, San Francisco, Calif., USA

Introduction

Research on the effects of dietary components on lipid and lipoprotein levels in blood plasma underlies recommendations for diets to treat hyperlipidemia and to reduce the concentration of atherogenic lipoproteins in westernized populations. Major emphasis has attached to the effects of cholesterol and fats in the diet, although recent research has been directed to other components as well. Formulas derived by *Keys* et al. [24] and *Hegsted* et al. [18] describe the average responses of plasma cholesterol levels in adult American populations to dietary cholesterol, saturated fats, and polyunsaturated fats. There is appreciable individual variability in response, especially to dietary cholesterol [34,36], which is poorly understood. The effects of dietary components upon whole body cholesterol metabolism have been carefully characterized [12], but this knowledge has done little to illuminate mechanisms by which dietary components affect plasma lipid and lipoprotein levels. Before research in this area can proceed beyond the descriptive stage, we need to understand the regulation of lipoprotein synthesis and catabolism and the pathways of fat and cholesterol transport in lipoproteins. Fortunately, recent discoveries have greatly increased our knowledge of these processes. In this brief review, I will describe relevant aspects of our present concepts of plasma lipid and

[1] Supported by grants from the US Public Health Service (Arteriosclerosis SCOR HL-14237 and HL-24696).

lipoprotein metabolism and the potential of some of the newer concepts to explain known effects of diet upon plasma lipid and lipoprotein levels.

Metabolism of Plasma Lipoproteins

Transport of Dietary Fat and Cholesterol

Absorbed fatty acids and monoglycerides are esterified to form triglycerides in intestinal mucosal cells and absorbed cholesterol is largely esterified as well [17]. These nonpolar lipids are packaged in the core of chylomicrons and secreted from the cell and transported by the lymphatic system into the blood. A form of apolipoprotein B apparently specific to the intestine (B–48) is essential to the secretory process [22, 31]. This protein, together with several 'A' apoproteins, is secreted with the chylomicron particle. During active fat absorption, the number of particles secreted increases appreciably, but large fat loads are accommodated mainly by an increase in the size of individual chylomicron particles [15]. After secretion, chylomicrons acquire other apoproteins from high-density lipoproteins (HDL) present in lymph and blood plasma (C and E apoproteins) [15]. One of the C apoproteins (apoC-II) permits lipoprotein lipase on the surface of blood capillaries to hydrolyze chylomicron triglycerides rapidly. As this occurs, the A apoproteins dissociate from the chylomicron, perhaps with some of the surface phospholipids, and are transferred to HDL [49]. With continuing hydrolysis, most of the C apoproteins are similarly transferred, leaving a shrunken particle (chylomicron remnant) which thereby becomes a poorer substrate for lipoprotein lipase. The remnants are delivered back into the blood, having lost most of their original complement of triglycerides, but retaining dietary cholesteryl esters. The remnant particles are rapidly absorbed to receptors on the surface of hepatic parenchymal cells, owing to the presence of apoprotein E, which, previously shielded by C apoproteins, now has high affinity for the receptor [16]. The bound remnant particles are rapidly taken up into liver by endocytosis and transferred to lysosomes where the lipid and protein components are hydrolyzed [17]. As these two steps in chylomicron metabolism occur in a few minutes, dietary cholesterol is rapidly transported from the intestine to the liver and contributes little directly to the level of cholesterol in blood plasma. The cholesterol, after lysosomal hydrolysis in the liver, can contribute to biliary cholesterol and (after oxidation) bile acids as well as to the synthesis of hepatogenous lipoproteins. Although only a small fraction of dietary triglyceride is delivered directly to the liver with remnant particles, a considerable fraction (up to one third) of the fatty acids released by lipoprotein lipase is transported to the liver as the albumin-fatty acid complex [17]. Thus, the liver receives not only the bulk of dietary cholesterol, but a considerable fraction of dietary fatty acids as well.

Endogenous Triglyceride Transport

The liver uses fatty acids preferentially as a fuel, accounting for about 20% of basal oxidative metabolism in resting humans. Fatty acids received by the liver in excess of this need are reesterified to form triglycerides that are packaged into lipoproteins secreted from the liver much as chylomicrons are secreted from the intestine [17]. Secretion of these very low-density lipoproteins (VLDL) depends upon an apolipoprotein B (B–100) differing from that produced by the intestine [31]. The C and E apolipoproteins are also secreted from the liver with VLDL particles, and more C apoproteins are acquired from HDL after secretion [50]. The initial catabolism of VLDL-triglycerides by lipoprotein lipase resembles that of chylomicrons, but in humans the process occurs more slowly. The resulting VLDL remnants are apparently taken up by the liver to some extent but mainly are further metabolized, by processes not yet clearly defined, whereby they lose most of the remaining triglycerides and virtually all proteins except apoB–100, to yield low-density lipoproteins (LDL) [17]. In mammals other than humans, much of the VLDL remnants are taken up by the liver and less are converted to LDL, accounting, in species such as rats, for the low plasma levels of LDL [17]. The basis for this species difference in metabolism of VLDL remnants is not known.

Endogenous Cholesterol Transport

In humans, as mentioned above, chylomicron cholesterol contributes little directly to plasma cholesterol. 75–80% of plasma cholesterol is present as cholesteryl esters, produced in the blood plasma by the action of lecithin-cholesterol acyltransferase (LCAT) which transfers a fatty acyl moiety from lecithin to cholesterol to yield mainly cholesteryl linoleate [10]. This enzyme exists in plasma as part of a complex (cholesteryl ester transfer complex), which is a minor component of plasma HDL [9]. The complex contains the enzyme, its cofactor protein (apoprotein A–I), and another protein (apoprotein D) that transfers the cholesteryl esters from the complex to other lipoproteins, principally VLDL or its remnants and

LDL [3].² Thus, the cholesteryl esters of these plasma lipoproteins are produced almost entirely in blood plasma and not during intestinal or hepatic secretion of triglyceride-rich lipoproteins.

A key question concerning plasma cholesterol metabolism is the source of the cholesterol substrate for LCAT. A priori, this cholesterol can be derived from the surface of plasma lipoproteins or from the plasma membrane of cells. In freshly obtained blood plasma incubated at body temperature, esterification of lipoprotein cholesterol continues for several hours, but eventually ceases as cholesteryl esters accumulate, mostly in LDL [7]. This indicates that plasma cholesterol can provide substrate for the enzyme in vivo. When fresh plasma is incubated with cells, some of the cholesterol esterified by the enzyme is drawn from the cells, presumably in the plasma membrane. Recent research shows that cellular and plasma cholesterol can compete for the enzyme, until the accumulation of cholesteryl esters prevents further movement from the transfer complex [6].

LCAT-derived cholesteryl esters are removed from the plasma by several routes. Some may be taken up by the liver with VLDL remnants, but the largest fraction is probably removed during the catabolism of LDL. LDL are catabolized by receptor-dependent adsorptive endocytosis in many cells, thereby providing a regulated supply of cholesterol to the cells [17]. The combined action of the cholesteryl ester transfer complex and the 'LDL receptor' thereby provides a mechanism to remove excess cholesterol from cells and utilize it in other cells, especially those that are actively synthesizing membranes or steroid hormones. Recent research has shown that LDL receptors are also present in the liver [51], which requires cholesterol for synthesis of bile acids. Like LDL receptors in other cells, these receptors in the liver interact with lipoproteins containing apoprotein B-100 and apoprotein E, the latter with much higher affinity [51]. The hepatic receptors are also subject to regulation. Thus, when bile acid synthesis rises during the administration of bile acid-binding resins, the number of functional receptors on hepatocytes increases [25], resulting in

---

² In rats, most of the cholesteryl esters of VLDL and some of those of LDL are secreted with VLDL from the liver. The activity of the enzyme responsible for synthesis of hepatic cholesteryl esters (acyl coenzyme A cholesterol acyltransferase) is high in rats but low in humans [17]. This fact and the high linoleate content of cholesteryl esters of human VLDL strongly suggest that most of these esters are transferred from the cholesteryl ester transfer complex to VLDL by a D apoprotein.

increased uptake and lysosomal degradation of LDL in the liver [25, 44].

In humans, an appreciable fraction, perhaps one to two thirds, of LDL is degraded by mechanisms other than that mediated by the LDL receptor [17, 39]. These mechanisms may include uptake by other receptors and receptor-independent uptake (bulk fluid endocytosis). The latter process is not known to be subject to metabolic regulation. The amount of LDL normally degraded by these mechanisms at various sites in the body is uncertain. Total endocytic uptake of LDL can account for degradation of about 60% of the cholesteryl esters produced by LCAT [17]. A small fraction of the remainder (perhaps about 10%) is probably removed during the catabolism of HDL, by poorly defined mechanisms. Some of the remainder may be catabolized in the liver during the uptake of VLDL or chylomicron remnants.[3] Finally, some cholesteryl esters may be removed from lipoproteins by mechanisms that do not involve endocytosis [5].

Formation and Catabolism of HDL

As described above, protein components of HDL are acquired during the catabolism of chylomicrons by lipoprotein lipase. In addition, some of these proteins are probably secreted directly from the liver, as nascent discoidal HDL particles [14]. As with other lipoproteins, the cholesteryl esters of HDL are produced by LCAT. Although HDL, like LDL, circulate in the blood for several days, they are continually modified by transfer of surface lipid and protein components from triglyceride-rich lipoproteins during the formation of remnant particles. Thus, during alimentary lipemia, the content of A apoproteins and phospholipids increases in HDL [49], increasing the average size of HDL particles. This is reflected in an increased concentration of the lighter species of HDL (HDL_2). The transferred phospholipids may provide some of the lecithin used by LCAT. In addition, recent research suggests that some of this lecithin may also be acted upon by a lipase located on hepatic endothelial cells to yield lysolecithin [21, 35]. The mechanisms of irreversible degradation of HDL are poorly defined, but degradation probably occurs in many cells of the body [17]. HDL other than the small component of the cholesteryl ester transfer complex evidently serve a number of functions in lipoprotein-lipid trans-

---

[3] Although chylomicrons and their remnants may accept cholesteryl esters from the transfer complex, the short life span of chylomicrons in the blood could limit such a process.

port: (a) they provide a reservoir permitting the reutilization of proteins, such as the C apoproteins that function in the metabolism of triglyceride-rich lipoproteins; (b) they accept surface lipids from triglyceride-rich lipoproteins during the formation of remnant particles; (c) they accept and transport some of the cholesteryl esters produced by LCAT. Another minor component of HDL appears to promote the transfer of cellular cholesterol to the transfer complex [6].

Effects of Dietary Lipids on Plasma Lipoproteins

Effects of Dietary Cholesterol on Plasma Cholesterol and Lipoprotein Levels

Mammals vary greatly in their response to dietary cholesterol. In rabbits, dietary cholesterol accumulates readily in large cholesteryl ester-rich particles that have a large complement of apoprotein E as well as apoprotein B, called β-VLDL [29]. The origin of these lipoproteins is uncertain. They may represent chylomicron remnants that are retained in the blood because hepatic receptors are saturated [26] or they may be secreted by the cholesteryl ester-loaded fatty liver that characterizes these animals. Similar β-VLDL accumulate in other mammals such as dogs and rats, provided that cholic acid is added to the diet and the animals are rendered hypothyroid [29], but cholesterol in blood plasma of these animals also accumulates in particles of somewhat greater density that also contain a large complement of apoprotein E, but no apoprotein B (HDL_c). In guinea pigs, both β-VLDL and LDL accumulate with cholesterol feeding, but in this species cholesterol-feeding evidently overwhelms the capacity of LCAT and lamellar particles accumulate that resemble those seen in genetically determined human LCAT deficiency [11], with accompanying accumulation of red blood cells and hemolytic anemia. The magnitude of the response to dietary cholesterol varies widely among subhuman primates, but in some, such as the rhesus monkey, large amounts of cholesteryl esters accumulate in enlarged LDL particles in which the protein component is principally apoprotein B [37].

In normal humans, the response to dietary cholesterol varies from nil to large. Over the usual range of intake, serum cholesterol levels increase an average of 0.12 mg/dl for each mg of cholesterol in 1,000 cal [32]. Sensitive persons seem to inhibit cholesterol synthesis less efficiently than those who are less sensitive [33, 34]. Most of the accumulated cholesterol is esterified and is found in LDL [1, 28, 33], although some may accumulate in HDL_2 and VLDL as well [33]. Although apolipoprotein E levels do not increase systematically in humans fed cholesterol-rich diets [1, 46], some humans fed dietary cholesterol may also accumulate particles that resemble HDL_c [30].

Several mechanisms may account for the lipoprotein-cholesterol response to dietary cholesterol. Perhaps the least likely is that dietary cholesterol molecules accumulate as such in blood plasma. In rats fed cholesterol-containing lipid mixtures, lymphatic chylomicrons are enriched in cholesteryl esters but not in cholesterol [8]. If this is so in humans, such esters are likely to be taken up predominantly by the liver with chylomicron remnants, or by endothelial cells [5], and then hydrolyzed. The cholesteryl esters accumulating in plasma lipoproteins are probably produced by LCAT (in which case they should be distinguishable by containing a preponderance of cholesteryl linoleate). The liver may be expected to compensate for accumulated dietary cholesterol by reducing de novo cholesterol synthesis, by excreting the cholesterol in the bile as such or after conversion to bile acids [28], or by secreting more cholesterol in VLDL or HDL. The latter would provide substrate for LCAT, leading to eventual accumulation of cholesteryl esters in the several lipoprotein classes that can accept them from the cholesteryl ester transfer complex. This cholesterol secreted from the liver, by competing at the level of the transfer complex with tissue cholesterol, would lead to the accumulation of cellular cholesterol. Direct accumulation of dietary cholesterol in hepatocytes and direct or indirect accumulation of dietary cholesterol in cells of peripheral tissues could down-regulate LDL receptors (as observed recently in blood mononuclear leukocytes of persons fed large amounts of cholesterol [33]). This would lead to an increased concentration of LDL and remnant particles, accompanied by increased plasma concentrations of apoprotein B [1, 33]. If the HDL that accumulate during cholesterol feeding contain appreciable amounts of apoprotein E, they might compete effectively with LDL for receptor-dependent uptake, further accentuating the accumulation of LDL. An additional possibility to be considered is that accumulation of LCAT-derived cholesteryl esters in lipoproteins reduces the affinity of the particles for cellular receptors.

It is evident that the extent to which cholesterol accumulates in hepatocytes is likely to be decisive in determining the hepatic responses (biliary excretion, down-regulation of receptors, secretion of cholesterol-enriched lipoproteins) to increased dietary cholesterol. These responses

Effects of Dietary Fats on Plasma Lipid and Lipoprotein Levels

may, in turn, be of major importance in determining the extent to which lipoprotein concentrations increase.

Low-fat diets are regularly accompanied by reduced concentration of plasma cholesterol, whereas plasma triglyceride levels tend to increase [12]. Studies in rats suggest that the latter change results from hepatic secretion of larger particles containing more triglycerides [52], but a role for reduced triglyceride catabolism by lipoprotein lipase has not been excluded. In humans fed diets containing < 5% of calories left from fat, the content of cholesterol is reduced in both LDL and HDL even when dietary cholesterol is held constant [27]. Reduction of phospholipids in LDL is comparable to that of cholesterol, but phospholipid is relatively preserved in HDL, indicating a change in composition of HDL particles [27]. In Japanese, who habitually ingest low-fat diets, HDL-cholesterol levels are at least as high as those in the USA, with no increase in the concentration of VLDL [23]. In vegetarians, cholesterol levels are reduced in both LDL and HDL, but reduction of HDL-cholesterol is less pronounced [38]. Reduced traffic of dietary fat could be accompanied by lower rates of synthesis of protein components of chylomicrons that contribute to HDL. The basis for the variability of HDL levels among populations ingesting low-fat diets is unclear.

In several recent studies the changes in plasma lipoprotein concentrations and composition that characterize the differing effects of saturated and polyunsaturated fats have been determined. In normolipidemic young women, changing the P/S ratio from 0.2 to 2 in a diet containing 43% of calories from fat (cholesterol intake not held constant) reduced the cholesterol content of VLDL and LDL [42]; the former were relatively enriched in protein with reduced content of cholesteryl esters. Little change occurred in lipid and protein components of HDL. In a similar study of hyperlipoproteinemic patients, cholesterol intake was held constant at 300 mg daily as the P/S ratio was altered [48]. In patients with isolated elevations of VLDL or combined elevations of VLDL and LDL, the polyunsaturated fat-rich diet reduced cholesterol levels in VLDL and LDL but not in HDL; in patients with isolated elevations of LDL, the cholesterol levels of all major lipoprotein fractions was reduced. In normal young men fed very polyunsaturated fat-rich diets (P/S ratio = 4.0), with cholesterol intake held constant at 400 mg daily, cholesterol concentrations fell by about 25% in all major lipoprotein fractions [40]. The composition of LDL (isolated between densities 1.019 and 1.063 g/ml) was systematically altered, with reduced content of cholesteryl esters and increased content of phospholipids.

Several mechanisms have been adduced to explain the hypocholesterolemic effect of polyunsaturated fatty acid-rich diets. As first proposed by *Spritz and Mishkel* [45], this effect might be explained by an altered composition of LDL, which normally carries about two thirds of the total serum cholesterol. Reduced content of cholesteryl esters in LDL has been observed in isolated LDL by others [40, 42], but this finding is not universal [13]. The concentration of the protein moiety of LDL (primarily apoprotein B-100) is also reduced by polyunsaturated fat-rich diets and its fractional catabolic rate has been found to be increased [40]. Intravenous infusion of a fat emulsion stabilized by egg lecithin was shown to increase the saturation of LDL-phospholipids and to reduce the fractional catabolic rate of apo-LDL [47], suggesting that the increased fluidity of the lipoprotein surface produced by feeding polyunsaturated fats may promote endocytic catabolism of LDL, perhaps by increasing the affinity of LDL for receptors that recognize apoprotein B-100. Polyunsaturated fat-rich diets also reduce the concentration of VLDL, the precursor of LDL [2]. Reduction of VLDL triglyceride levels in such diets occurs in the absence of evidence for increased triglyceride catabolism [2, 48], suggesting that the synthesis of the protein moiety of LDL might be reduced, but in one study this has not been found to be the case [40]. The reduced concentration of HDL-cholesterol produced by polyunsaturated fat-rich diets is accompanied by reduced concentration of apoprotein A–I, with comparable reduction in the synthetic rate of this protein [41]. One possibility raised by this observation is that the content of chylomicron proteins is affected by the composition of dietary fatty acids.

As with dietary cholesterol, dietary fat composition evidently can affect serum lipoprotein levels by several mechanisms. Although biliary sterol and bile acid excretion are increased by polyunsaturated fat-rich diets in some persons (especially those with hypertriglyceridemia) [12], the effects of such diets on body steroid turnover and balance seem to bear no necessary relationship to the effects on plasma lipoproteins.

Effects of Therapeutic Diets on Plasma Lipoprotein Levels

It is evident that the content and composition of dietary fat as well as cholesterol can affect the levels of HDL as well as those of LDL and VLDL. Hence, diets usually recommended to reduce the levels of the potentially atherogenic lipoproteins, LDL and VLDL, may reduce HDL levels as well. In some recent short-term studies, diets restricted in cholesterol and saturated fats and augmented in polyunsaturated fats have been found to reduce the concentration of cholesterol in all major lipoprotein fractions [4, 46]. However, in long-term trials, diets with reduced cholesterol and fat content and a moderate increase in P/S ratios (ca 1/1) have regularly reduced VLDL and LDL levels, without affecting HDL levels [19, 20]. In these studies, alterations in body weight and cigarette use may have influenced the results. However, it cannot be assumed that effects observed in trials lasting 2–4 weeks will be sustained. Nor can it be assumed that changes in HDL produced by drastic modifications of dietary fats and cholesterol will be observed with diets containing about 300 mg cholesterol daily and a P/S ratio near unity.

The components of therapeutic diets that can influence serum lipid and lipoprotein levels are not limited to fats and cholesterol, as exemplified by the recent demonstration that diets containing soy protein as the sole protein source lead to substantial reductions of LDL-cholesterol, with no change in HDL-cholesterol [43]. In future studies, it will be necessary not only to isolate the effects of individual dietary components, but to characterize in detail their influences on lipoprotein concentration and composition and then to conduct appropriate studies of lipoprotein-lipid and protein metabolism. Hepatic lipoprotein receptors have major roles in the catabolism of remnant particles and LDL. The possibility that these receptors are subject to dietary regulation offers particularly exciting prospects for research in experimental animals and humans, with due regard for species differences in responses. Much more research is needed to clarify the mechanisms by which dietary components influence the formation and catabolism of HDL, and the formation and transport of plasma cholesteryl esters. Future research on the regulation of plasma lipid and lipoprotein metabolism by dietary components is likely to find practical application in the formulation of diets that minimize atherogenesis.

References

1 Applebaum-Bowden, D.; Hazzard, W.R.; Cain, J.; Cheung, M.C.; Kushwaha, R.S.; Albers, J.J.: Short-term egg yolk feeding in humans. Increase in apolipoprotein B and low density lipoprotein cholesterol. Atherosclerosis *33:* 385–396 (1979).
2 Chait, A.; Onitiri, A.; Nicoll, A.; Rabaya, E.; Davies, J.; Lewis, B.: Reduction of serum

triglyceride levels by polyunsaturated fat. Studies on the mode of action and on very low density lipoprotein composition. Atherosclerosis 20: 347–364 (1974).

3 Chajek, T.; Fielding, C. J.: Isolation and characterization of a human serum cholesteryl ester transfer protein. Proc. natn. Acad. Sci. USA 75: 3445–3449 (1978).

4 Ernst, N.; Fisher, M.; Bowen, P.; Schaefer, E. J.; Levy, R. I: Changes in plasma lipids and lipoproteins after a modified fat diet. Lancet ii: 111–113 (1980).

5 Fielding, C. J.: Metabolism of cholesterol-rich chylomicrons. Mechanism of binding and uptake of cholesteryl esters by the vascular bed of the perfused rat heart. J. clin Invest. 62: 141–151 (1979).

6 Fielding, C. J.; Fielding P. E.: Evidence for a lipoprotein carrier in human plasma catalyzing sterol efflux from cultured fibroblasts, and its relationship to lecithin: cholesterol acyltransferase. Proc. natn. Acad. Sci. USA 78: 3911–3914 (1981).

7 Fielding, C. J.; Fielding P. E.: Regulation of human plasma lecithin: cholesteryl acyltransferase activity by lipoprotein acceptor cholesteryl ester content. J. biol. Chem. 256: 2102–2104 (1981).

8 Fielding, C. J.; Renston, J. P.; Fielding, P. E.: Metabolism of cholesterol-enriched chylomicrons. Catabolism of triglycerides by lipoprotein lipase of perfused heart and adipose tissue. J. Lipid Res. 19: 705–711 (1978).

9 Fielding, P. E.; Fielding, C. J.: A cholesteryl ester transfer complex in human plasma. Proc. natn. Acad. Sci. USA 77: 3327–3330 (1980).

10 Glomset, J. A.: The plasma lecithin: cholesterol acyltransferase reaction. J. Lipid Res. 9: 155–167 (1968).

11 Guo, L. S. S.; Meng, M.; Hamilton, R. L.; Ostwald, R.: Changes in the plasma lipoprotein-apoproteins of guinea pigs in response to dietary cholesterol. Biochemistry, N.Y. 16: 5807–5812 (1977).

12 Grundy, S. M.: Dietary fats and sterols; in Levy, Rifkind, Dennis, Ernst, Nutrition, lipids and coronary heart disease, pp. 89–118 (Raven Press, New York 1979).

13 Grundy, S. M.: Personal communication.

14 Hamilton, R. L.; Williams, M. C.; Fielding, C. J.; Havel, R. J.: Discoidal bilayer structure of nascent high density lipoproteins from perfused rat liver. J. clin. Invest. 58: 667–680 (1976).

15 Havel, R. J.: Origin of HDL; in Gotto, Miller, Oliver, High density lipoproteins and atherosclerosis, pp. 21–35 (Elsevier/North-Holland, Amsterdam 1978).

16 Havel, R. J.: Lipoprotein biosynthesis and metabolism. Ann. N.Y. Acad. Sci. 348: 16–27 (1980).

17 Havel, R. J.; Goldstein, J. L.; Brown, M. S.: Lipoproteins and lipid transport; in Bondy, Rosenberg, Metabolic control and disease; 8th ed., pp. 393–494 (Saunders, Philadelphia 1980).

18 Hegsted, D. M.; McGandy, R. B.; Myer, M. L.; Stare, F. J.: Quantitative effects of dietary fat on serum cholesterol in man. Am. J. clin. Nutr. 17: 281–295 (1965).

19 Hjermann, J.; Enger, S.-C.; Helgeland, A.; Holme, I.; Leren, P.; Trygg, K.: The effect of dietary changes on high density lipoprotein cholesterol. The Oslo study. Am. J. Med. 66: 105–109 (1979).

20 Hulley, S. B.; Cohen, R.; Widdowson, G.: Plasma high density lipoprotein concentration. Influence of risk factor intervention. J. Am. med. Ass. 238: 2269–2271 (1977).

21 Jansen, H.; Van Tol, A.; Hülsmann, W. C.: On the metabolic function of heparin-releasable liver lipase. Biochem. biophys. Res. Commun. 92: 53–59 (1980).

22 Kane, J. P.; Hardman, D.; Paulus, H. E.: Heterogeneity of apolipoprotein B.: isolation of a new species from human chylomicrons. Proc. natn. Acad. Sci. USA 77: 2463–2469 (1980).
23 Kano, Y.; Irie, N.; Homma, Y.; Tsushima, M.; Takeuchi, J.; Nakaya, N.; Goto, Y.: High density lipoprotein cholesterol levels in the Japanese. Atherosclerosis 36: 173–181 (1980).
24 Keys, A.; Anderson, J. T.; Grande, F.: Prediction of serum cholesterol responses in man to change in fats in the diet (1957). Lancet ii: 959–966 (1965).
25 Kovanen, P. T.; Bilheimer, D. W.; Goldstein, J. L.; Jaramillo, J. J.; Brown, M. S.: Regulatory role for hepatic low density receptors in vivo in the dog. Proc. natn. Acad. Sci. USA 78: 1194–1198 (1981).
26 Kovanen, P., T.; Brown, M. S.; Basu, S. K.; Bilheimer, D. W.; Goldstein, J. L.: Saturation and suppression of hepatic lipoprotein receptors: a mechanism for the hypercholesterolemia of cholesterol-fed rabbits. Proc. natn. Acad. Sci. USA 78: 1396–1400 (1981).
27 Kroes, J.; Havel, R. J.; Kane, J. P.: Unpublished data.
28 Lin, D. S.; Connor, W. E.: The long-term effects of dietary cholesterol upon the plasma lipids, lipoproteins, cholesterol absorption, and the sterol balance in man: the demonstration of feedback inhibition of cholesterol biosynthesis and increased bile acid excretion. J. Lipid Res. 21: 1042–1052 (1980).
29 Mahley, R. W.: Dietary fat, cholesterol and accelerated atherosclerosis; in Paoletti, Gotto, Atherosclerosis reviews, vol. 5, pp. 1–35 (Raven Press, New York 1979).
30 Mahley, R. W.; Innerarity, T. L.; Bersot, T. P.; Lipson, A.; Margolis, S.: Alterations in human high-density lipoproteins, with or without increased plasma-cholesterol, induced by diets high in cholesterol. Lancet ii: 807–809 (1978).
31 Malloy, M. J.; Kane, J. P.; Hardman, D. A.; Hamilton, R. L.; Dalal, K. B.: Normotriglyceridemic abetalipoproteinemia: absence of the B-100 apolipoprotein. J. clin. Invest. 67: 1441–1450 (1980).
32 Mattson, F. N.; Erickson, R. A.; Klegman, A. M.: Effect of dietary cholesterol on serum cholesterol in man. Am. J. clin. Nutr. 25: 589–594 (1972).
33 Mistry, P.; Miller, N. E.,; Laker, M.; Hazzard, W. R.; Lewis, B.: Individual variation in the effects of dietary cholesterol on plasma lipoproteins and cellular cholesterol homeostasis in man. Studies of low density lipoprotein receptor activity and 3-hydroxy-3-methylglutaryl coenzyme A reductase activity in blood mononuclear cells. J. clin. Invest. 67: 493–502 (1981).
34 Nestel, P. J.; Poyser, A.: Changes in cholesterol synthesis and excretion when cholesterol intake is increased. Metabolism 25: 1591–1599 (1976).
35 Nikkilä, E. A.; Kuusi, T.; Harno, K.; Tikkanen, M.; Tashkinen, M. R.: Lipoprotein lipase and hepatic endothelial lipase are key enzymes in the metabolism of plasma high density lipoproteins, particularly of HDL_2; in Gotto, Smith, Allen, Atherosclerosis V, p. 387 (Springer, New York 1980).
36 Quintao, E.; Grundy, S. M.; Ahrens, E. H., Jr.: Effects of dietary cholesterol and the regulation of total body cholesterol in man. J. Lipid Res. 12: 233–247 (1971).
37 Rudel, L. L.; Shah, R.; Green, D. G.: Studies of the atherogenic dyslipoproteinemia induced by dietary cholesterol in rhesus monkeys (Macaca mulatta). J. Lipid Res. 20: 55–65 (1979).

38 Sacks, F.M.; Castelli, W.P.; Donner, A.; Kass, E.H.: Plasma lipids and lipoproteins in vegetarians and controls. New Engl. J. Med. 292: 1148–1151 (1975).
39 Shepherd, J.; Bicker, S.; Lorimer, A.R.; Packard, C.J.: Receptor mediated low-density lipoprotein catabolism in man. J. Lipid Res. 20: 999–1006 (1979).
40 Shepherd, J.; Packard, C.J.; Grundy, S.M.; Yeshurun, D.; Gotto, A.M., Jr.; Taunton, O.D.: Effects of saturated and polyunsaturated fat diets on the chemical composition and metabolism of low density lipoproteins in man. J. Lipid Res. 21: 91–99 (1980).
41 Shepherd, J.; Packard, C.J.; Patsch, J.R.; Gotto, A.M., Jr.; Taunton, O.D.: Effects of dietary polyunsaturated and saturated fat on the properties of high density lipoproteins and the metabolism of apolipoprotein A–I. J. clin. Invest. 61: 1582–1592 (1978).
42 Shepherd, J.; Stewart, J.M.; Clark, J.G.; Carr, K.: Sequential changes in plasma lipoproteins and body fat composition during polyunsaturated fat feeding in man. Br. J. Nutr. 44: 265–271 (1980).
43 Sirtori, C.R.; Agradi, E.; Conti, F.; Mantero, O.; Gatti, E.: Soybean-protein diet in the treatment of type II hyperlipoproteinemia. Lancet i: 275–277 (1977).
44 Slater, H.W.; Packard, C.J.; Bicker, S.; Shepherd, J.: Effects of cholestryramine on receptor-mediated plasma clearance and tissue uptake of human low-density lipoproteins in the rabbit. J. biol. Chem. 255: 10210–10213 (1980).
45 Spritz, N.; Mishkel, M.A.: Effects of dietary fat on plasma lipids and lipoproteins: an hypothesis for the lipid lowering effects of polyunsaturated fatty acids. J. clin. Invest. 48: 78–86 (1969).
46 Tan, M.H.; Dickenson, M.A.; Albers, J.J.; Havel, R.J.; Cheung, M.C.; Vigne, J.-L.: The effect of a high cholesterol and saturated fat diet on serum high density lipoprotein-cholesterol, apoprotein A–I, and apoprotein E levels in normolipidemic humans. Am. J. clin. Nutr. 33: 2259–2265 (1980).
47 Thompson, G.R.; Jadhav, A.; Nava, M.; Gotto, A.M.: Effect of intravenous phospholipid on low density lipoprotein turnover in man. Eur. J. clin. Invest. 6: 241–248 (1976).
48 Vessby, B.; Gustufsson, I.-B.; Boberg, J.; Karlström, B.; Lithell, H.; Werner, J.: Substituting polyunsaturated for saturated fat as a single change in a Swedish diet: effects on serum lipoprotein metabolism and glucose tolerance in patients with hyperlipoproteinemia. Eur. J. clin. Invest. 10: 193–202 (1980).
49 Vigne, J.-L.; Havel, R.J.: Metabolism of apolipoprotein A–I of chylomicrons in rats and humans. Can. J. Biochem. 59: 613–618 (1981).
50 Windler, E.; Chao, Y.-s.; Havel, R.J.: Determinants of hepatic uptake of triglyceride-rich lipoproteins and their remnants in the rat. J. biol. Chem. 255: 5475–5480 (1980).
51 Windler, E.E.T.; Kovanen, P.T.; Chao, Y.-s.; Brown, M.S.; Havel, R.J.; Goldstein, J.L.: The estradiol-stimulated lipoprotein receptor of rat liver: a binding site that mediates the uptake of rat lipoproteins containing apoproteins B and E. J. biol. Chem. 255: 10464–10471 (1980).
52 Witztum, J.L.; Schonfeld, G.: Carbohydrate diet-induced changes in very low density lipoprotein composition and structure. Diabetes 27: 1215–1229 (1978).

R.J. Havel, Cardiovascular Research Institute and Department of Medicine, University of California, San Francisco, CA 94143 (USA)

Studies on Diet and Plasma Lipids in Naples

E. Farinaro, P. Rubba, A. Postiglione, G. Riccardi, M. Mancini

Center for Arteriosclerosis and Metabolic Disease, Semeiotica Medica, 2nd Medical School, University of Naples, Naples, Italy

Many population studies have shown low levels of plasma cholesterol in the Mediterranean area, including Naples [8, 10]. It has been suggested that this might be related to features of the habitual diet in this region [8, 10]. In particular, some studies suggest that the low consumption of saturated fat and cholesterol is the main determinant of low levels of plasma cholesterol.

Plasma lipid concentrations do not show marked differences in newborns of different geographical areas [9, 11]. However, differences among groups from different populations become apparent later in life when children are exposed to different environments. The increase in plasma lipid concentration after birth is mainly due to an increase in low-density lipoproteins (LDL) and very low-density lipoproteins (VLDL). An increase in serum lipids is already detectable after the first 6 months of life [7]. Among many environmental factors, diet is most likely to play a central role in producing this change. If risk of coronary heart disease starts in the early years of life, it is not surprising that much interest has been recently devoted to the identification and correction of risk factors in childhood [2, 13, 14]. In this regard, we have joined an international program on the prevention of chronic diseases at an early age organized by the American Health Foundation. This program involves the early detection of children at high risk and nonpharmacological control efforts [4, 14].

Findings from this study have shown that young people living in the Mediterranean area have lower levels of serum lipids than others studied in the United States and northern Europe (table I). These differences might be related to the diet consumed in southern Italy which is traditionally low in saturated fats and cholesterol and rich in complex carbohydrates.

Table I. Serum lipid concentration in school age children – Mugnano (Naples), Italy, 1979

		Cholesterol, mg/dl	Triglycerides, mg/dl
Boys	n	215	158
	M ± SD	148 ± 26	67 ± 22
Girls	n	171	154
	M ± SD	152 ± 25	78 ± 28
All	n	386	312
	M ± SD	150 ± 25	73 ± 25

Table II. Calories available per person per day from specific food sources in four European countries (FAO, 1977)

	Italy	Switz.	UK	Sweden
Butter	34	130	164	190
Dairy	214	600	528	656
Meat	118	348	509	351
Grains + starchy vegetables	1,484	1,067	1,045	1,021
Eggs	36	40	59	46
Oils + nuts	320	314	120	337

Comparison of different eating patterns gives further support to the idea that dietary lipids are major factors responsible for population differences in plasma cholesterol levels.

Recent data from the United Nations' Food and Agriculture Organization (FAO) [6] on the availability of major nutrients indicate that consumption of dietary fat in Italy is still very low, while foods rich in carbohydrates are preferred in Italy as compared to other European countries. Data on major nutrients, of course, give only limited information on dietary patterns in different countries and ought to be integrated with information on specific items of consumption.

Table II shows that little butter, dairy products, or meat are consumed in Italy while grains and starchy vegetables are major sources of nutrients. The limitation of these data is that they report differences in the availability of items, from which consumption can only be inferred.

Table III. Dietary survey among healthy Neapolitans (alcohol excluded)

Population	Year	Total cal	Fat, %	C %	PT, %
Working class	1957	2,900	25	63	12
SIP telephone company	1980	2,227	36	49	15

C = carbohydrates; PT = protein.

Table IV. Consumption of different food items in a sample of adult Neapolitan population (201 males, 234 females)

Item	Percent of subjects consuming the item			
	daily	weekly	monthly	never
Wine	87	–	–	13
Bread	98	2	–	–
Pasta, rice	87	13	–	–
Meat	2	73	2	23
Butter	41	–	–	59
Egg	1	70	22	7
Olive oil	95	–	–	5

To evaluate in greater detail the characteristics of the habitual diet in Naples, we investigated the major nutrient composition of the diet and specific food items consumed by individuals employed in the Naples Telephone Company using the 7-day food record. As shown in table III, daily intake of dietary fat, mostly unsaturated fatty acids, is about 36% of total calories. It must be noted that, when compared to a previous survey conducted in 1957 [5], the proportion of dietary fat to total daily calories appears to be higher today than in the late 1950s.

Preliminary data from another study carried out by dietary interview of a free-living population sample in the Naples area are shown in table IV [3]. They indicate that animal fat is not consumed at all by the great majority of Neapolitans while use of fresh fruit, vegetables, beans, olive oil, pasta, and bread is very common. In addition, drinking wine during the meal is a common habit in the Neapolitan male population, with an

Table V. Composition of a representative 2,200 kcal diet

Diets	A		B		C	
	g/day	kcal %	g/day	kcal %	g/day	kcal %
Protein	98	17	98	17	121	21
Fat, total	76	30	74	30	93	37
Saturated	17		18		26	
Monounsaturated	50		47		54	
Polyunsaturated	9		9		13	
Cholesterol	<0.250		<0.250		<0.250	
Carbohydrate, total	318	53	321	53	252	42
Simple	111		112		108	
Complex	207		209		144	
Plant fiber, total	16		54		20	

Table VI. Blood glucose (mg/dl) and glycosuria (g/24 h) at the end of the three dietary periods (M ± SD)

	Diets		
	A	B	C
Postprandial (2 h) blood glucose	167.6*** ± 52.08	130.1 ± 48.8	159.6* ± 59.8
Mean daily blood glucose	145.1* ± 36.4	114.4 ± 42.7	137.0* ± 27.8
Glycosuria	13.1** ± 12.8	4.0 ± 5.3	10.1* ± 8.5

Significance vs B: * $p < 0.05$; ** $p < 0.02$; *** $p < 0.005$.

average daily consumption of 40 g [6]. Both studies thus show that the Neapolitan diet is low in animal fat and relatively high in complex carbohydrates including nondigestible fibers.

The effects of a diet rich in fiber have been studied recently by our group and others [1, 12]. The data indicate that this diet induces both a

Table VII. Lipoprotein composition at the end of the three dietary periods (M ± SD)

Diets	Cholesterol, mg/dl				Triglycerides, mg/dl	
	total	VLDL	LDL	HDL	total	VLDL
A	227.0* ± 50.2	25.1 ±13.1	140.1* ± 40.9	52.1 ±13.9	142.5* ± 62.8	83.2* ±58.4
B	179.9 ± 25.1	20.1 ± 8.9	110.0 ± 27.0	50.2 ±13.9	133.6 ± 52.2	77.0 ±39.8
C	215.0*** ± 37.0	18.1 ±10.0	128.9** ± 29.0	56.0*** ±15.8	122.1 ± 47.8	69.0 ±31.0

Significance vs B: * $p < 0.05$; ** $p < 0.01$; *** $p < 0.001$.

reduction of plasma lipids and an improvement of glucose tolerance. In our experiment with a fiber-rich diet, diabetic patients were put on three different diets of an identical, fixed caloric content as shown in table V, and blood glucose and serum lipoprotein analyses were performed.

All diets were restricted in monosaccharides and disaccharides and were composed only of foodstuffs normally available at the food stores. The first diet (A) contained a usual amount of plant fiber and was designed to resemble as closely as possible the usual Italian diet in the caloric distribution of nutrients. This diet was taken for 10 days each, in a random order. Diet B was identical in composition to diet A except for the amount of plant fiber which was increased 3- to 4-fold. The plant fiber was provided by: starchy vegetables (i.e. beans, peas, chickpeas, lentils, 45%); wholemeal bread (12%); other vegetables such as broccoli, cabbage, eggplant, artichokes, cauliflower, bean sprouts, mushrooms, and fennel (37%); and fresh fruit (6%). Diet C was the classical low carbohydrate diet used for the care of diabetic patients. The three diets were identical in their polyunsaturated/saturated (P/S) ratios and in the amount of soluble carbohydrates.

Both 2-hour postprandial glucose and mean daily glucose levels were significantly lower with diet B than with either of the other two diets (table VI), as were total and LDL cholesterol levels. Total and VLDL triglycerides levels with diet B were significantly lower than those with diet A, but they were almost identical to those with diet C. The concentration of high-density lipoprotein cholesterol was not affected by dietary fiber, but was significantly increased by the low carbohydrate diet (table VII). A high-fiber, normal carbohydrate diet therefore improves blood glucose control and decreases the concentration of atherogenic lipoproteins in diabetic patients. This effect seems to be independent of the amount on carbohydrate available in the diet.

References

1 Anderson, J.W.; Kyllen, W.: High carbohydrate, high-fiber diets for insulin treated men with diabetes and hyperlipidemia. Lancet *ii:* 128–189 (1979).
2 Ball, K.; Brook, C. (eds): The prevention of coronary heart disease in childhood. Post-grad. med. J. *54:* 629 (1978).
3 Farinaro, E.; Mancini, M.: Unpublished data.
4 Farinaro, E.; Panico, S.; Mancini, M.: The Know Your Body Program in Italy. Prev. Med. *10:* 187–194 (1981).
5 Fidanza, F.; Mancini, M.; Cioffi, L.: Quadro lipidemico ed emocoagulazione nell'uomo in rapporto all'alimentazione ed alla attività fisica. Minerva med. Roma *51:* 1183–1198 (1960).
6 Food and Agriculture Organization, United Nations: Food Balance Sheets 1975–77, average and per capita food supplies.
7 Glueck, C.J.; Tsang, R.; Balister, R.; et al.: Plasma and dietary cholesterol in infancy: effect of early low or moderate dietary cholesterol intake on subsequent response to increased dietary cholesterol. Metabolism *21:* 1181–1193 (1972).
8 Keys, A.: Coronary heart disease in seven countries. Circulation *41:* suppl. 1 (1970).
9 Kwiterovich, P.O.; Rifkind, B.M.; Levy, R.I. (eds): Pediatric aspect of hyperlipoproteinemia in hyperlipidemia. Diagnosis and therapy (Grune & Stratton, New York 1977).
10 Lewis, B.; Chait, A.; Sigurdsson, G.; Mancini, M.; Farinaro, E.; Oriente, P.; Carlson, L.A.; Ericson, M.: Serum lipoproteins in four European communities: a quantitative comparison. Eur. J. clin. Invest. *8:* 165–173 (1978).
11 Plasma lipid and lipoprotein levels at birth in Naples. 32nd Meet. of the Eur. Atherosclerosis Group, 1978, p. 32.
12 Rivellese, A.; Riccardi, G.; Giacco, A.; Pacioni, D.; Genovese, S.; Mattioli, P.L.; Mancini, M.: Effect of dietary fibre on glucose control and serum lipoproteins in diabetic patients. Lancet *ii:* 447–450 (1980).
13 Stamler, J.: Lectures on preventive cardiology (Grune & Stratton, New York 1967).
14 Williams, C.; Arnold, C.B.; Wynder, E.L.: Primary prevention of chronic disease beginning in childhood. The Know Your Body Program design of study. Prev. Med. *6:* 344–347 (1977).

E. Farinaro, MD, Center for Arteriosclerosis and Metabolic Diseases, Semeiotica Medica, 2nd Medical School, University of Naples, I-80131 Naples (Italy)

Metabolism of Lipoproteins in the Postprandial Phase

Giovannella Baggio, Carlo Gabelli, Renato Fellin, Maria Rosa Baiocchi, Scipione Martini, Goretta Baldo, Enzo Manzato, Gaetano Crepaldi

Department of Internal Medicine, Division of Gerontology and Metabolic Diseases, University of Padova, Padova, Italy

Introduction

The role played by the lipoproteins in the process of atherogenesis is long known. The increase in low-density lipoproteins (LDL) and very low-density lipoproteins (VLDL) is widely recognized as one of the main risk factors in atherosclerotic cardiovascular diseases. High-density lipoproteins (HDL) instead seem to be not only negatively correlated with ischemic heart disease but act as protective factors as well [11].

Nevertheless, frank atherosclerotic processes have been observed in patients where risk factors are absent and whose lipid and lipoprotein pattern is absolutely normal. All epidemiological studies are conducted on fasting subjects. For many years now, it has been demonstrated that the lipoprotein classes are not separated statically from each other, but rather continuously exchange both their protein and lipid portions [12]. Some enzymes act as the regulators of these processes or part of them, especially the lipoprotein-lipases and lecithin-cholesterol acyltransferase (LCAT) which are activated or inhibited by the apolipoproteins [18]. The postprandial phase is particularly interesting, since it is a moment of activation of all the processes of formation, catabolism, and interchange of the different lipoprotein fractions [2]. However, there are very few reports regarding lipoprotein metabolism during this stage and the results are not always concordant. This is probably due to the wide variability in the postprandial state which is highly conditioned by the processes of intestinal absorption [5] as well as the difficulty in standardizing precise methods that may be applied to epidemiological studies.

On the other hand, the study of lipoprotein metabolism during the postprandial phase is extremely important not only for a better comprehension of the metabolic processes themselves, but also for understanding if and how the process of atherogenesis is correlated with them.

In the 1950s *Havel* demonstrated that, following fat ingestion, the increase in the triglycerides (TG) was attributable to the increase in VLDL with Sf > 10, while cholesterol increase was insignificant if not nil.

Nonetheless, the growing interest during the 1970s in HDL has led to more profound study of their origin as well as their significance in relation to the clearance of TG-rich lipoproteins [23]. The postprandial phase thus represents a privileged moment for the study of these relationships.

Unlike the chylomicrons and VLDL, HDL are never secreted as such into the circulation, but appear to be formed within the plasmatic circulation from precursors that derive from the intestine (chylomicrons) and the liver (VLDL) if not also from all the other lipoprotein classes and the cellular membranes [4, 10]. Thus, the processes that regulate the HDL level would depend on lipase activity [13].

In 1973, attention was called to the fact that HDL without doubt furnished components (particularly C peptides) that were important for the removal of chylomicrons during alimentary lipemia, and yet in turn received surface constituents during chylomicron catabolism [6].

During the lipoprotein lipase-mediated processes of TG-rich lipoprotein catabolism, apolipoprotein A-I, apoC peptides, phospholipids, and free cholesterol leave the surface shell and appear within the HDL density range [21]. These HDL, termed 'nascent', lose their disc shape and become spherical through LCAT action. The difference between HDL derived from chylomicrons and HDL derived from VLDL is that the latter contain less apoA-I [20].

The distinction between the subfraction HDL_2 (d = 1.063–1.120 g/ml) and HDL_3 (d = 1.120–1.210 g/ml) is very important. It appears that HDL_3 are produced first and their pool also seems more stable. HDL_2 instead undergo more fluctuations directly in relationship to lipase activity [10, 23].

In vitro studies have shown that incubation of isolated lipoprotein lipase with VLDL and HDL_3 leads to the formation of HDL_2 as well as intermediate-density lipoproteins (IDL) and LDL [15].

In fact, HDL_2 are more concentrated in females, and increase after physical exercise or nicotinic acid therapy, while they decrease in the postmenopausal age [19].

Table I. Cholesterol, triglycerides, and phospholipids (mg/dl, $\overline{m} \pm$ SE) of whole serum (WS), VLDL, LDL, and HDL in 6 males (♂) and 6 females (♀) at 0 time, and at 4.5 and 9 h postprandial

	0 h		4.5 h		9 h	
	♂	♀	♂	♀	♂	♀
Cholesterol						
WS	187.2±15.4	183.7±11.1	184.6±13.0	180.0±10.8	187.8±14.9	181.2±11.5
VLDL	9.5± 3.2	6.3± 1.4	21.7± 4.8*	13.2± 3.4	10.8± 3.3	5.3± 1.3
LDL	127.7±15.5	115.2±11.9	118.2±13.3	112.5±12.2	128.5±14.7	115.8±12.6
HDL	50.2± 3.9	62.2± 6.0	44.7± 4.3	56.5± 6.7	48.6± 5.0	60.0± 5.9
Triglycerides						
WS	78.8±12.8	72.0± 8.0	189.3±33.1*	138.7±29.2*	104.2±13.5	78.3±10.7
VLDL	28.2± 6.3	17.7± 5.3	118.7±29.4*	65.3±25.3	35.8± 9.4	20.6± 5.7
LDL	34.2± 5.9	32.7± 3.5	50.4± 8.2	48.2± 7.6	48.6± 5.7	35.9± 5.8
HDL	16.8± 2.0	21.5± 1.0	20.2± 1.3	25.2± 2.3	19.7± 1.7	21.8± 1.3
Phospholipids						
WS	183.0±11.8	209.3±10.1	203.0±13.4	224.8± 8.1	209.8±11.5	228.5±11.4
LDL	75.2±14.0	97.8±10.3	89.7±15.5	93.4± 9.6	83.8±11.7	98.9±11.1
HDL	99.5± 8.9	105.8± 9.5	97.7± 9.0	117.8± 8.1	117.7± 7.3	122.2± 7.0

* $p < 0.05$ (4.5 vs 0).

There are numerous observations regarding the changes and the composition of lipoproteins, in particular chylomicrons and VLDL, during postprandial phase [8, 16]. Reports regarding HDL and their relationships with the TG-rich lipoproteins are instead less frequent. This study, therefore, concerns the relationship between chylomicrons and VLDL and HDL_2 and HDL_3 following a standard Italian meal in humans. Some preliminary data regarding the behavior of plasma lipids following a fat load will also be presented.

Materials and Methods

6 male and 6 female healthy volunteers were studied in the outpatient clinic. The average age in each group was 27.8 and 25.8 years, respectively. All subjects had normal body weight and normal blood lipid levels (table I).

Following an overnight fast, a blood sample was collected at 11 a.m. At 12:30 p.m., a meal containing 1,500 cal and consisting of 40% fat, 40% carbohydrate, and 20% protein was administered. Cholesterol content (CH) was about 500 mg, and alcohol content, in the form of wine, about 7.5 g. The polyunsaturated/saturated fatty acids ratio (P/S) was 2.6. This meal reflects the alimentary habits of Northern Italy. At 5:30 and 10:00 p.m., blood samples were taken. During the study, the volunteers were required to avoid all stressful psychological activity, as well as not to smoke or take drugs (contraceptives included).

Blood specimens were collected in 0.01% ethylenediamine tetraacetic acid (EDTA). Plasma was separated immediately in a low-speed centrifuge and stored at 4 °C. VLDL were isolated by ultracentrifugation for 22 h at 120,000 g at 10 °C, using a Beckman L5-65B ultracentrifuge and the Ti50 rotor at 1.006 g/ml hydrated density. Floating VLDL were removed by the tube slicing technique. Another plasma aliquot was precipitated with 5% (v/v) sodium-heparin (Liquemin 5000, Hoffmann-La Roche AG, Grenzach) and 5% 1 M $MnCl_2$, kept at 4 °C for 16 h, and then centrifuged for 30 min at 4,000 rpm at room temperature. The filtrate containing HDL was ultracentrifuged for 40 h at 145,000 g at 10 °C and 1.22 g/ml density. The top fraction was ultracentrifuged again under the same conditions at 1.21 g/ml density and the resulting top fraction containing plasma protein-free HDL was centrifuged for 30 h at 145,000 g at 10 °C and 1.11 g/ml density. Thus, HDL_2 were obtained in the supernatant, and HDL_3 in the infranatant [7].

Whole plasma and all the fractions were assayed for free and esterified CH (*Röschlau's* enzymatic method [17]), TG (*Wahlefeld's* enzymatic method [22]), phospholipids (*Zilversmit's* colorimetric test [25]), and proteins (according to *Lowry* [9]).

Preliminary data about TG level after a fat load are presented. 3 males and 3 females (mean age 22.3 ± 1 year), under the same conditions as in the first experiment, were given a fat meal: 1.5 g fat/kg body weight (CH = 3.5%, TG = 96.5%, P/S = 0.016).

Results

The basal levels of CH, TG, and phospholipids in plasma as well as VLDL and LDL were very similar in both sexes. However, HDL levels, at 0 time, were higher in the females but, most likely due to the low number of subjects studied, the difference was not significant in any of the three parameters examined (table I). Although basal levels were very close, the TG showed a significant increase during the postprandial phase in both sexes, but it was more marked in males (table I; fig. 1 on the left). This increase is determined by an accumulation of VLDL (chylomicrons and VLDL) which only in males reached significant levels at 4.5 h compared to 0 h. The TG-LDL increase was slight and similar for both sexes. CH levels in the postprandial phase (table I, fig. 1 on the right) did not undergo significant variations in serum in toto; they increased in VLDL while they tended to decrease in LDL in both males and females. The VLDL-CH increase was significant compared to basal values only in males. In LDL

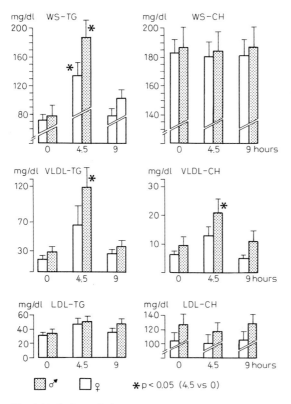

Fig. 1. Variations of triglycerides (TG) and cholesterol (CH) of whole serum (WS) in 6 males and 6 females at 0 time, 4.5, and 9 h postprandial.

the decrease in the CH/TG ratio may be an indirect index of the increase in LDL_1, or IDL, already demonstrated in this phase [3].

The most interesting results, however, emerge from the study of HDL_2 and HDL_3 behavior in the two sexes (table II). Females had significantly higher HDL_2 levels compared to males, in agreement with *Shepherd* et al [19]. This significance, which was already observed at basal time, increased in the postprandial phase. In fact, at 4.5 h, HDL_2 showed a marked increase (fig. 2) which was significant compared to basal values; at 9 h postprandial, despite a decrease, HDL_2 levels were still significantly higher

Table II. Cholesterol (CH), phospholipids (PL), and proteins (PR) (mg/dl, $\overline{m} \pm SE$) of HDL$_2$ and HDL$_3$ in 6 males (♂) and 6 females (♀) at 0 time, and at 4.5 and 9 h postprandial

0 h			4.5 h			9 h		
CH	PL	PR	CH	PL	PR	CH	PL	PR
HDL$_2$								
♂ 7.5± 0.9	13.0± 2.2	17.6± 2.1	8.6± 0.9	14.2± 1.9	21.1± 1.4	9.1± 1.2	16.5± 1.6	23.6± 1.6△
♀ 12.2± 0.8*	15.5± 2.8	25.4± 2.8*	18.7± 2.7*△	24.9± 3.1*△	39.3± 5.4*△	16.0± 2.0*△	22.4± 2.2*△	36.0± 3.6*△
HDL$_3$								
♂ 31.4± 2.8	45.9± 3.8	131.7± 7.4	28.2± 2.7	45.0± 4.3	122.8± 7.9	30.3± 3.1	54.1± 2.8	130.5± 7.2
♀ 34.6± 2.6	49.2± 4.6	144.0± 5.7	27.6± 3.0	41.8± 4.3	125.6± 8.9	32.5± 1.6	47.2± 3.9	135.3± 8.4

* $p < 0.05$ (♀ vs ♂).
△ $p < 0.05$ (4.5 vs 0; 9 vs 0)

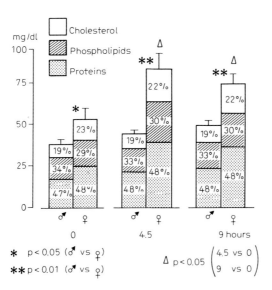

* $p < 0.05$ (♂ vs ♀)
** $p < 0.01$ (♂ vs ♀)
△ $p < 0.05$ (4.5 vs 0 / 9 vs 0)

Fig. 2. Variations of HDL$_2$ mass, expressed as a sum of proteins, phospholipids, and cholesterol at 0 time, 4.5, and 9 h postprandial in 6 males and 6 females. Percentage of the different constituents is expressed.

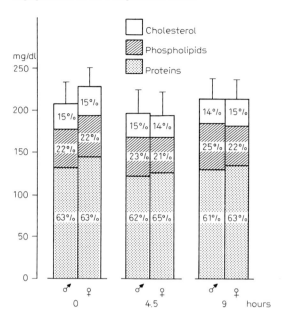

Fig. 3. Variations of HDL$_3$ mass, expressed as a sum of proteins, phospholipids, and cholesterol at 0 time, 4.5, and 9 h postprandial in 6 males and 6 females. Percentage of the 3 different constituents is expressed.

compared to basal values. This phenomenon was less pronounced in the males where HDL$_2$ showed a slight but continuous increase up to 9 h postprandial.

CH, phospholipids, and protein, expressed as the percent of the total lipoprotein mass of HDL$_2$ and HDL$_3$, did not exhibit meaningful variations during the postprandial phase. However, it was noted that HDL$_2$ in males, both at fasting and in postprandial phase, were richer percentagewise in phospholipids and poorer in CH compared to HDL$_2$ in females. HDL$_3$ had almost the same percent composition in both sexes (fig. 3).

From these results it is evident that the HDL$_2$/HDL$_3$ ratio is the parameter that undergoes the major modification during the postprandial phase (fig. 4). This ratio in females undergoes a statistically significant increase compared to both basal levels and levels observed in males, and exhibits a maximum increase at 4.5 h postprandial.

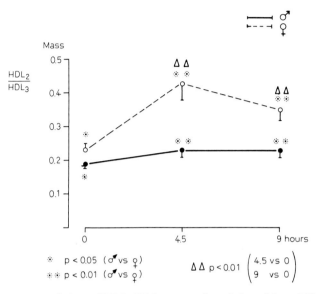

Fig. 4. Variations of HDL_2/HDL_3 mass ratio at 0 time, 4.5, and 9 h postprandial in 6 males and 6 females.

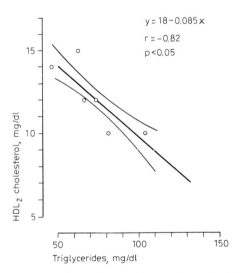

Fig. 5. Correlation between HDL_2 cholesterol and TG of whole serum in the 6 females at 0 time.

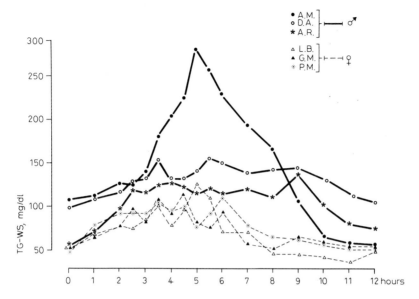

Fig. 6. Behavior of TG in whole serum (TG-WS) in 3 males and 3 females during 12 h after a fat load.

In the search for a relationship between HDL_2 and TG, it emerges that, despite the limited number of cases, a statistically significant negative correlation existed between $CH\text{-}HDL_2$ and plasma TG in females at $t = 0$ (fig. 5).

Extremely interesting data were obtained from a preliminary study conducted on 6 volunteers (3 males and 3 females) following a fat load. All the subjects were absolutely not lipemic and fasting, but after a fat meal, TG show behavior patterns that differ from subject to subject (fig. 6). In fact, 1 male volunteer showed an abnormal increase in TG which went from 110 mg/dl at basal time to almost 300 mg/dl at 5 h postprandial; the other 2 male volunteers did not show peaks of this kind but maintained higher TG levels compared to the females. The females showed a more uniform tendency. At 7–8 h postprandial, TG tended to decrease in all subjects; in some of the volunteers and most markedly in the subject with anomalous TG behavior, TG levels fell to lower than basal values at 12 h postprandial.

Discussion

The most important observation that emerges from the data presented in this study is that the lipid and lipoprotein levels differ in both sexes not only at fasting but also during the postprandial phase. This difference had not yet been taken into consideration by workers in the field.

The fraction which is already higher at time 0 and which undergoes more meaningful variations is the HDL_2 fraction (fig. 2), which in females rises significantly at 4.5 h postprandial, unlike the HDL_3 fraction which seems to diminish slightly (fig. 3). This phenomenon is also present in males, but it is much less pronounced (fig. 4).

It appears that not only is there a correlation between HDL and lipoprotein lipase levels, as demonstrated by *Nikkilä,* but also between HDL_2 and TG levels (fig. 4).

The HDL_2 level and/or the HDL_2/HDL_3 ratio conditions the clearance of the TG-rich lipoproteins. This observation seems confirmed not only by the data presented here, but also by *Patsch and Gotto* [14] who demonstrated, in males only, a negative correlation between the height of the TG peak following a fat load and HDL_2, using the zonal ultracentrifuge.

An observation not yet reported in the literature up to the present consists of the slight difference in HDL_2 composition in males compared to females detectable both at fasting and during the postprandial phase at equal protein concentration: HDL_2 in females appears to have a lesser phospholipid (29 vs 34%) and a greater CH (23 vs 19%) content.

A preliminary report (fig. 5) of a study still in progress in our laboratory demonstrates that among 6 apparently healthy and normolipemic persons at fasting and after a fat meal, the 3 male subjects show a clear tendency to develop a more intense lipemia after the fat meal than the females. 1 of the males in particular showed frankly anomalous behavior, reaching values near 300 mg/dl and returning to lower than basal values at 12 h postprandial.

Therefore, the postprandial phase may represent a physiological moment during which the female regulates HDL_2 at higher levels, compared to the male. The reasons for this behavior are not known. Since

HDL_2 formation appears related to chylomicron metabolism [6, 20, 21], it may be hypothesized that the greater increase in HDL_2 in fertile females is due to a more pronounced capacity to remove TG-rich particles. It may not be excluded that transformation of HDL_3 into HDL_2 contributes to the increase in the latter. In vitro studies seem to favor this possibility [15].

It has been recently suggested that the putative protection against atherosclerosis in premenopausal women would be related to their higher HDL_2 subfraction [1]. Therefore, we consider it highly interesting to evaluate the HDL_2/HDL_3 ratio in relationship to the formation and catabolism of chylomicrons in the postprandial phase in order to detect possible alterations in subjects with signs of premature atherosclerosis. This hypothesis is now gaining experimental evidence in animals as well as in humans [24].

References

1 Cheung, M. C.; Albers, J. J.: Distribution of cholesterol and apolipoproteins A-I and A-II in human' high density lipoprotein subfractions separated by CsCl equilibrium gradient centrifugation: evidence for HDL subpopulation with differing A-I/A-II. J. Lipid Res. *20:* 200–207 (1979).
2 Eisenberg, S.; Levy, R. I.: Lipoprotein metabolism. Adv. Lipid Res. *13:* 1–89 (1975).
3 Fellin, R.; Agostini, B.; Rost, W.; Seidel, D.: Isolation and analysis of human plasma lipoproteins accumulating post-prandial in an intermediate density fraction (d. = 1.006–1.019 g/ml). Clin. chim. Acta *54:* 325–333 (1974).
4 Hamilton, R. L.; Williams, M. C.; Fielding, C. J.; et al.: Discoidal bilayer structure of nascent high density lipoproteins from perfused rat liver. J. clin. Invest. *58:* 667–680 (1978).
5 Havel, R. J.; Goldstein, J. L.; Brown, M. S.: Lipoproteins and lipid transport; in Bondy, Rosenberg, Metabolic control and disease; 8th ed., pp. 393–494 (Saunders, Philadelphia 1980).
6 Havel, R. J.; Kane, J. P.; Kashyap, M. L.: Interchange of apolipoproteins between chylomicrons and high density lipoproteins during alimentary lipemia in man. J. clin. Invest. *52:* 32–38 (1973).
7 Kostner, G. M.; Holasek, A.: The separation of human serum high density lipoproteins by hydroxyapatite column chromatography. Biochim. biophys. Acta *488:* 417–431 (1977).
8 Lewis, B.; Chait, A.; February, A. W.; Mattock, M.: Functional overlap between 'chylomicra' and 'very low density lipoproteins' of human plasma during alimentary lipaemia. Atherosclerosis *17:* 455–462 (1973).
9 Lowry, O. H.; Rosebrough, N. J.; Farr, A. L.; Randall, R. J.: Protein measurement with the Folin phenol reagent. J. biol. Chem. *193:* 265–278 (1951).

10 Miller, G.J.: High density lipoproteins and atherosclerosis. A. Rev. Med. *31:* 97–108 (1980).
11 Miller, G.J.; Miller, N.E.: Plasma-high-density-lipoprotein concentration and development in ischaemic heart disease. Lancet *i:* 16–19 (1975).
12 Nikkilä, E.: Studies on the lipid-protein relationships and pathological sera and the effect of heparin on serum lipoproteins. Scand. J. clin. Lab. Invest. *5:* suppl. 8, pp. 1–101 (1953).
13 Nikkilä, E.A.: Metabolic and endocrine control of plasma high density lipoprotein concentration. Relation to catabolism of triglyceride-rich lipoproteins; in Gotto, Miller, Oliver, High density lipoproteins and atherosclerosis, pp. 177–192 (Elsevier/North Holland, Amsterdam 1978).
14 Patsch, J.R.; Gotto, A.M., Jr.: Separation and analysis of HDL subclasses by zonal ultracentrifugation; in Lippel, Report of the high density lipoprotein methodology workshop. NIH Publ. No. 79–1961, pp. 310–324 (NIH Washington 1979).
15 Patsch, J.R.; Gotto, A.M., Jr.; Olivecrona, R.; Eisenberg, S.: Formation of high density lipoprotein$_2$-like particles during lipolysis of very low density lipoproteins in vitro. Proc. nat. Acad. Sci. USA *75:* 4519–4523 (1978).
16 Redgrave, T.G.; Carlson, L.A.: Changes in plasma very low density and low density lipoprotein content, composition, and size after a fatty meal in normo- and hypertriglyceridemic man. J. Lipid Res. *20:* 217–228 (1979).
17 Röschlau, P.: Enzymatische Bestimmung des Gesamt-Cholesterins in Serum. Chem. klin. Biochim. *12:* 226–232 (1974).
18 Schaefer, E.J.; Eisenberg, G.; Levy, R.I.: Lipoprotein apoprotein metabolism. J. Lipid Res. *19:* 667–687 (1978).
19 Shepherd, J.; Packard, C.J.; Patsch, J.R.; Gotto, A.M., Jr.; Taunton, O.D.: Metabolism of apolipoproteins A-I and A-II and its influence on high density lipoproteins subfraction distribution in males and females. Eur. J. clin. Invest. *8:* 115–120 (1978).
20 Smith, L.C.; Pownall, H.J.; Gotto, A.M.: The plasma lipoproteins: structure and metabolism. A. Rev. Biochem. *47:* 751–777 (1978).
21 Tall, A.R.; Small, D.M.: Plasma high density lipoproteins. New Engl. J. Med. *299:* 1232–1236 (1978).
22 Wahlefeld, A.M.: Triglycerides determination after enzymatic hydrolysis; in Bergmeyer, Methods of enzymatic analysis; trans. from 3rd Ger. ed.; vol. 4, pp. 1831–1835 (Verlag Chemie, New York 1974).
23 Witztum, J.; Schonfeld, G.: High density lipoproteins. Diabetes *28:* 376–386 (1979).
24 Zilversmit, D.B.: Atherogenesis: a post-prandial phenomenon. Circulation *60:* 473–485 (1979).
25 Zilversmit, D.B.; Davis, A.K.: Microdetermination of plasma phospholipids by trichloroacetic. J. Lab. clin. Med. *35:* 155–163 (1950).

Dr. Giovannella Baggio, Department of Internal Medicine, Division of Gerontology and Metabolic Diseases, Policlinico, Via Giustiniani 2, I-35100 Padova (Italy)

Relationships of Body Weight and Fatness to Lipoprotein Components. Variations following Hypocaloric Diet

P. Avogaro, G. Bittolo Bon, F. Belussi, G. Cazzolato

General Regional Hospital, National Council for Research, Preventive Medicine, Sub-Project Atherosclerosis, Unit of Venice, Venice, Italy

Introduction

Nutrients can modify plasma lipids and/or lipoprotein concentrations. These variations are obtained through a complex mechanism involving the kind of food and its quantity, the synthesis and excretion of cholesterol and bile acids, the hepatic synthesis of very low-density lipoproteins (VLDL), and the hepatic uptake of 'remnants'. Studies of this mechanism, however, have not yet achieved satisfactory results and many observed correlations are obscured by the interference of too many different risk factors. The relationship between obesity and coronary heart disease (CHD) appears to be weak when the many covariables are ruled out. Previous research has focused mostly on variations induced by diet on cholesterol and triglycerides (TG). Information on variations in lipoproteins is scanty and there is no available information concerning apolipoproteins. This brief review summarizes certain data on the relationship between body weight, fatness, and plasma lipids, lipoproteins, and apolipoproteins as well as on some variations induced in various lipoprotein components by a hypocaloric diet.

Body Weight, Lipoproteins, and Apolipoproteins

There is a weak correlation between obesity and lipid and/or lipoprotein plasma levels [3, 4, 10, 13, 26]. TG and VLDL are usually higher in obese subjects than in normal subjects [3, 4, 10, 13, 26].

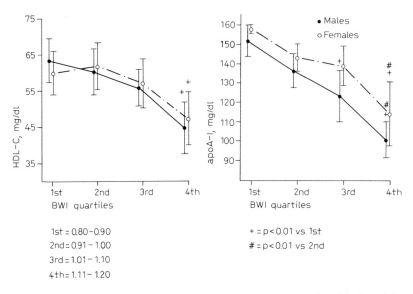

Fig. 1. Values of plasma HDL-C and apoA-I in the various quartiles of body weight index.

It has been shown that obesity is associated with reduced levels of high-density lipoprotein cholesterol (HDL-C) [14, 19, 27]. It is not clear whether this is related to an expanded adipose tissue mass, a higher carbohydrate intake [24, 27] or reduced physical activity, or whether the insulin resistance of obesity and/or reduced lipoprotein-lipase activity of adipose tissue play a role [20]. Epidemiological data from the Framingham study [14] and the Tromsö study [18], however, do not reveal any close relationship between HDL-C and relative body weight. This problem has been investigated recently by *Albrink* et al. [2]. It has been stressed that the negative relationship between fatness and plasma level of HDL is entirely due to the subclass HDL_2 and its subfractions HDL_{2b} and HDL_{2a}, while a weak positive correlation exists between HDL_3 and TG. The basis for the relationship of HDL_2 with fatness is not known. A reduced activity of lipoprotein lipase due to the insulin resistance of large adipose cells in acquired obesity may be a link. The reduced plasma levels of HDL_2 in obese subjects may, therefore, be a sign of impairment in the removal of VLDL and chylomicrons.

We have stressed an inverse correlation between the body weight index (BWI) and plasma levels of HDL-C and apolipoprotein A-I (apoA-I)

Table I. Plasma concentration of plasma lipids, lipoproteins, and apolipoproteins A-I and B in 30 obese subjects and 30 controls

	BWI	C mg/dl	TG mg/dl	apoB mg/dl	apoA mg/dl	VLDL-C mg/dl	VLDL-TG mg/dl	VLDL-apoB mg/dl	LDL-C mg/dl	LDL-apoB mg/dl	HDL-C mg/dl
Controls											
Mean	1.02	258.2	159.6	119.1	143.8	31.3	88.7	15.9	164.6	103.2	62.3
SEM	0.02	8.9	13.2	5.3	4.3	4.9	7.1	1.2	6.8	3.7	1.8
Obese subjects											
Mean	1.53*	256.9	162.4	129.2	112.3*	30.2	87.2	16.8	180.5	112.4	46.2*
SEM	0.04	9.2	13.3	4.9	2.6	5.0	6.8	1.2	6.5	3.8	2.1

* $p < 0.01$ vs controls. BWI = Body weight index; C = cholesterol; TG = triglycerides.

[8]. We studied 120 subjects (60 males and 60 females) with a BWI between 0.8 and 1.2. Lower levels of HDL and apoA-I were found in men than in women, although the difference was not significant. A high negative correlation was found between BWI, HDL-C, and apoA-I in both males and females, whereas no correlation was found between BWI and apoB.

When the values of HDL-C and apoA-I were considered according to quartiles of BWI in both sexes, significant variations were found between the mean values of the lowest and the highest quartile (fig. 1). These data are even more significant since they were obtained from 'normal' people with a BWI within normal limits, excluding both obese and very lean subjects. Recently, emphasis has been placed on the role of HDL-C and apoA-I as a biochemical marker of human atherosclerosis [5, 19]. Because the difference in concentration of HDL-C between normal and atherosclerotic patients is only a few milligrams, the relevant influence of body weight on these parameters must be emphasized.

Lipoproteins and Apolipoproteins A-I and B in Obese Subjects

We undertook a study [7] to ascertain the plasma levels of the major lipoprotein classes and the two major apolipoproteins, apoB and apoA-I, in obese people. 30 obese subjects were included in the study (BWI

Table II. Ratios between some plasma lipids and some apolipoproteins in 30 subjects and 30 controls

	C / HDL-C	LDL-C / HDL-C	apoB / apoA-I	LDL-C / LDL-apoB	HDL-C / apoA-I
Controls					
Mean	4.14	2.64	0.83	1.59	0.43
SEM	0.27	0.19	0.06	0.13	0.03
Obese subjects					
Mean	5.56*	3.91*	1.15*	1.61	0.41
SEM	0.29	0.26	0.16	0.13	0.02

* $p < 0.01$ vs controls.

between 1.30 and 1.75; mean, 1.53). Their plasma lipid levels were matched with a control series (BWI 0.82 to 1.11; mean, 1.02). For controls, we selected subjects who had plasma levels of cholesterol and TG close to the baseline lipid levels of the obese subjects. Before inclusion in the study, both obese subjects and controls followed a typical Italian diet (50% carbohydrate, 35% fat, 15% protein; cholesterol 400–600 mg; P/S 0.8:1). In comparing the obese subjects and the controls, we found that the only significant variations were that obese subjects had lower values of HDL-C and apoA-I and higher ratio values of total cholesterol (TC)/HDL-C, low-density lipoprotein cholesterol (LDL-C)/HDL-C, and apoB/apoA-I (tables I, II).

Lipoproteins and Apoliproteins A-I and B in Obese Subjects Fed a Hypocaloric Diet

Information on the variations in lipoproteins during weight loss is rather scanty [8, 20, 26, 27]. No information is available on the behavior of apolipoproteins. Previous research has not ascertained any variations of β-lipoproteins in subjects with normal lipid plasma levels following fasting up to 1 week [26] or in patients with phenotype II following a hypocaloric, hypocholesterolemic diet [12]. In patients with phenotypes III, IV, and V, an increase in β-lipoproteins concomitant with a reduction in triglycerides

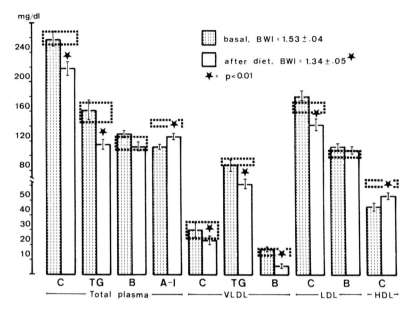

Fig. 2. Behavior of lipids and apoA-I and apoB in the total plasma and in the major lipoprotein classes following a hypocaloric diet (1,000 kcal/day) in obese people. The shaded area represents the range of values recorded in controls.

has been recorded following negative caloric dietary treatment [12]. *Wilson and Lees* [27] have recorded reduced levels of plasma VLDL and LDL while HDL increased following weight loss. Unchanged levels of HDL-C were recorded by *Nestel and Miller* [20], while specific activity following an injection of labeled cholesterol increased in HDL. In other studies [11], a decrease of HDL-C concentration was recorded in obese subjects during active weight loss. When body weight, however, was stabilized at a reduced level, HDL-C increased substantially above the baseline values; in the same patients both VLDL-TG and LDL-C decreased. Even in normal subjects, given a 400-kcal diet for a few days, a decrease of HDL-C was recorded while total TG and VLDL-TG did not change and LDL-TG increased [25].

In our study [7], 30 obese subjects were fed a hypocaloric, hypocholesterolemic diet (1,000 kcal; 45% carbohydrate, 35% fat, 20% protein; cholesterol 300 mg/day; P/S 1). A second analysis was performed at the point when a major weight loss was no longer obtainable and the body weight had been stable for some days.

Table III. Behavior of the various ratios in obese subjects (30) following the hypocaloric diet

	C / HDL-C	LDL-C / HDL-C	apoB / apoA-I	LDL-C / LDL-apoB	HDL-C / apoA-I
Baseline					
Mean	5.56	3.91	1.15	1.61	0.41
SEM	0.29	0.26	0.16	0.13	0.02
After diet					
Mean	4.03*	2.62*	0.89*	1.34*	0.42
SEM	0.28	0.18	0.05	0.12	0.02

* $p < 0.01$ vs baseline.

Following dietary treatment, the BWI decreased significantly (between 5 and 18%) in all subjects (fig. 2). A significant decrease in TC, VLDL-C, LDL-C, total TG, VLDL-TG, and VLDL-apoB, and in the ratios TC/HDL-C, LDL-C/HDL-C, apoB/apoA-I, and LDL-C/LDL-apoB was recorded following dietary treatment, while the HDL-C/apoA-I ratio remained unchanged; a significant increase in both HDL-C and apoA-I was also observed (fig. 2, table III). Even when VLDL-TG in obese subjects following a hypocaloric diet reached normal levels, the values of apoA-I and HDL-C were still significantly lower than in the controls. There may be another reason, therefore, to explain the decreased levels of apoA-I and HDL-C in obese subjects, a reason unrelated to the levels of VLDL-TG.

The relationship between the degree of weight loss and variations in the different parameters was ascertained (table IV). A significant relationship between the degree of weight decrease and the increase in plasma levels of apoA-I and HDL-C has been stressed. No other correlations were observed.

Our findings in obese people following a negative caloric diet are rather complex. TG and their major carrier, the VLDL, show a uniform tendency to decrease; VLDL-C and VLDL-TG exhibit the same behavior. These data support previous findings [27]. An accelerated secretion of TG mediated by hyperinsulinemia in obesity has been suggested by *Robertson* et al. [23]. This may explain how, by dietary manipulation, a reversal of

Table IV. Relationship between the decrease in body weight index (BWI) and the various parameters analyzed following the dietary period

△ BWI vs	
△ TC	R = –0.09
△ TG	R = 0.26
△ apoB	R = 0.10
△ apoA-I	R = 0.37*
△ VLDL-C	R = 0.08
△ VLDL-TG	R = 0.20
△ VLDL-apoB	R = 0.13
△ LDL-C	R = –0.04
△ LDL-apoB	R = 0.02
△ HDL-C	R = 0.49**

* $p < 0.05$; ** $p < 0.01$. TC = Total cholesterol; TG = triglycerides.

the metabolic derangement in obese people is followed by a decrease in the secretion of TG and VLDL. ApoB-VLDL, one of the major peptides of this lipoprotein class, also undergoes a uniform decrease, according to the findings of *Wilson and Lees* [27], which differ from the data of *Walker* et al. [26] and *Fredrickson* [12]. Both total cholesterol and LDL-C show a significant reduction, while total apoB and LDL-apoB mimic the same trend, but not significantly.

ApoA-I and HDL-C, at the end of the dietary treatment, reached values significantly higher than the baseline values. The metabolic variation induced by the diet is probably better explained by the two ratios TC/HDL-C and LDL-C/HDL-C, both indicating a major mobilization of cholesterol in the HDL. This increase reflects the mobilization of cholesterol from adipose tissue into plasma and provides 'direct evidence that HDL is the principal acceptor of adipocyte cholesterol in human plasma'.

The variations shown by both apoB and apoA-I need consideration. ApoB is synthesized in the intestinal mucosal cells [16, 17]. The source of apoA-I is more uncertain. It seems, however, that the larger part originates as such from the liver, but that a substantial amount of apoA-I enters the blood from the intestine in chylomicrons.

The shortage of dietary fats and protein experienced by our patients can explain the decrease in VLDL-apoB. The significant increase of apoA-

I following dietary treatment agrees with the known inverse relationship between VLDL-TG and HDL-C [19, 21]. Moreover, previous research has stressed that a carbohydrate-enriched diet decreases both HDL-C and apoA-I [24].

It is relevant, however, that even if VLDL-TG in obese people following dietary treatment are lower than in controls, the values of apoA-I and HDL-C are still significantly lower than in normal individuals. There may be other unexplained reasons, therefore, that cause in obese subjects the decreased levels of apoA-I and HDL-C that are unrelated to the levels of VLDL-TG. This finding may mean that only apoA-I and HDL-C values are directly correlated to body weight (and to adipose tissue size), while variations in VLDL-C and/or apoB are more dependent on variations in food quantity and/or quality.

Nutrition plays an outstanding role among the various factors having an impact on atherogenesis. However, the multifactorial nature of the disease, the differences among national cultures, and the heterogeneity in design of various experimental trials have not led to any firm conclusions. Moreover, it has not been soundly established whether the frequency of cardiovascular morbidity and/or mortality may be reduced by variations in diet [22]. In recent years some of the data have suggested that the historical lipid parameters such as cholesterol and TG may be of limited help in establishing a firm relationship between plasma lipid levels and atherogenesis. A special role has been claimed for HDL [6, 19] and for the ratio LDL/HDL [15].

Moreover, the protein part of lipoproteins plays an unexpected role [1, 5]. The limited review presented here stresses the real complexity of these problems. The nutritional status of the individual has a close relationship with some of the newer parameters and particularly with HDL-C and apoA-I. The small amount of HDL-C present in the human plasma and the small differences existing between normal and atherosclerotic patients [5, 14] emphasize the need for special attention to avoid the interference of body weight on HDL-C. This is especially true for epidemiological surveys. In this respect, the greater plasma levels of apoA-I and its close positive relationship with HDL-C makes apoA-I a better and more useful parameter than HDL-C. Another relevant finding coming from our experience is found in the behavior of apoA-I peculiar to obese people even when body weight and plasma TG are returned to normal values. The low level of this apolipoprotein appears, therefore, to be a constitutional trait which is only partly modified by the environment.

References

1 Alaupovic, P.: Apolipoproteins and lipoproteins. Atherosclerosis *13:* 141–146 (1971).
2 Albrink, M.J.; Krauss, R.M.; Lindgren, F.T.; Von der Groeben J.: Intercorrelations among plasma high density lipoprotein, obesity and triglycerides in a normal population. Lipids *15:* 668–676 (1980).
3 Albrink M.J.; Meigs, J.W.: Interrelationship between skinfold thickness, serum lipids and blood sugar in normal men. Am. J. clin. Nutr. *15:* 255–261 (1964).
4 Albrink, M.J.; Meigs, J.W.; Granoff, M.A.: Weight gain and serum triglycerides in normal men. New Engl. J. Med. *266:* 484–489 (1962).
5 Avogaro, P.; Bittolo Bon, G.; Cazzolato G.; Quinci, G.B.: Are apolipoproteins better discriminators than lipids for atherosclerosis? Lancet *i:* 901–903 (1979).
6 Avogaro, P.; Cazzolato, G.: Familial hyper-HDL-(α)-cholesterolemia. Atherosclerosis *22:* 63–77 (1975).
7 Avogaro, P.; Cazzolato, G.; Bittolo Bon, G.; Quinci, G.B.: Variations of plasma lipoproteins and apolipoproteins B and AI in obese subjects fed with hypocaloric diet. Obesity/Bariatric Med. *8:* 158–161 (1979).
8 Avogaro, P.; Cazzolato, G.; Bittolo Bon, G.; Quinci G.B.; Chinello, M.: HDL-cholesterol, apolipoproteins A-I and B, age and index body weight. Atherosclerosis *31:* 85–91 (1978).
9 Avogaro, P.; Cazzolato, G.; Pais, M.: High-density lipoprotein and atherosclerosis. Lancet *i:* 691–692 (1975).
10 Avogaro, P.; Crepaldi, G.; Enzi G.; Tiengo, A.; Association of hyperlipaemia, diabetes mellitus and mild obesity. Acta diabet. lat. *4:* 572–590 (1967).
11 Contaldo, F.; Strazzullo, P.; Postiglione, A.; Riccardi, G.; Patti, L.; Di Biase, G.; Mancini M.: Plasma high density lipoprotein in severe obesity after stable weight loss. Atherosclerosis *37:* 163–167 (1980).
12 Fredrickson, D.S.: The role of lipids in acute myocardial infarction. Circulation *40:* suppl. IV, pp. 99–111 (1969).
13 Gofman, J.W.; Jones, M.B.: Obesity, fat metabolism and cardiovascular disease. Circulation *5:* 514–518 (1952).
14 Gordon T.; Castelli, W.P.; Hjortland, M.C.; Kannell, W.B.; Dawber, T.R.: High density lipoproteins as a protective factor against coronary heart disease. Am. J. Med. *62:* 707–714 (1977).
15 Kannel, W.B.; Castelli, V.P.: Is the serum total cholesterol an anachronism? Lancet *ii:* 950–951 (1979).
16 Margolis, S.; Capuzzi, D.: Serum lipoproteins synthesis and metabolism; in Nelson, Blood lipids and lipoproteins: quantitation, composition and metabolism, pp. 825–880 (Wiley-Interscience, New York 1972).
17 Marsh, J.B.: The incorporation of aminoacids into soluble lipoproteins by cell free preparations from rat liver. J. biol. Chem. *238:* 1752–1756 (1963).
18 Miller N.E.; Forde, O.H.; Thelle, D.S.; Mjos, O.D.: The Tromsö Heart Study. High density lipoprotein and coronary heart disease: a prospective case-control study. Lancet *i:* 965–968 (1977).
19 Miller, C.J.; Miller N.E.: Plasma high density lipoprotein concentration and development of ischaemic heart disease. Lancet *i:* 16–19 (1975).

20 Nestel, P.J.; Miller, N.E.: Mobilization of adipose tissue cholesterol in high density lipoprotein during weight reduction in man; in Gotto, Miller, Oliver, High density lipoproteins and atherosclerosis, pp. 51–54 (Elsevier, Amsterdam 1978).
21 Nikkilä, E.A.: Metabolic and endocrine control of plasma high density lipoprotein concentration; in Gotto, Miller, Oliver, High density lipoproteins and atherosclerosis, pp. 177–192 (Elsevier, Amsterdam 1978).
22 Oliver, M.F.: Diet and coronary heart disease. Br. med. Bull. 37: 49–58 (1981).
23 Robertson, R.P.; Gavereski, D.J.; Henderson, J.D.; et al.: Accelerated triglyceride secretion. A metabolic consequence of obesity. J. clin. Invest. 52: 1620–1626 (1973).
24 Schonfeld, G.; Weidman, W.; Witztum, J.L.; et al.: Alterations in levels and interrelations of plasma apolipoproteins induced by diet. Metabolism 25: 261–275 (1976).
25 Taskinen, M.-R.; Nikkilä, E.A.: Effects of caloric restriction on lipid metabolism in man. Changes of tissue lipoprotein lipase activities and of serum lipoproteins. Atherosclerosis 32: 289–299 (1979).
26 Walker, W.J.; Weiner, N.; Milch, L.J.: Differential effect of dietary fat and weight reduction on serum levels of beta-lipoprotein. Circulation 15: 31–37 (1957).
27 Wilson, D.E.; Lees, R.S.: Metabolic relationship among the plasma lipoproteins. Reciprocal changes in the very-low and low-density lipoproteins in man. J. clin. Invest. 51: 1051–1057 (1972).

P. Avogaro, M.D., Ph.D., General Regional Hospital, National Council for Research, Preventive Medicine, Sub-Project Atherosclerosis Unit of Venice, I-30100 Venice (Italy)

Dietary Influences on Lipids and Lipoprotein Levels in Animals and Atherosclerosis[1]

David Kritchevsky

Wistar Institute of Anatomy and Biology, Philadelphia, Pa., USA

There are periodic reviews of the effects of dietary factors (specific or general) on lipid metabolism in animals. The reviews are usually slanted towards effects on atherosclerosis. An exhaustive review is beyond the scope of this presentation which will focus on aspects of diet which are of greatest interest today; atherosclerosis will not be its central theme.

Lipids

Cholesterol added to a standard laboratory ration or to a semipurified diet is hypercholesterolemic for rabbits and chickens and some species of primates. Rats can be rendered hypercholesterolemic only if the diet contains both cholesterol and bile acid. Obviously, the effect of cholesterol on blood lipids is a function of the animal's own homeostatic control mechanisms which may include its normal low-density lipoprotein (LDL)/high-density lipoprotein (HDL) cholesterol ratio [20]. The animal's capacity for cholesterol absorption limits a dose-response relationship. *Scebat* et al. [66] found that rabbits fed 0.5 or 1.0 g of cholesterol responded with similar levels of cholesterolemia and atherosclerosis within 3 months; when fed 0.25 g/day, their cholesterol levels were half of those observed in the other groups. The form in which cholesterol is administered may be a factor. Thus, rabbits fed 2% cholesterol as an amorphous solid, crystalline material, or suspended in corn oil, exhibited

---

[1] Supported in part by Research Career Award HL-00734 and by Grants HL-03299 and HL-05209 from the National Institutes of Health.

serum cholesterol levels of 543, 1,119, and 1,060 mg/dl, respectively. Average atherosclerosis on the three regimens was 1.23, 1.53, and 1.31, respectively [30].

Recently, *Imai* et al. [18] have shown that fresh (recrystallized, sharp melting point) cholesterol is less angiotoxic for rabbits than is aged (yellowed, rancid) cholesterol. The principal oxidation products were 25-hydroxycholesterol and 3β, 5α, 6β-cholestanetriol. The oxidation of cholesterol in air has been recognized for over 70 years [8]. The findings of *Imai* et al. [18] suggest that the oxidation state of cholesterol in the diet can influence its effects on vascular tissue and probably on cholesterolemia but, to date, there are few data on this point.

The extent of unsaturation of dietary fat also affects levels of cholesterolemia [20, 22]. While there is generally a straight line relationship between iodine value of a fat and severity of atherosclerosis this correspondence does not always hold for cholesterol level. Rabbits fed 2% cholesterol and 6% coconut oil (iodine value 9), lard (iodine value 63), or corn oil (iodine value 124) exhibited cholesterol levels (mg/dl) of 2,827, 2,245, and 1,908, respectively [33].

One fat which is inordinately atherogenic despite its iodine value and cholesterolemic effect is peanut oil. Peanut oil is more atherogenic than would be predicted for rats [12], rabbits [45], and rhesus monkeys [72, 77]. The peanut oil effect is even seen in rabbits fed a semipurified, cholesterol-free diet [39]. Peanut oil differs from most other vegetable fats in that it contains up to 6% of arachidic, behenic, and lignoceric acids, all of which are present in the sn-3 position of the triglyceride [61]. We have shown [45] that a mixed fat whose fatty acid composition resembles that of peanut oil minus the long-chain saturated fatty acids is no more atherogenic than corn oil. Introduction of these fatty acids into the synthetic fat by interesterification results in a new fat whose composition is identical with that of peanut oil but whose atherogenic effects are considerably lower [46]. Interesterification (randomization) of peanut oil significantly reduces its atherogenicity but has no influence on its cholesterolemic effect (table I). The earlier studies were carried out using North American peanut oil. Peanut oil obtained from Africa or South America is also atherogenic and the effect on atherosclerosis and serum cholesterol levels is unrelated to its linoleic acid content [41]. North American peanut oil was slightly less atherogenic than oil from either of the other sources. The data obtained in this series of experiments suggest that the structure of some fats may be as important as fatty acid spectrum and iodine value.

Table I. Influence of peanut oil and other fats on cholesterolemia and atherosclerosis in rabbits. After Kritchevsky et al. [41, 45, 46]

Fat	Survival	Serum cholesterol mg/dl	Average atherosclerosis[1]
Series 1			
Coconut oil	44/45	1,360	1.91
Peanut oil	98/106	1,483	1.62
PGF[2]	73/76	1,650	1.34
PGF + LC[3]	31/31	1,723	1.17
Peanut oil/R[4]	31/31	1,833	1.18
Corn oil	100/106	1,548	1.29
Series 2			
Peanut oil (US)	29/34	2,404	1.59
Peanut oil (Africa)	31/34	2,473	1.75
Peanut oil (South America)	18/27	2,567	1.88
Corn oil	27/34	1,917	1.36

[1] (Arch + thoracic) ÷ 2. Graded on 0–4 basis.
[2] PGF = Blend of 55% olive oil, 35% safflower oil and 10% cottonseed oil.
[3] PGF interesterified with arachidic and behenic glycerides. Fatty acid composition identical to that of peanut oil. LC = Long-chain triglycerides.
[4] Randomized peanut oil.

Trans fatty acids, which are formed during hydrogenation of fat, have been impugned as being especially cholesterolemic and atherogenic [49]. Weigensberg and co-workers [56, 73–75] carried out experiments in rabbits fed cholesterol and elaidic acid, trielaidin, or linoleic acid containing 9 and 12 trans double bonds. The trans fat was more cholesterolemic than its cis counterpart but not more atherogenic (table II). When rabbits were fed a cholesterol-free, semipurified diet containing trans fat it was found to be more cholesterolemic than control fat but to have no effect on the activity of a number of hepatic enzymes [65]. Liver homogenates from vervet monkeys fed trans fat convert less radioactive acetate to cholesterol than do controls [24]. In this property trans fat resembles saturated fat which inhibits hepatic cholesterogenesis (compared to unsaturated fat) in rats [5, 35, 51], and in rhesus [71], cebus [3], and cynomolgus [3] monkeys. Jackson et al. [19] fed pigs diets containing trans fat or tallow. There were no significant differences between the plasma lipoproteins of any of the groups although trans fatty acids were found in the lipoproteins of the pigs fed trans fat. Houtsmiller [14] reviewed the biological effects of trans fats

Table II. Influence of *trans* fatty acids on cholesterolemia and atherosclerosis in rabbits. After *McMillan* et al. [56] and *Weigensberg* et al. [75]

Vehicle[1]	Number of animals	Serum cholesterol mg/dl	Aorta sudanophilia, % ± SD	F/E[2]
Oleic acid	33	2,413 ± 1,290	32 ± 29	0.82
Elaidic acid	30	3,830 ± 1,326	37 ± 28	0.69
Corn oil	18	838 ± 646	10 ± 14	0.28
Olive oil (OO)	32	1,437 ± 911	23 ± 26	1.22
Elaidinized OO	36	2,446 ± 1,474	30 ± 27	1.01

[1] Rabbits fed 1 g cholesterol and 6 g vehicle for 84 days.
[2] Ratio of free to esterified cholesterol.

and concluded that the biochemical properties of *trans* monoenoic acids are intermediate between those of saturated and monounsaturated fats.

Protein

Interest in the effects of protein on lipid metabolism dates back to 1909 [17], but until recently there have been few concerted research efforts in this area. *Meeker and Kesten* [57, 58] fed rabbits diets containing 38% casein or 39% soy protein plus 60–250 mg of cholesterol per day. Soy protein, in their hands, was more cholesterolemic but less atherogenic. Comparison of different levels of casein or soy protein on cholesterolemia in rats showed the former to be more cholesterolemic [60]. The relative hypocholesterolemic effect of soy protein persisted in chickens even when they were germ-free [29].

Recently, *Carroll* and co-workers [1, 13, 16] have revived interest in the effects of proteins on lipid metabolism. Rabbits were fed diets containing 30% protein for 28 days, and it was shown that, generally, animal protein was more cholesterolemic than vegetable protein but there was wide individual variation (table III). Dilution of the casein with soy protein decreased its cholesterolemic effect so that cholesterol levels of rabbits fed casein-soy 3:1 exhibited cholesterol levels 42% lower than those of rabbits fed 100% casein. Cholesterol levels of rabbits fed casein-soy 1:1 had cholesterol levels identical with those of rabbits fed 100% soy protein. Partial

Table III. Effect of proteins on plasma cholesterol in rabbits. After *Hamilton and Carroll* [13]

Diet[1]	Number	Plasma cholesterol, mg/dl ± SEM	
		total	% ester
Commercial	22	70 ± 5	63
SP casein[2]	20	200 ± 22	64
Whole egg	4	235 ± 89	45
Skim milk	6	230 ± 40	71
Lactalbumin	5	215 ± 69	57
Beef	5	160 ± 60	62
Pork	6	110 ± 17	68
Egg white	6	105 ± 28	67
Wheat gluten	6	80 ± 21	69
Peanut protein	6	80 ± 10	62
Peanut meal	4	75 ± 27	67
Soy	6	25 ± 5	56

[1] Rabbits fed 27% protein (equivalent) for 28 days.
[2] SP = Semipurified.

enzymatic hydrolyzates were less cholesterolemic than casein or soy protein but amino acid mixtures corresponding to the respective proteins were not. We have found [37] that beef protein is as cholesterolemic and atherogenic as casein but that beef-soy 1:1 behaves like soy protein (table IV).

We have examined the effects of casein and soy protein on lipoprotein distribution and atherosclerosis in rabbits. We have confirmed the hypocholesterolemic property of soy and have further shown that addition of single amino acids to soy protein or casein can bring about changes in their effects [4, 44]. On the premise that the lysine/arginine ratio in a protein can affect its atherogenicity, we added lysine to soy protein (lysine/arginine = 0.90) and arginine to casein (lysine/arginine = 2.00). Casein was 94% more cholesterolemic and 144% more atherogenic than soy protein. Addition of arginine to casein did not affect its effects on serum cholesterol but reduced atherogenicity by 25%. Addition of lysine to soy protein enhanced cholesterolemia by 53% and atherosclerosis by 64%. The addition of arginine to casein reduced the levels of very low-density lipoproteins (VLDL) and intermediate density lipoproteins (IDL) by 52 and 32%, respectively, and increased LDL and HDL levels by 41 and 35%, respectively. Addition of

Table IV. Effect of protein on cholesterolemia and atherosclerosis in rabbits. After Kritchevsky et al. [37]

Protein[1]	Number	Serum cholesterol mg/dl ± SEM	HDL cholesterol % ± SEM	Average atherosclerosis[2]
Beef	12/12	185 ± 24	20.1 ± 24	1.02
Soy	9/12	37 ± 4	38.8 ± 5.1	0.50
Beef-soy 1:1	11/12	61 ± 6	43.4 ± 4.2	0.55
Casein	11/12	200 ± 18	29.6 ± 3.6	1.09

[1] 25% protein, 40% sucrose, 14% tallow. Fed 8 months.
[2] (Arch + thoracic) ÷ 2. Graded on a 0–4 scale.

Table V. Influence of specific amino acids on cholesterolemic effects of casein or soy protein. After Czarnecki and Kritchevsky [4], and Kritchevsky et al. [44]

	Group[1]			
	casein (C)	C + arginine	soy (S)	S + lysine
Number	20/31	20/31	25/31	25/31
Serum lipids, mg/dl				
Cholesterol	241	238	124	190
Triglycerides	105	123	66	74
Lipoproteins, %				
VLDL	2.3	0.9	1.1	2.2
IDL	14.4	7.8	7.7	17.8
LDL	31.9	35.8	30.0	30.2
HDL	51.4	55.5	61.2	49.8
Atherosclerosis[2]				
Arch	1.63	1.30	0.70	1.10
Thoracic	1.05	0.73	0.40	0.70

[1] Diets fed 8 months.
[2] Graded on a 0–4 scale.

lysine to soy protein increased VLDL and IDL levels by 56 and 84%, respectively, and decreased LDL and HDL levels by 20 and 25%, respectively (table V).

Quantity of dietary protein is also a factor in determining its effects on cholesterol levels and atherosclerosis. *Lofland* et al. [53] found high levels (25%) of soy protein to be somewhat more cholesterolemic and signifi-

Table VI. Influence of proteins and fats on cholesterolemia (mg/dl ± SEM) in white carneau pigeons. After *Lofland* et al. [54]

	Protein			
	casein		wheat gluten	
Fat	8%	30%	8%	30%
Butter	472 ± 22	399 ± 47	419 ± 23	398 ± 22
Corn oil	345 ± 41	404 ± 31	492 ± 32	676 ± 96
Crisco	577 ± 98	643 ± 56	495 ± 32	377 ± 37
Margarine	375 ± 24	242 ± 21	442 ± 32	387 ± 32

cantly more atherogenic than low levels (5%). In a later study [54] they compared wheat gluten with casein-lactalbumen (85:15). Low levels of protein (8% of calories) were generally less cholesterolemic than high levels (table VI).

Strong and McGill [70] fed baboons diets high or low in casein (20 or 8% of calories), high or low in cholesterol, and containing saturated or unsaturated fat. In three of four dietary sets the high-protein diets were more cholesterolemic and sudanophilic.

The data regarding effects of vegetable protein or protein quantity are not unanimous. *Charkoff* et al. [2] examined spontaneous aortic lesions in three groups of chickens fed 7.3% protein, 14.5% protein (pair-fed with 7.3% diet), and 14.5% protein fed ad libitum. Serum cholesterol levels (mg/dl) for the three groups were 140, 108, and 121, respectively. Atherosclerosis in the abdominal aorta was most severe in animals fed the low-protein regimen. *Polcak* et al. [62] fed rabbits cholesterol plus 10 or 15% beef protein and found lower serum and aortic cholesterol levels in rabbits fed 15% protein.

The mechanism by which vegetable protein lowers cholesterol involves enhanced steroid excretion [10] and increased oxidation of cholesterol [15].

Carbohydrate

We have fed rabbits a semipurified, cholesterol-free diet containing 40% carbohydrate in which the source of carbohydrate has varied (table VII). In one study [31], we compared the effects of glucose, sucrose,

Table VII. Carbohydrate effects on cholesterolemia and atherosclerosis in rabbits. After Kritchevsky et al. [40]

Carbohydrate[1]	Number	Serum lipids, mg/dl ± SEM		Average atherosclerosis
		cholesterol	triglycerides	
Glucose	7	451 ± 102	92 ± 15	0.85
Fructose	4	922 ± 231	116 ± 51	1.50
Sucrose	9	520 ± 119	248 ± 72	1.45
Lactose	4	329 ± 144	107 ± 21	0.50
Starch	6	532 ± 152	254 ± 102	1.35

[1] Diets contained 40% carbohydrate, fed 40 weeks.

starch, and hydrolyzed starch. Starch was the most cholesterolemic and atherogenic carbohydrate and glucose was the least. In a second experiment [40] we compared glucose, fructose, sucrose, lactose, and starch. Fructose was the most cholesterolemic and atherogenic carbohydrate in this study and lactose gave the lowest cholesterol levels and least severe atherosclerosis. When fed with cholesterol, lactose is very atherogenic for rabbits [76]. Sucrose is more cholesterolemic than glucose when fed with cholesterol to rabbits [11, 63] or chickens [28, 29].

Lang and Barthel [50] fed a diet containing 66% sucrose or dextrin and 0.5% cholesterol to three species of monkey. Dextrin was more cholesterolemic and atherogenic in *Macaca mulatta,* and was more atherogenic but not more cholesterolemic in *Cebus albifrons.* The two carbohydrate regimens gave similar results when fed to *Macaca arctoides.* We have fed baboons [27] and vervet monkeys [25] semipurified diets similar to those we fed to rabbits. Sucrose, glucose, fructose, or starch gave similar cholesterol levels in the baboons but fructose gave the most extensive aortic sudanophilia. Fructose, sucrose, and glucose were compared in vervet monkeys, and fructose and sucrose gave higher cholesterol levels than did glucose. Fructose feeding led to much more severe aortic atherosclerosis than seen in the monkeys fed the other two sugars. When the semipurified diet, augmented with 0.1% cholesterol, was fed to baboons the extent of aortic sudanophilia was increased and atherosclerosis was observed [26]. Atherosclerosis was observed in 5 of 6 baboons fed lactose plus 0.1% cholesterol. In the absence of dietary cholesterol lactose was mildly sudanophilic (table VIII).

Table VIII. Influence of carbohydrates + 0.1% cholesterol on lipids of baboons. After Kritchevsky et al. [26]

Carbohydrate[1]	Serum lipids, mg/dl ±SEM[2]		Lipoproteins α/(β + pre-β)	Aortic sudanophilic, % ± SEM
	cholesterol	triglycerides		
Fructose	164 ± 10	128 ± 14	0.59	11.3 ± 4.2
Sucrose	164 ± 12	106 ± 14	0.69	10.4 ± 5.4
Starch	178 ± 23	119 ± 25	0.72	21.3 ± 8.9
Glucose	173 ± 24	120 ± 18	0.72	17.2 ± 10.3
Lactose	190 ± 16	155 ± 37	0.82	65.8 ± 13.6
Control	101 ± 4	120 ± 27	1.13	1.4 ± 0.4

[1] 40% carbohydrate. Diets fed 17 months.
[2] Terminal levels.

Fiber

Over the years there have been reports pro and con concerning the atherogenicity for rabbits of saturated fat in a cholesterol-free diet. A summary of the available literature in 1964 [21] revealed that saturated fat was atherogenic only when fed as part of a semipurified diet; when added to commercial ration it had no effect. Inasmuch as the dietary fat was the same, it was concluded that some other component of the diet, most probably the fiber, was exercising a protective effect. Experiments designed to test this point [34, 36] involved the addition of hydrogenated coconut oil to the defatted residue of commercial ration, showing that the diet was not atherogenic. Clearly, then, the fiber was inhibiting the atherogenic effect of the fat. *Moore* [59] fed rabbits a semipurified diet in which the fat (20%) was butter oil and the fiber (19%) was wheat straw, cellulose, cellophane, or cellophane-peat (14:5). The diet containing cellophane was significantly more cholesterolemic and atherogenic than that containing wheat straw or cellophane plus peat. Pectin is noteworthy in that it reduces cholesterolemia and inhibits atherogenesis in rabbits [6], chickens [9], and swine [7].

Fiber is a generic term covering a number of substances, mostly carbohydrate in nature, which are not affected by mammalian enzymes. The four major classes of fiber are cellulose, hemicellulose, pectin, and lignin, and their structures and physiological effects can be diverse. When rats are

Table IX. Interaction of fiber with animal and vegetable protein. After Kritchevsky et al. [47]

	Fiber[1]		
	cellulose	wheat straw	alfalfa
Casein			
Serum lipids, mg/dl ± SEM			
Cholesterol	402 ± 40	375 ± 42	193 ± 34
Triglycerides	164 ± 45	94 ± 19	60 ± 8
Average atherosclerosis[2]	1.50	1.03	0.63
Soy protein			
Serum lipids, mg/dl ± SEM			
Cholesterol	248 ± 44	245 ± 35	159 ± 20
Triglycerides	41 ± 8	66 ± 9	62 ± 17
Average atherosclerosis[2]	1.25	0.91	0.73

[1] Proteins fed as 25% of diet: fiber as 15%. Fed for 10 months.
[2] (Arch + thoracic) ÷ 2. Graded on a 0–4 scale.

fed a fiber-free diet containing cholesterol their serum cholesterol is not greatly altered but liver cholesterol may increase 5-fold. Addition of pectin or vegetable gums to this diet will result in a significant reduction in liver cholesterol but cellulose, agar, or alginic acid will increase it. These findings have been reviewed recently [23, 69]. Bran has no effect on serum lipids of rats [67] or cynomolgus monkeys [55].

Hamilton and Carroll [13] compared the effects of a variety of types of crude fiber and of alfalfa, wheat, and oats on plasma cholesterol levels in rabbits. Alfalfa, ground wheat, and ground oats all reduced cholesterol levels compared to the controls who were fed a semipurified diet. Cellulose at levels of 20 or 40% increased cholesterol levels. The mode of action of fiber may involve steroid excretion and/or bile acid binding. *Portman and Murphy* [64] found that rats fed a stock diet excreted more neutral and acidic steroids than rats fed a semipurified diet. Pectin increases bile acid excretion in rats [52], and alfalfa increases neutral steroid excretion in rats [42]. Rabbits fed a semipurified diet exhibit a lower ratio of biliary primary to secondary bile acids than do rabbits fed stock diets [38]; the same effect has been observed in baboons [27]. It has been suggested that semipurified diets reduce the conversion of cholesterol to bile acids [43, 48]. Different types of fiber have been shown to bind bile acids in vitro [32, 68] and they may exert a similar effect in vivo.

Table X. Influence of carbohydrate (60%) on plasma cholesterol in rabbits. After *Hamilton and Carroll* [13]

Diet	Number	Plasma cholesterol	
		mg/dl ± SEM	% ester
Control	22	70 ± 5	63
Dextrose	20	200 ± 22	64
Sucrose	6	185 ± 41	66
Lactose	6	135 ± 24	67
Wheat starch	5	260 ± 66	67
Corn starch	6	185 ± 44	68
Rice starch	6	140 ± 19	62
Potato starch	5	95 ± 24	68

Interaction among dietary components can affect cholesterolemia and atherosclerosis. Thus, casein is more cholesterolemic than soy protein in diets in which the fiber is cellulose. When the fiber is wheat straw, cholesterolemia is unaffected but atherosclerosis is inhibited. Substitution of alfalfa for cellulose reduces both cholesterolemia and atherosclerosis (table IX) [47]. *Hamilton and Carroll* [13] fed rabbits a semipurified diet containing casein (27%) and dextrose (60%) and after 28 days the plasma cholesterol levels were 200±22 mg/dl. Replacement of the dextrose by sucrose did not affect plasma cholesterol (185±41 mg/dl) but, when the carbohydrate was lactose, cholesterol levels fell to 135±24 mg/dl. Potato starch reduced levels of cholesterol to the normal range (95±24 mg/dl) but wheat starch raised them (260±66 mg/dl) (table X).

Investigation of dietary effects on lipids and lipoproteins is a dynamic area of research. There is still much left to be clarified. The influence of diet on apolipoprotein metabolism has not been fully explored and there is still much to be learned about dietary interactions. New findings will provide new clues to dietary treatment of lipidemias.

References

1   Carroll, K.K.; Hamilton, R.M.G.: Effects of dietary protein and carbohydrate on plasma cholesterol levels in relation to atherosclerosis. J. Food. Sci. *40:* 18–23 (1975).

2   Chaikoff, I.L.; Nichols, C.W., Jr.; Gaffey, W.; et al.: The effect of dietary protein level on the development of naturally occurring aortic arteriosclerosis in the chicken. J. Atheroscler. Res. *1:* 461–469 (1961).
3   Corey, J.E.; Hayes, K.C.: Effect of diet on hepatic and intestinal lipogenesis in squirrel, cebus and cynomolgus monkeys. Atherosclerosis *20:* 405–416 (1974).
4   Czarnecki, S.K.; Kritchevsky, D.: The effect of dietary proteins on lipoprotein metabolism and atherosclerosis in rabbits. J. Am. Oil Chem. Soc. *56:* 388A (1979).
5   Dupont, J.: Synthesis of cholesterol and total lipid by male and female rats fed beef tallow or corn oil. Lipids *1:* 409–414 (1966).
6   Ershoff, B.H.: Effects of pectin N.F. and other complex carbohydrates on hypercholesterolemia and atherosclerosis. Expl. Med. Surg. *21:* 108–112 (1963).
7   Fausch, H.D.; Anderson, T.A.: Influence of citrus pectin feeding on lipid metabolism and body composition of swine. J. Nutr. *85:* 145–149 (1965).
8   Fiorith, J.A.; Sims, R.J.: Autoxidation products of cholesterol. J. Am. Oil Chem. Soc. *44:* 221–224 (1967).
9   Fisher, H.; Soller, W.G.; Griminger, P.: The retardation by pectin of cholesterol-induced atherosclerosis in the fowl. J. Atheroscler. Res. *6:* 292–298 (1966).
10   Fumagalli, R.; Paoletti, R.; Howard, A.N.: Hypocholesterolaemic effect of soya. Life Sci. *22:* 947–952 (1965).
11   Grant, W.C.; Fahrenbach, M.J.: Effect of dietary sucrose and glucose on plasma cholesterol in chicks and rabbits. Proc. Soc. exp. Biol. Med. *100:* 250–252 (1959).
12   Gresham, G.A.; Howard, A.N.: The independent production of atherosclerosis and thrombosis in the rat. Br. J. exp. Path. *41:* 395–402 (1960).
13   Hamilton, R.M.G.; Carroll, K.K.: Plasma cholesterol levels in rabbits fed low fat, low cholesterol diets. Effects of dietary proteins, carbohydrates and fibre from different sources. Atherosclerosis *24:* 47–62 (1976).
14   Houtsmiller, U.M.T.: Biochemical aspects of fatty acids with trans double bonds. Fette Seifen Anstr-Mittel *80:* 162–170 (1978).
15   Huff, M.W.; Carroll, K.K.: Effects of dietary protein on turnover, oxidation and absorption of cholesterol and on steroid excretion in rabbits. J. Lipid Res. *21:* 546–558 (1980).
16   Huff, M.W.; Hamilton, R.M.G.; Carroll, K.K.: Plasma cholesterol levels in rabbits fed low fat, cholesterol-free, semipurified diets: effects of dietary proteins, protein hydrolysates and amino acid mixtures. Atherosclerosis *28:* 187–195 (1977).
17   Ignatowski, A.: Über die Wirkung des tierischen Eiweisses auf die Aorta und die parenchymatosen Organe der Kaninchen. Virchows Arch. path. Anat. Physiol. Klin. Med. *198:* 248–270 (1909).
18   Imai, H.; Werthessen, N.T.; Taylor, C.B.; et al.: Angiotoxicity and arteriosclerosis due to contaminants of USP-grade cholesterol. Archs. Pathol. Lab. Med. *100:* 565–572 (1976).
19   Jackson, R.L.; Morrisett, J.D.; Pownall, H.J.; et al.: Influence of dietary *trans* fatty acids on swine lipoprotein composition and structure. J. Lipid Res. *18:* 182–190 (1977).
20   Kritchevsky, D.: Experimental atherosclerosis; in Paoletti, Lipid pharmacology, pp. 63–130 (Academic Press, New York 1964).
21   Kritchevsky, D.: Experimental atherosclerosis in rabbits fed cholesterol-free diets. J. Atheroscler. Res. *4:* 103–105 (1964).

22 Kritchevsky, D.: Role of cholesterol vehicle in experimental atherosclerosis. Am. J. clin. Nutr. *23:* 1105–1110 (1970).
23 Kritchevsky, D.: Atherosclerosis and dietary fiber; in Paoletti, Gotto, Atherosclerosis reviews, vol. 2, pp. 179–186 (Raven Press, New York 1977).
24 Kritchevsky, D.: Unpublished data.
25 Kritchevsky, D.; Davidson, L. M.; Kim, H. K.; et al.: Influence of semipurified diets on atherosclerosis in African green monkeys. Exp. molec. Path. *26:* 28–51 (1977).
26 Kritchevsky, D.; Davidson, L. M.; Kim, H. K.; et al.: Influence of type of carbohydrate on atherosclerosis in baboons fed semipurified diet plus 0.1% cholesterol. Am. J. clin. Nutr. *33:* 1869–1887 (1980).
27 Kritchevsky, D.; Davidson, L. M.; Shapiro, I. L.; et al.: Lipid metabolism and experimental atherosclerosis in baboons: influence of cholesterol-free semi-synthetic diets. Am. J. clin. Nutr. *27:* 29–50 (1974).
28 Kritchevsky, D.; Grant, W. C.; Fahrenbach, M. J.; et al.: Effect of dietary carbohydrate on the metabolism of cholesterol-4-C^{14} in chickens. Archs. Biochem. Biophys. *75:* 142–147 (1958).
29 Kritchevsky, D.; Kolman, R. R.; Guttmacher, R. M.; et al.: Influence of dietary carbohydrate and protein on serum and liver cholesterol in germ-free chickens. Archs Biochem. Biophys. *85:* 444–451 (1959).
30 Kritchevsky, D.; Marcucci, A. M.; Sallata, P.; et al.: Comparison of amorphous and crystalline cholesterol in establishment of atherosclerosis in rabbits. Med. Exp. *19:* 185–193 (1969).
31 Kritchevsky, D.; Sallata, P.; Tepper, S. A.: Experimental atherosclerosis in rabbits fed cholesterol-free diets. 2. Influence of various carbohydrates. J. Atheroscler. Res. *8:* 697–703 (1968).
32 Kritchevsky, D.; Story, J. A.: Binding bile salts in vitro by non-nutritive fiber. J. Nutr. *104:* 458–462 (1974).
33 Kritchevsky, D.; Tepper, S. A.: Cholesterol vehicle in experimental atherosclerosis. VIII. Influence of naturally occurring saturated fats. Med. Pharmacol. Exp. *12:* 315–320 (1965).
34 Kritchevsky, D.; Tepper, S. A.: Factors affecting atherosclerosis in rabbits fed cholesterol-free diets. Life Sci. *4:* 1467–1471 (1965).
35 Kritchevsky, D.; Tepper, S. A.: Influence of medium-chain triglyceride (MCT) on cholesterol metabolism in rats. J. Nutr. *86:* 67–72 (1965).
36 Kritchevsky, D.; Tepper, S. A.: Experimental atherosclerosis in rabbits fed cholesterol-free diets: influence of chow components. J. Atheroscler. Res. *8:* 357–369 (1968).
37 Kritchevsky, D.; Tepper, S. A.; Czarnecki, S. K.; et al.: Experimental atherosclerosis in rabbits fed cholesterol-free diets. 9. Beef protein and textured vegetable protein. Atherosclerosis *39:* 169–175 (1981).
38 Kritchevsky, D.; Tepper, S. A.; Kim, H. K.; et al.: Experimental atheroclerosis in rabbits fed cholesterol-free diets. 4. Investigation into the source of cholesteremia. Exp. molec. Path. *22:* 11–19 (1975).
39 Kritchevsky, D.; Tepper, S. A.; Kim, H. K.; et al.: Experimental atherosclerosis in rabbits fed cholesterol-free diets. 5. Comparison of peanut, corn, butter and coconut oils. Exp. molec. Path. *24:* 375–391 (1976).
40 Kritchevsky, D.; Tepper, S. A.; Kitagawa, M.: Experimental atherosclerosis in rabbits

fed cholesterol-free diets. 3. Comparison of fructose and lactose with other carbohydrates. Nutr. Rep. int. 7: 193–202 (1973).

41 Kritchevsky, D.; Tepper, S.A.; Scott, D.A.; et al.: Cholesterol vehicle in experimental atherosclerosis. 18. Comparison of North American, African and South American peanut oils. Atherosclerosis 38: 291–299 (1981).

42 Kritchevsky, D.; Tepper, S.A.; Story, J.A.: Isocaloric, isogravic diets in rats. III. Effect of non-nutritive fiber (alfalfa or cellulose) on cholesterol metabolism. Nutr. Rep. int. 9: 301–308 (1974).

43 Kritchevsky, D.; Tepper, S.A.; Story, J.A.: Nonnutritive fiber and lipid metabolism. J. Food Sci. 40: 8–11 (1975).

44 Kritchevsky, D.; Tepper, S.A.; Story, J.A.: Influence of soy protein and casein on atherosclerosis in rabbits. Fed. Proc. 37: 747 (1978).

45 Kritchevsky, D.; Tepper, S.A.; Vesselinovitch, D.; et al.: Cholesterol vehicle in experimental atherosclerosis. 11. Peanut oil. Atherosclerosis 14: 53–64 (1971).

46 Kritchevsky, D.; Tepper, S.A.; Vesselinovitch, D.; et al.: Cholesterol vehicle in experimental atherosclerosis. 13. Randomized peanut oil. Atherosclerosis 17: 225–243 (1973).

47 Kritchevsky, D.; Tepper, S.A.; Williams, D.E.; et al.: Experimental atherosclerosis in rabbits fed cholesterol-free diets. 7. Interaction of animal or vegetable protein with fiber. Atherosclerosis 26: 397–403 (1977).

48 Kyd, P.A.; Bouchier, I.A.D.: Cholesterol metabolism in rabbits with oleic acid-induced cholelithiasis. Proc. Soc. exp. Biol. Med. 141: 846–849 (1972).

49 Kummerow, F.A.: Current studies on relation of fat to health. J. Am. Oil Chem. Soc. 51: 255–259 (1974).

50 Lang, C.M.; Barthel, C.H.: Effects of simple and complex carbohydrates on serum lipids and atherosclerosis in nonhuman primates. Am. J. clin. Nutr. 25: 470–475 (1972).

51 Leveille, G.A.; Pardini, R.S.; Tillotson, J.A.: Influence of medium-chain triglycerides on lipid metabolism in the rat. Lipids 2: 287–294 (1967).

52 Leveille, G.A.; Sauberlich, H.F.: Mechanism of the cholesterol-depressing effect of pectin in the cholesterol-fed rat. J. Nutr. 88: 209–214 (1966).

53 Lofland, H.B.; Clarkson, T.B.; Goodman, H.O.; et al.: Atherosclerosis and thiamine deficiency in pigeons in relation to soybean protein. J. Atheroscler. Res. 2: 123–130 (1962).

54 Lofland, H.B.; Clarkson, T.B.; Rhyne, L.; et al.: Interrelated effects of dietary fats and proteins on atherosclerosis in the pigeon. J. Atheroscler. Res. 6: 395–403 (1966).

55 Malinow, M.R.; McLaughlin, P.; Papworth, L.; et al.: Effect of bran and cholestyramine on plasma lipids in monkeys. Am. J. Clin. Nutr. 29: 905–911 (1976).

56 McMillan, G.C.; Silver, M.D.; Weigensberg, B.I.: Elaidinized olive oil and cholesterol atherosclerosis. Archs. Path. 76: 106–112 (1963).

57 Meeker, D.R.; Kesten, H.D.: Experimental atherosclerosis and high protein diets. Proc. Soc. exp. Biol. Med. 45: 543–545 (1940).

58 Meeker, D.R.; Kesten, H.D.: Effect of high protein diets on experimental atherosclerosis of rabbits. Archs. Path. 31: 147–162 (1941).

59 Moore, J.H.: The effect of the type of roughage in the diet on plasma cholesterol levels and aortic atherosis in rabbits. Br. J. Nutr. 21: 207–215 (1967).

60 Moyer, A.W.; Kritchevsky, D.; Logan, J.B.; et al.: Dietary protein and serum cholesterol in rats. Proc. Soc. exp. Biol. Med. *92:* 736–737 (1956).
61 Myher, J.J.; Marai, L.; Kuksis, A.; et al.: Acylglycerol structure of peanut oils of different atherogenic potential. Lipids *12:* 775–785 (1977).
62 Polcak, J.; Melichar, F.; Sevelova, D.; et al.: The effect of a meat-enriched diet on the development of experimental atherosclerosis in rabbits. J. Atheroscler. Res. *5:* 174–180 (1965).
63 Pollak, O.J.: Cholesteremia of rabbits fed carbohydrate diets. J. Am. Geriat. Soc. *9:* 349–358 (1961).
64 Portman, O.W.; Murphy, P.: Excretion of bile acids and hydroxysterols by rats. Archs. Biochem. Biophys. *76:* 367–376 (1958).
65 Ruttenberg, H.; Little, N.A.; Davidson, L.M.; et al.: Influence of dietary *trans* fatty acid on atherosclerosis in rabbits. Fed. Proc. *39:* 1039 (1980).
66 Scebat, L.; Renais, J.; Lenegre, J.: Experimental atherosclerosis in the rabbit. Preliminary study. Revue Athéroscler. *3:* 14–26 (1961).
67 Story, J.A.; Czarnecki, S.K.; Baldino, A.; et al.: Effect of components of fiber on dietary cholesterol in the rat. Fed. Proc. *36:* 1134 (1977).
68 Story, J.A.; Kritchevsky, D.: Comparison of the binding of various bile acids and bile salts in vitro by several types of fiber. J. Nutr. *106:* 1292 (1976).
69 Story, J.A.; Kritchevsky, D.: Nutrients with special functions: dietary fiber; in Alfin-Slater, Kritchevsky, Human nutrition – a comprehensive treatise, vol. 3A, pp. 259–279 (Plenum Press, New York 1980).
70 Strong, J.P.; McGill, H.C., Jr.: Diet and experimental atherosclerosis in baboons. Am. J. Path. *50:* 669–690 (1967).
71 Van Bruggen, J.T.; Elwood, J.C.; Marco, A.; et al.: In vitro studies on oxidative and lipid metabolism in the immature macaque monkey. J. Atheroscler. Res. *2:* 388–399 (1964).
72 Vesselinovitch, D.; Getz, G.S.; Hughes, R.H.; et al.: Atherosclerosis in the rhesus monkey fed three food fats. Atherosclerosis *20:* 303–321 (1974).
73 Weigensberg, B.I.; McMillan, G.C.: Lipids in rabbits fed elaidinized olive oil and cholesterol. Exp. molec. Path. *3:* 201–214 (1964).
74 Weigensberg, B.I.; McMillan, G.C.: Serum and aortic lipids in rabbits fed cholesterol and linoleic acid stereoisomers. J. Nutr. *83:* 314–324 (1964).
75 Weigensberg, B.I.; McMillan, G.C.; Ritchie, A.C.: Elaidic acid: effect on experimental atherosclerosis. Archs. Path. *72:* 358–366 (1961).
76 Wells, W.W.; Anderson, S.C.: The increased severity of atherosclerosis in rabbits on a lactose-containing diet. J. Nutr. *68:* 541–549 (1959).
77 Wissler, R.W.; Vesselinovitch, D.; Getz, G.S.; et al.: Aortic lesions and blood lipids in rhesus monkeys fed three food fats. Fed. Proc. *26:* 371 (1967).

D. Kritchevsky, PhD, Associate Director, Wistar Institute of Anatomy and Biology, 36th Street at Spruce, Philadelphia, PA 19104 (USA)

Nutrients and Lipoproteins: Dietary Profiles of Population Groups and Experimental Models

A. Rabbi

Istituto Nazionale della Nutrizione, Rome, Italy

Introduction

A widely debated and still unsolved problem like the relationship between diet and cardiovascular diseases lends itself to, or one might say, suggests some general considerations. Some of these considerations may appear to be useless or not new at all. Still, when faced with a problem that, for a long time now and in spite of substantial research efforts, is still hazy, the questions arise: Why is the problem still obscure? How can we explain that the relationship is still poorly understood when so many other and often much more sophisticated questions have been resolved thanks to the formidable biochemical and biomedical developments during the last decades?

In cases like this the explanation is often found in the fact that the process is multifactorial and the analysis is extremely complex. One might also think, however, that the philosophy at the basis of the methodological approach is not the most suitable one. Does such a doubt apply to the question of the relationship between dietary intake and atherosclerosis?

We think that the question is properly posed. In fact, the lengthy debate on the subject and the large number of new hypotheses that have been advanced was described by *Ahrens* [1] as an 'unfinished business'. If one acknowledges that the century-old debate between the advocates of 'Virchow's infiltration hypothesis and those of 'Rokitansky's incrustation hypothesis is still going on, then one must say that much work remains to be done [5].

One of the key problems today is finding new experimental models for our investigations, models that are no longer in the wake of a rich but not very illuminating tradition. As one example, we can note that the available data on the effects of diet on high-density lipoprotein (HDL) cholesterol are still insufficient and inadequate to indicate whether a specific diet can raise plasma HDL cholesterol [3]. Perhaps it will be possible in the future to assess the value of the experience of *Anitschkow and Chalatow* [2] with an experimental model that 70 years later is still applicable to research on atherosclerosis; the same applies to the early observations by *Ignatowski* [7].

We might ask why experimental hyperlipidemic or hyperproteinemic diets bring about different responses, in terms of the onset of atherosclerosis, in a number of different experimental animal species. The same question arises about the influence of hormones on the atherosclerotic process. We know, for instance, that experiments on different animals (dogs, cats, rats, and rabbits) demonstrate the involvement of the hypothalamus in the onset of hypercholesterolemia due to a hyperlipidemic diet [4]. We still have to determine the 'actual weight' of individual factors that contribute to the complex onset of the atherosclerotic process in man.

Certainly the answer to our question of the relationship of diet to the development of atherosclerosis, which appears to depend on more than one cause, can only be obtained by a multidisciplinary approach. For a long time in the future we may still have to consider a number of possible solutions, and this will certainly contribute to a sense of frustration among biologists who usually base their deductions on specific relationships that link causes to effects and structures to functions.

Dietary Profiles

Epidemiological studies, in their prospective, retrospective, and intervention variants, represent an approach for obtaining evidence on the nature and causality of the relationship between dietary intakes of specific foods and nutrients and the development of coronary heart disease (CHD). While significant relationships have been reported in the literature between habitual intakes of fat or cholesterol and levels of serum cholesterol or mortality from CHD, when cross-cultural data have been obtained there has been a general failure to demonstrate unequivocal associations within culturally homogeneous groups. Intervention studies have often generated conflicting and inconsistent results that are difficult to interpret.

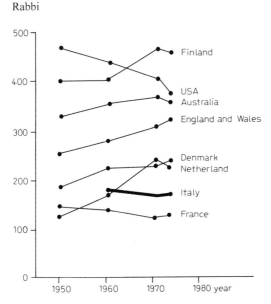

Fig. 1. Trends in standardized mortality rates in the specified countries for all forms of heart disease per 100,000 population (men, age group 35-64). Based on *Ovacov and Bystrova* [9].

A vast literature is available on the subject, and the role of environmentally confounding variables, the faults and limits of experimental design, and methodological errors have been reviewed extensively and critically.

Among the industrially developed countries, Italy occupies a somewhat privileged position in terms of CHD mortality rates. The data shown in figure 1 indicate that, in 1960, the standardized mortality rates for males aged 35-64 years were lower for Italy than in most other countries. The rate appears to be fairly constant throughout the examined years, or show a slight decline. It is too soon to advance any hypothesis on the continuation of such a decline and its causes remain obscure. However, atherosclerotic disease is a chronic process which requires time to develop; that is, a delay of undetermined length occurs before one can observe any change in CHD mortality rates in any given community as a consequence of a modification in its risk factors.

If we consider what was and what has become the dietary profile of the Italian population, we have legitimate grounds to suspect that we may be moving toward an epidemic of atherosclerosis. The exposure of the Italian population to the dietary risk factor is quite recent, dating, as shown in

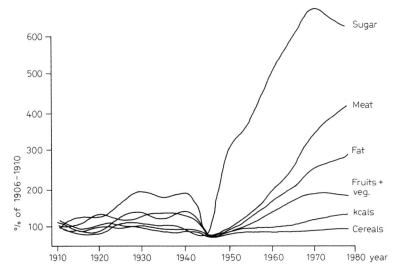

Fig. 2. Relative changes with time in consumption of several food groups in Italy. Reprinted from *Mariani* [8].

figure 2, from the late 1950s after the end of World War II; we can expect, after a due time lag, that there could well be evidence of a rise in the CHD mortality rate.

Figure 2 shows the profound modifications that have occurred both in the quantity and the quality of the food comprising the average Italian diet and its fat moiety. The values are derived from food disappearance data, and have all the limitations inherent in these types of data. The changes, however, have been impressive, particularly in the introduction into the average diet of energy and nutrient-dense foods which have displaced vegetable staples such as wheat products. These foods and nutrients, namely meat, eggs, total fats, and saturated fatty acids, are notoriously associated with the atherosclerotic process.

Even if such a dietary evolution is the expression of a general improvement in the quality of life, indicating that larger sectors of the population have progressively gained access to a diet which meets their nutrient requirements and which at the same time provides the symbolic and social gratification deriving from the use of prestigious foods, legitimate concern focuses on the possible untoward effects on the state of health of the community that such a diet may have. Serum cholesterol levels have been traditionally low in Italy, although differences have been reported within

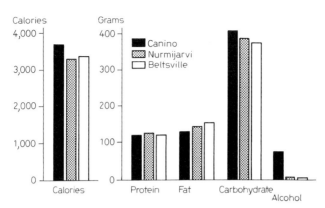

Fig. 3. Average daily intakes. Each bar represents the mean intake for 7 consecutive days per subject. n = 168 in Canino, 203 in Nurmijarvi, and 147 in Beltsville. Reprinted from *Iacono* et al. [6].

the population, suggesting higher values among the most affluent and urban populations. Recent findings appear to substantiate these concerns, inasmuch as an increase of about 25% has been reported in the serum cholesterol level of Neapolitan men in the last 15 years. We do not know if the time lag linking a progressive rise in serum cholesterol to an increase in CHD mortality rates will expire soon, but we feel that the issue is of extreme importance and some action should be promoted to monitor and assess the risk to the Italian population, in order to initiate a preventive campaign before it is too late.

Within the framework of an epidemiological approach, the Italian National Institute of Nutrition has participated in an international study with American scientists [6]. The aim of the study was to investigate the role of dietary fats on atherogenic and thrombogenic indexes in free-living populations characterized by contrasting dietary habits and CHD mortality rates.

A pilot study was conducted in the rural areas of Italy (Canino), Finland (Nurmijarvi), and the USA (Beltsville, Md.). The dietary intakes of selected groups of middle-aged, clinically healthy farmers (fig. 3–5) were assessed during 7 consecutive days by the precise weighing method. At the end of the dietary study, blood samples were obtained from the subjects for measuring the quantity and quality of scrum, red blood cells, and platelet lipids, as well as for tests on the function of platelets.

The results indicate that the structure of the diet was reflected in the composition of serum lipids, red blood cells, and platelets (fig. 6–9), and

Nutrients and Lipoproteins

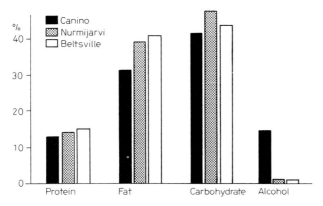

Fig. 4. Percent of energy derived from different sources in the diets. Reprinted from *Iacono* et al. [6].

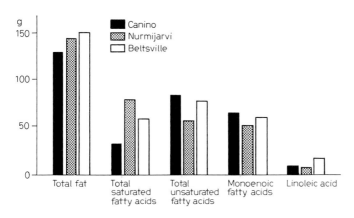

Fig. 5. Fatty acid composition of diets. Reprinted from *Iacono* et al. [6].

that the higher reactivity of the platelets to aggregating agents differed in the three localities as expected, suggesting an easier clotting response in the subjects exposed to a diet with higher saturated fats and a lower content of polyunsaturated fatty acids. A second stage of this study is currently under preparation. The project will consist of intervention programs aimed at assessing the reversibility of thrombogenic and atherogenic indexes in free-living communities by short-term dietary manipulation.

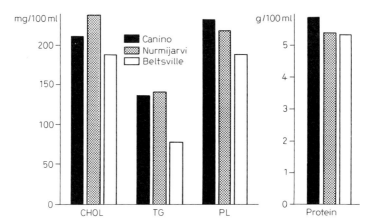

Fig. 6. Lipid composition of plasma. Each bar represents the mean of 20 subjects in Canino, 21 subjects in Nurmijarvi, and 21 subjects in Beltsville. Reprinted from *Iacono* et al. [6].

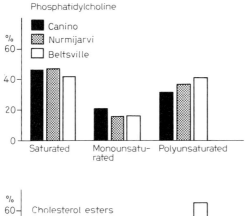

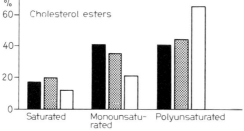

Fig. 7. Major fatty acid groups in phosphatidylcholine and cholesterol esters of plasma. Reprinted from *Iacono* et al. [6].

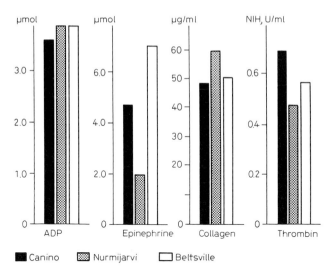

Fig. 8. Threshold concentrations for aggregation of platelets. Reprinted from *Iacono* et al. [6].

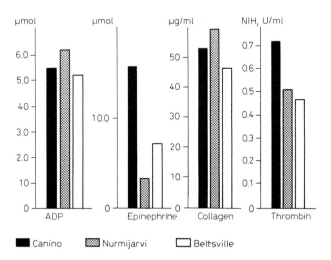

Fig. 9. Concentrations of aggregating agents required for release of equal amounts of ^{14}C-serotonin from platelets. Reprinted from *Iacono* et al. [6].

Experimental Models

The use of animal models for research on atherosclerosis has developed along two lines in an attempt to find answers to two sets of problems: pathogenesis and treatment. The wide field of studies aimed at assessing the role of diets to prevent or treat atherosclerosis belongs to the second group. Animal models are used to obtain information on the behavior of blood cholesterol and other connected biochemical parameters. Even if the etiology of coronary disorders is still a matter for speculation, high blood cholesterol is certainly one of the 10 risk factors involved.

In the early period of research, the rabbit was the experimental animal of choice; since then many other species have been investigated, including monkeys. Selection of the most suitable animal model must obviously take into account the specific problem that is being tackled. Other factors involved in the selection are availability, cost, and the ease with which the animal adjusts to both the environment and dietary habits.

An extensive literature survey shows that, in order to assess the effects of dietary intake on the metabolism of cholesterol, the rat model can afford information as reliable and as transferable to man as pig or even monkey models. Rats are easy to treat and to manipulate, their environment can be easily controlled, and it is possible to operate on selected strains and on sample sizes that permit very accurate statistical manipulation.

For these reasons we have used rats to assess the effect of a fava bean protein concentrate developed in our laboratories by 'air flow classification'. This research work is part of a larger project promoted by the Italian National Research Council. The project seeks to promote a rational use of vegetable protein sources, particularly legumes.

Although in recent years there has been renewed interest in research on the hypocholesterolemic effects of vegetable proteins, this research, except for a few Indian examples, has mainly concentrated on soya bean protein concentrates. Still, from the work done by *Carroll* [3] on rabbits it appears that, at least on this animal model, fava beans have the best effect in lowering blood cholesterol.

We have conducted three sets of experiments in our Institute to study the way in which fava bean protein concentrates lower blood cholesterol that had been raised previously by dietary means, and the ability of the concentrates to prevent cholesterol levels from increasing. The hypercholesterolemic diets had a semisynthetic basis with casein proteins. In the first set of experiments the fat content consisted of hydrogenated peanut oil, 15%; cholesterol, 2%; and carbohydrates given as starch. The second and third sets of experiments had a fat content consisting of coconut oil, 25%; cholesterol, 1%; and sucrose, 48%. An isocaloric fava bean concentrate was added to replace either 100 or 50% of the casein proteins in the basic diet.

First Set of Experiments. The presence of a high concentration of highly unsaturated lipids and cholesterol in the diet did not necessarily raise the levels of blood cholesterol and triglycerides in rats even if these parameters were very high in the liver. Total or partial replacement of casein with fava bean concentrate brought about a moderate but significant reduction of blood cholesterol. The reduced energy consumption, even if the diet is hyperlipidemic and hypercholesterolemic, is by itself sufficient to avoid storage of cholesterol and triglycerides in the liver and reduces the concentration of circulating triglycerides.

Second Set of Experiments. The 25% coconut oil and 48% sucrose diet induced a substantial increase in blood cholesterol with fluctuations which stabilized after 8 weeks at three times the original level. The increase particularly affected very low-density lipoprotein (VLDL) cholesterol but was also noticeable in low-density lipoprotein (LDL) cholesterol. The introduction in the diet of fava bean concentrate to replace 50% of the casein proteins substantially lowered the level of blood cholesterol, particularly VLDL and LDL. A crossover experiment induced a modulation of VLDL and LDL cholesterol in the expected direction.

Third Set of Experiments. Fava bean concentrates partly prevented the induction of hypercholesterolemia caused by the 25% coconut oil diet. The effect was still mainly at the VLDL and LDL level. There was a concomitant marked increase (approximately 10-fold) in the fecal excretion of cholic acid, indicating that the hypocholesterolemic effect is due to the drainage of cholic acid from the intestine, which implies an enhanced cholesterol catabolism and a decrease of its level in the circulation.

The conclusions drawn from the experiments reported here are that fava bean concentrates play a decisive hypocholesterolemic role, particularly in cases of diet-induced hypercholesterolemia. Under our experimental conditions, we observed no change in the cholesterol concentration at the aortic level. It is also true that rats are very resistant to atherosclerosis.

Many assumptions could be made as to which factors were responsible for the cholesterol-lowering effect of the fava bean concentrates. At the moment, however, nothing can be said for sure. The hypothesis that proteins per se might be responsible for the effect is very appealing. Since the fava bean concentrate that was used included only 60% protein, we cannot exclude other factors that may play a role: fibers, saponins, and isoflavons, the last of which are in fact almost always present in legumes. Research is continuing in this field.

References

1 Ahrens, E. H., Jr.: Dietary fats and coronary heart disease: unfinished business. Lancet *ii:* 1345–1348 (1979).
2 Anitschkow, N.; Chalatow, S.: Über experimentelle Cholesterinsteatose und ihre Bedeutung für die Entstehung einiger pathologischer Prozesse. Central bl. allg. Path. path. Anat., Jena *24:* 1–9 (1913).
3 Carroll, K. K.: Dietary protein in relation to plasma cholesterol levels and atherosclerosis. Nutr. Rev. *36:* 1–5 (1978).
4 Chernukh, A. M.; Chernysheva, G. V.; Lebedeva, L. N.: The effect of the hypothalamus on the lipid metabolism and ATPase activity of the cardiac mitochondria of rabbits in the early stages of atherogenesis. Proc. 2nd US-USSR Joint Symp. on Myocardial Metabolism, Sochi 1975. DHEW Publ. No.(NIH) 77–924, pp.105–119 (US Department of Health, Education, and Welfare, Public Health Service, National Institutes of Health, Washington 1977).
5 Gryglewski, R. J.: Prostacyclin and atherosclerosis. Trends Pharmacol. *1:* 164–166 (1980).
6 Iacono, J. M.; et al.: Pilot epidemiological studies in thrombosis; in Chandler, Eurenius, McMillan, Nelson, Schwartz, Wessler, Workshop on the thrombotic process in atherogenesis, Reston, pp. 309–328 (Plenum Publishing, New York 1978).
7 Ignatowski, A. I.: Influence of animal food on the organism of rabbits. Archs Méd. exp. Anat. *20:* 1–20 (1908).
8 Mariani, A.: Nutritional needs and the situation in the European Community; in Symposium on Nutrition Food Technology and Nutritional Information, London 19–20 March 1980; Commission of the European Communities, Report Eur 7085EN (1981).
9 Ovacov, V. K.; Bystrova, V. A.: Present trends in mortality in the age group 35–64 in selected developed countries between 1950–73, Wld Hlth Stat. Quart. *31:* 208 (1978).

A. Rabbi, MD, Istituto Nazionale della Nutrizione, Via Ardeatina 546, I-00179 Rome (Italy)

Diet and Hypertension

Role of Nutrition in Hypertension and Its Control – Experimental Aspects

Harriet P. Dustan

Cardiovascular Research and Training Center, The University of Alabama in Birmingham, School of Medicine, Birmingham, Ala., USA

Introduction

It is now clear that nutrition plays a role in hypertension but the magnitude of that role is, as yet, undetermined. Three aspects are known or suspected as being important: an excess of calories leading to obesity, an excess of dietary sodium, and a relative deficiency of dietary potassium. In regard to caloric excess, the evidence seems clear that obesity and/or weight gain is a substantial risk factor for development of hypertension in man. For sodium, the evidence is less sure because although cross-cultural studies have shown that hypertension is rare in societies with low salt intake and frequent in those with high salt intake, within-population studies have not shown a correlation of sodium intake and arterial pressure. However, the ease with which experimental hypertension can be produced by salt feeding strongly argues that some fraction of essential hypertension is salt-sensitive. As far as potassium is concerned, there is but little information available. Potassium feeding mitigates salt-dependent experimental hypertension and hypertension in blacks is associated with a lesser urinary potassium than found in whites. These findings suggest that the sodium/potassium ratio may be an important factor in arterial pressure control.

Not all obese people are hypertensive and salt feeding of laboratory animals has a variable effect on arterial pressure. These two facts indicate that there are other factors which are necessary for the pressor potential to be expressed. It is now becoming clear that genetic makeup is one of these factors.

The purpose of this discussion is to review the evidence linking the three nutritional aspects – obesity, salt intake and potassium intake – to hypertension.

Caloric Excess – Obesity

Obesity and Hypertension

The evidence relating obesity to hypertension seems now well-established. The most recent evidence comes from the Hypertension Detection and Follow-up Program [21] which found that about 60% of the over 10,000 enrollees were at least 20% overweight. This statistic is in contrast to the current findings for the general US population with 14% of men and 24% women, aged 20–74 years, being more than 20% overweight.

The importance of obesity to hypertension has been a subject of epidemiologic interest for several decades and, in 1969, *Chiang* et al. [5] reviewed the available information and concluded that obesity and/or weight gain acts as an environmental stress causing hypertension in genetically predisposed individuals.

One of the earliest studies of this phenomenon is that of *Short and Johnson* [57] which investigated the relationship between excess weight and the prevalence of hypertension. Their findings led them to conclude that being overweight exerts a positive influence on the development of hypertension. Since that time much additional evidence has been found. In 1946, *Levy* et al. [30] in a study of 22,741 US Army officers reported that sustained hypertension developed more than twice as frequently in overweight men as in men who were of normal weight. *Padmavati and Gupta* [45] studied the relationship of obesity to hypertension in 1,132 people of low socioeconomic status and 224 affluent people in Delhi, India. Hypertension, defined as a blood pressure greater than 140/90 mm Hg, was found in 25% of the well-to-do but in only 0.17% of the poor. The former were heavier and had a rise in body weight and arterial pressure with age while the latter weighed less and had only a slight rise in weight with age; arterial pressure increased only slightly with age and that rise was found only in women.

Lowe [31] reported on weight and arterial pressure relationships in over 5,000 employees of a British utility company. He found that within each decade, from 20 to 69 years, pressure increased with increasing body weight and arm circumference. *Epstein* et al. [15] from the Tecumseh, Michigan study found a positive correlation between relative weight and pressure that was highest for the age group of 30–39 years.

Weight gain during early adult life and/or adolescent obesity seems to be a potent risk factor for subsequent development of hypertension. This evidence comes from several independent investigations. In the Framingham population, *Kannel* et al. [23] found that relative weight, weight gain during the years of follow-up, and skinfold thickness were positively correlated with existing levels of arterial pressure and the subsequent development of hypertension. For those individuals who were normotensive in the early years of the study, the risk of becoming hypertensive with the passage of time was eight times greater for those who were 20% overweight than it was for those 10% underweight. *Abraham* et al. [1] reported that people who had not been obese as children but gained weight during young adulthood had a higher incidence of hypertension and cardiovascular-renal disease later in life than did those who did not gain weight.

There are four important longitudinal studies that clearly show the importance of weight gain to the development of hypertension. *Stamler* [59] followed employees of a Chicago utility company for 20 years and found a strong relationship between the onset of hypertension during middle age and increase in body weight. *Paffenbarger* et al. [46] reviewed the postgraduate health histories of more than 7,000 men who had attended the University of Pennsylvania between 1931 and 1940. They found that in-college predictors of subsequent development of hypertension were overweight, high normal arterial pressure, increased heart rate, and positive family history for hypertension. *Oberman* et al. [41] reported in a 24-year follow-up study of 1,056 World War II aviators that weight gain was a risk factor for subsequent development of hypertension. The Evans County, Georgia, study has also yielded evidence concerning weight gain and hypertension in young adults. *Johnson* et al. [22] did a 7-year resurvey of individuals who were between 15 and 29 years of age at the original examination. Hypertension was more frequent at resurvey and arterial pressure correlated with body mass index in white men and women and black women. *Heyden* et al. [18] studied a hypertensive subset of this group and found that weight gain was associated with worsening of hypertension except in black males.

The evidence cited above is but a fraction of that which has shown the relationship of obesity to hypertension. The evidence is compelling and led *Tyroler* et al. [63] to estimate from the Evans County experience that the control of obesity could prevent 48% of hypertension in white people and 28% in black.

There is little evidence from animal studies that can tell us with certainty whether the relationship of obesity and/or weight gain to hypertension is a human characteristic. Apparently it is not, because in 1939 *Wood and Cash* [69] reported the effects of weight gain on arterial pressure of normotensive and renal hypertensive dogs who gained from 16 to 45% of control weight with a high-fat diet. The normotensive dogs showed some rise in pressure while those with renal hypertension became severely hypertensive. Considering how much we have learned from animal models of hypertension, it is surprising that there is a dearth of information about the model, i.e. obesity, that bears the closest relationship to hypertension in man.

Body Weight and Blood Pressure

Why obesity predisposes to hypertension is unknown but it is now well-recognized that body weight and arterial pressure (within the normal range) are positively correlated. This information comes from studies of children as well as adults, from ethnically homogeneous as well as inhomogeneous populations, and from widely separated geographic areas.

Chiang et al. [5] found in their 1969 review that the correlation of body weight with arterial pressure was clearest in populations in which obesity is common and arterial pressure rises with age. However, subsequent epidemiologic studies have established the relationship of body weight to arterial pressure in a number of populations that are not obese and in which hypertension is rare.

Two studies are from island societies of the South Pacific. *Page* et al. [47] measured arterial pressure and body weight of six groups living in the Solomon Islands. In contrast to findings in acculturated peoples, weight correlated negatively with age in all the groups while pressure and weight correlated positively in women of two of the societies. The populations of Rarotonga and Pukapuka were studied by *Prior* et al. [50]. In both populations, arterial pressure and body weight were positively associated. The Rarotongans were taller and heavier than the Pukapukans and showed a rise in pressure with age which was absent in the Pukapukans.

Whyte [66] compared body weight and arterial pressure characteristics of Australian men and natives of New Guinea of two age groups – 29–39 years and 40–59 years. In the New Guineans, pressure did not rise with age nor was it correlated with weight or skinfold thickness. In contrast, the Australians showed a rise in pressure with age as well as an increase in body fat. In the younger Australians, systolic and diastolic pressures were positively correlated with body weight and indices of fatness. The Delhi, India study of *Padmavati and Gupta* [45] referred to earlier, also showed a positive correlation of body weight and blood pressure.

From South America come two studies. *Glanville and Geerdink* [16] examined natives of Surinam on the northeast coast of South America. They found that among men, systolic pressure was negatively correlated with age and positively correlated with stature while diastolic pressure correlated positively with weight as it did also in the women. The Multinational Andean Genetic and Health Program [34] studied three groups of Chileans living at different altitudes to explore reasons for the supposed immunity against hypertension of high-altitude living. Three groups were studied which lived at three different altitudes: coastal, < 300 m; sierran, approximately 3,000 m; and altiplano, > 4,000 m. Among all three groups, arterial pressure and body weight were positively correlated and the lack of hypertension at high altitudes was found to be a function of lower body weight.

Children also have the same characteristic relationship as adults between arterial pressure and body weight. This was evident in the Andean study [34] and in several recent reports contained in a supplement to *Hypertension,* 'High Blood Pressure in the Young' [20]. *Prineas* et al. [49] found that, among 9,977 children aged 6–9 years, weight was the best single measure of body size for predicting blood pressure. *Katz* et al. [24] carried out a longitudinal study of black children selected from the Philadelphia Collaborative Perinatal Project and found that weight percentile groupings provided the best discrimination for systolic arterial pressure at these ages. They concluded that 'blood pressure variation is so closely associated with growth and maturation that these factors must be taken into account when assessing blood pressure in childhood and adolescence'. Five other reports in this symposium [14, 19, 28, 58, 65] also showed a strong positive correlation between arterial pressure and body weight or other indices of obesity.

The report of the Task Force on Blood Pressure Control in Children [53] collated the arterial pressure values for children from three studies according to age. From age 6 years to 18 years arterial pressure rose in boys and girls. Pressure correlated positively with both height and weight in both groups.

Chiang et al. [5] make the point clearly that the relationship of arterial pressure to body weight is not a function of obesity per se but of body size. There is no explanation for this association although a recent report by *Magrini* [33] sheds some light on it. He carried out hemodynamic studies in unanesthetized puppies monthly from age 1 month to maturity at 8–9 months. As the puppies grew, arterial pressure and total peripheral resistance rose while cardiac output (related to body weight) fell. Weight gain and increase in peripheral resistance were positively correlated as were arterial pressure and the fall in cardiac output. Thus, it is apparent from all these studies that with increasing body size and/or maturation changes in vascular resistance and flow take place that account for the increase in arterial pressure.

Mechanisms of Obesity Hypertension

Obesity changes metabolic, hemodynamic, volume, and hormonal functions – any one of which, or some abnormal interrelationship among them, could produce hypertension. However, there is no evidence to date that normotensive obese individuals differ from obese hypertensive patients in any of these abnormalities nor in their interrelationships.

Alexander [2] did hemodynamic and volume studies in obese subjects weighing over 300 lbs. He found an increase in cardiac output that was proportional to the degree of overweight. About two thirds of his subjects were hypertensive: they had normal vascular resistance while those that were normotensive had decreased vascular resistance. Blood volume was also increased and correlated with the amount of weight over the ideal value. *Whyte* [67] has suggested that the elevated cardiac output of obese individuals accounts for the hypertension, but *Mujais* et al. [40] failed to find hemodynamic differences among non-obese and obese hypertensives that would explain the hypertension. When cardiac output and vascular resistance were indexed according to body weight, no differences in these two functions were found between the two groups.

Caloric excess is necessary for the development of most obesity and since an increased food intake is usually equivalent to an increased sodium intake, *Dahl* et al. [10] proposed that obesity hypertension is salt-dependent. They studied a group of obese hypertensive patients during weight reduction with varied intakes of sodium chloride and found that arterial pressure fell only when salt was restricted, regardless of the degree of weight loss. Two recent studies contrast with that experience. *Reisin* et al. [51] reported that weight reduction of obese hypertensives reduced arterial pressure in spite of an unrestricted sodium intake. *Tuck* et al. [62] studied obese hypertensives during weight reduction with sodium intakes of either 120 or 40 mEq/day. All patients lost weight and arterial pressure fell without regard to the sodium intake.

In both spontaneous [55] and experimental obesity in man [42], the secretion rate, metabolic clearance, and excretion of 17-OH corticoid metabolites of cortisol are increased and such increased steroid production could work with other factors to increase arterial pressure. However, this seems unlikely because the increases are proportional to body surface area, and total and ultrafilterable plasma cortisol and urinary free cortisol are not elevated.

Obesity influences the production of thyroid hormones and, since triiodothyronine (T_3) influences the number of β-adrenergic receptors in animals [68], the possibility has been raised that the increased T_3 that occurs with overfeeding [42] may increase receptors and produce increased reponsiveness to normal amounts of circulating pressor substances.

Although a wide variety of changes occur with obesity, no studies have systematically looked for differences between non-obese and obese hypertensives (except that of *Mujais* et al. [40]) and between normotensive and hypertensive obese subjects. Thus, there is no understanding of the hypertension of obesity.

Salt Intake

Population Studies

Generally speaking, epidemiologic studies have shown that populations with high dietary sodium have a substantial prevalence of hypertension while, in those with a low salt intake, hypertension is rare and pressure does not rise with age. *Prior* et al. [50] concluded that a higher sodium intake accounted for the higher arterial pressures found among the natives of Rarotonga as compared with those of Pukapuka. *Page* et al. [47] concluded that the absence of hypertension in six Solomon Island societies reflected their relatively low sodium intake. *Oliver* et al. [43] studied a primitive tribe, the Yanomamo Indians of northern Brazil and found that they had an extremely low sodium intake (<1 mEq/day as estimated from urinary sodium) and no rise in arterial pressure with age; the authors concluded that the normal arterial pressure was because of the low salt intake. *Truswell* et al. [61] found that the Kung bushmen of northern Botswana of Africa also had no age-correlated rise in blood pressure and suggested that this was because of a low salt intake. For Japan *Sasaki* [54] has reported a positive correlation between stroke death rate and salt intake. *Page* et al. [48] has studied a nomadic tribe in Iran that has a dietary sodium similar to industrialized populations (i.e. >150 mEq/day), a similar prevalence of hypertension and an age-related rise in arterial pressure.

Dahl [7] was impressed that hypertension in industrialized societies results from high salt intake, and his graph showing the positive relationship between habitual sodium intake and prevalence of hypertension in several populations has done much to solidify the opinion that culturally determined high salt intake is responsible for the high prevalence of hypertension.

As important as sodium excess may be as a determinant of arterial pressure in certain circumstances, it should be stated that primitive low-salt cultures differ from industrialized societies in many ways other than sodium intake. As emphasized by the Task Force on Hypertension at the National Heart, Lung, and Blood Institute [52], primitive peoples are ethnically more homogeneous, are smaller, more physically active, and have a higher potassium intake than industrialized societies. Accordingly, it is currently impossible to determine the importance of a high salt intake in the prevalence of hypertension in industrialized populations. This difficulty is compounded by the lack of relationship between arterial pressure and salt intake in within-population studies [11, 38].

The therapeutic efficacy of dietary salt restriction in hypertensive patients has been taken as evidence to support the hypothesis that a high salt intake causes hypertension. In the 1940s, it was shown by *Kempner* [26] and several other investigators [6, 12, 64] that rigid sodium restriction can reduce arterial pressure to normal levels in a substantial percentage of hypertensive patients. However, this finding does not establish conclusively that dietary salt caused their hypertension in the first place because from experimental models of hypertension it has been learned that mechanisms for maintaining hypertension can be completely different than those that initiate it.

Experimental salt-dependent hypertension is a useful model to consider when reviewing the possible role of sodium in human hypertension.

Experimental Salt-Dependent Hypertension

Salt-dependent hypertension is a frequently used experimental model. Although most widely studied in the rat, it has been produced in a variety of animal species and none has been found resistant to the hypertension-producing potential of salt although there is much intra-species variation in responsiveness. Methods of production vary but each supplies more salt than can be promptly excreted so that a positive sodium balance results. Thus, it can be produced by high salt feeding alone or by using deoxycorticosterone acetate injections or implants along with 0.9% saline as drinking fluid. Unilateral nephrectomy is often used because it hastens the development of hypertension, and subtotal nephrectomy that leaves but a fraction of functioning renal mass promotes the process yet more. High salt diets as used by *Meneely* et al. [37] produce hypertension slower than does the DOCA-salt combination [56] suggesting that the steroid interference with renal sodium excretion is an important factor; to date, there has been no evidence to show that DOCA causes hypertension through mechanisms other than sodium retention.

Dahl and Schackow [9] showed that development of salt-dependent hypertension is genetically determined when they bred the salt-sensitive (S) and salt-resistant (R) strains of rats. They also found that age is an important factor in the response to salt loading, with young rats developing more severe hypertension than older rats [8].

Salt-Dependent Hypertension in Man

It is not easy to produce salt-dependent hypertension in normotensive people. *Kirkendall* et al. [27] gave three levels of sodium intake to normal control subjects – 10, 210, and 410 mEq/day – and found that arterial pressure was unaffected even at the highest intake although inulin space and total exchangeable sodium rose progressively. *Luft* et al. [32] used intakes of 10, 300, 600, 800, 1,200, and 1,500 mEq/day for 3-day periods each in normotensive controls and found, in most subjects, small progressive rises in arterial pressure although, in some, pressure did not change. Markedly positive sodium balance equivalent to several liters of extracellular fluid occurred (from 850 to 1,400 mEq).

Among hypertensives the case is clearly different than in normals. Salt restriction reduces arterial pressure in many, though not all, patients. *Dustan* et al. [13] described volume-dependent hypertension as that which disappears in response to a 9-mEq sodium diet, and they found this type in 7 of 14 essential hypertensives. *Kawasaki* et al. [25] designated as salt-sensitive hypertension arterial pressure that rose when sodium intake was changed from 9 to 249 mEq/day, regardless of the response to the previous 9-mEq sodium diet. Neither group of investigators found any changes in renin, volume, or steroid factors to account for the different arterial pressure response to variations in sodium intake.

In a sense, primary aldosteronism is like the DOCA-salt model of experimental hypertension. The hypertension varies in severity [60] and is salt-dependent [3].

Another useful comparison between clinical and experimental models is hypertension of endstage kidney disease in men and subtotal nephrectomy-salt hypertension in animals. Hypertension is frequent in chronic uremia, and this contrasts sharply with the 15–20% prevalence in the overall US population. The largest proportion of the hypertension of uremia is salt- and water-dependent. Because some uremic or anephric patients never develop hypertension regardless of the amount of sodium and water overload, it is obvious that additional factors are needed to elevate arterial pressure.

Mechanisms of Salt-Dependent Hypertension

Hemodynamic studies in man and animals have shown a marked heterogeneity of responses to salt loading. Thus, in salt-sensitive hypertension in pigs [39] and dogs [4] and in anephric man [44], salt loading raises arterial pressure by increasing mainly cardiac output, increasing mainly peripheral resistance, or through long-term, whole body autoregulations. These changes point to organ differences in responses to salt and these could be genetically determined.

In summary, it is clear that salt is a potent pressor substance in some, but not all, circumstances. The necessary ingredients for salt-dependent hypertension seem to be a salt load, a reduced renal sodium excretory ability, and a genetic predisposition for hypertension.

Potassium Intake

Available Evidence

Very little is known about the importance of potassium intake in arterial pressure regulation. *Meneely* et al. [35] found that the life-shortening effect of high salt feeding in the rat could be ameliorated by feeding a high potassium intake as well and that this was associated with a somewhat lower blood pressure and total exchangeable sodium [36].

There is little evidence in man, also. *Langford and Watson* [29] reported a positive correlation between diastolic arterial pressure and the sodium/potassium ratio in black and white school girls. *Grim* et al. [17]

found that black people in Evans County, Georgia, in comparison with whites, had higher pressures, greater prevalence of hypertension, and the same dietary sodium but a lower dietary potassium intake. From these fragmentary data no conclusion can be drawn; however, the effects of potassium supplementation in animals with a high salt intake are clear enough to warrant further study.

References

1. Abraham, S.; Collins, G.; Nordsieck, M.: Relationship of childhood weight status to morbidity in adults. HSMHA Hlth Rep. 86: 273–284 (1971).
2. Alexander, J.K.: Obesity and the circulation. Mod. Concepts cardiovasc. Dis. 32: 799–803 (1963).
3. Bravo, E.L.; Dustan, H.P.; Tarazi, R.C.: Spironolactone as a nonspecific treatment for primary aldosteronism. Circulation 48: 491–498 (1973).
4. Bravo, E.L.; Tarazi, R.C.; Dustan, H.P.: Multifactorial analysis of chronic hypertension induced by electrolyte-active steroids in trained, unanesthetized dogs. Circulation Res. 1977: suppl I, pp. 140–145.
5. Chiang, B.N.; Perlman, L.V.; Epstein, F.H.: Overweight and hypertension. Circulation 39: 403–421 (1969).
6. Corcoran, A.C.; Taylor, R.D.; Page, I.H.: Controlled observations on effect of low sodium dietotherapy in essential hypertension. Circulation 3: 1 (1951).
7. Dahl, L.K.: Possible role of salt intake in the development of essential hypertension; in Bock, Cottier, Essential hypertension, p. 53 (Springer, Berlin 1960).
8. Dahl, L.K.; Knudsen, K.D.; Heine, M.A.; Leitl, G.J.: Effects of chronic excess salt ingestion: modification of experimental hypertension in the rat by variations in the diet. Circulation Res. 22: 11–18 (1968).
9. Dahl, L.K.; Schackow, E.: Effects of chronic excess salt ingestion: experimental hypertension in the rat. Can. med. Ass. J. 90: 155–160 (1964).
10. Dahl, L.; Silver, L.; Christie, R.: Role of salt in the fall of blood pressure accompanying reduction of obesity. New Engl. J. Med. 258: 1186–1192 (1958).
11. Dawber, T.R.; Kannel, W.B.; Kagan, A.; Donalbedian, R.K.; McNamara, P.; Pearson, G.: Environmental factors in hypertension; in Stamler, Stamler, Pullman, The epidemiology of hypertension, pp. 255–288 (Grune & Stratton, New York 1967).
12. Dole, V.P.; Dahl, L.K.; Cotzias, G.C.; Eder, H.A.; Krebs, M.E.: Dietary treatment of hypertension. Clinical and metabolic studies of patients on the rice diet. J. clin. Invest. 29: 1189–1206 (1950).
13. Dustan, H.P.; Bravo, E.L.; Tarazi, R.C.: Volume-dependent essential and steroid hypertension. Am. J. Cardiol. 31: 606–615 (1973).

14 Ellison, R.C.; Sosenko, J.M.; Harper, G.P.; Gibbons, L.; Pratter, F.E.; Miettinen, O.S.: Obesity, sodium intake, and blood pressure in adolescents. Hypertension 2: suppl. I, 78–82 (1980).

15 Epstein, S.H.; Francis, R., Jr.; Hayner, N.S.; Johnson, B.C.; Kjelsberg, M.O.; Napier, J.A.; Ostrander, L.D., Jr.; Payne, M.W.; Dodge, H.J.: Prevalence of chronic disease and distribution of selected physiological variables in a total community, Tecumseh, Michigan. Am. J. Epidemiol. 81: 307–323 (1965).

16 Glanville, E.V.; Geerdink, R.A.: Blood pressure of Amerindians of Surinam. Am. J. phys. Anthrop. 37: 251–254 (1972).

17 Grim, C.E.; Luft, F.C.; Miller, J.Z.; Meneely, G.R.; Battarbee, H.D.; Hames, C.G.; Dahl, L.K.: Racial differences in blood pressure in Evans County, Georgia: relationship to sodium and potassium intake and plasma renin activity. J. chron. Dis. 33: 87–94 (1980).

18 Heyden, S.; Bartel, A.G.; Hames, C.G.: Elevated blood pressure levels in adolescents, Evans County, Ga. Seven-year follow-up of 30 patients and 30 controls. J. Am. med. Ass. 209: 1683–1689 (1969).

19 Higgins, M.W.; Keller, J.B.; Metzner, H.L.; Moore, F.E.; Ostrander, L.D., Jr.: Studies of blood pressure in Tecumseh, Michigan. II. Antecedents in childhood of high blood pressure in young adults. Hypertension 2: suppl. I, 117–123 (1980).

20 Kotchen, T.A.; Havlik, R.J. (ed.): High blood pressure in the young. Hypertension 2: suppl., July/August (1980).

21 Hypertension Detection and Follow-up Program Cooperative Group: Race, education and prevalence of hypertension. Am. J. Epidemiol. 106: 351–361 (1977).

22 Johnson, A.L.; Cornoni, J.C.; Cassel, J.C.; Tyroler, H.A.; Hayden, S.; Hames, C.G.: Influence of race, sex and weight on blood pressure behavior in young adults. Am. J. Cardiol. 35: 523–530 (1975).

23 Kannel, W.; Brand, N.; Skinner, J.; Dawber, T.; McNamara, P.: Relation of adiposity to blood pressure and development of hypertension: the Framingham Study. Ann. intern. Med. 67: 48–59 (1967).

24 Katz, S.H.; Hediger, M.L.; Schall, J.I.; Bowers, E.J.; Barker, W.F.; Aurand, S.; Eveleth, P.B.; Gruskin, A.B.; Parks, J.S.: Blood pressure, growth and maturation from childhood through adolescence. Mixed longitudinal analyses of the Philadelphia blood pressure project. Hypertension 2: suppl. I, 55–69 (1980).

25 Kawasaki, T.; Delea, C.S.; Bartter, F.C.; Smith, H.: The effect of high-sodium intakes on blood pressure and other related variables in human subjects with idiopathic hypertension. Am. J. Med. 64: 193–198 (1978).

26 Kempner, W.: Treatment of hypertensive vascular disease with rice diet. Am. J. Med. 4: 545–577 (1948).

27 Kirkendall, W.M.; Connor, W.E.; Abboud, F.; Rastogi, S.P.; Anderson, T.A.; Fry, M.: The effect of dietary sodium on the blood pressure of normotensive man; in Genest, J.; Koiw, E. (eds.) Hypertension '72, pp. 360–373 (Springer, Berlin 1972).

28 Kuller, L.H.; Crook, M.; Almes, M.J.; Detre, K.; Reese, G.; Rutan, G.: Dormont high school (Pittsburgh, Pennsylvania) blood pressure study. Hypertension 2: suppl. I, 103–116 (1980).

29 Langford, H.G.; Watson, R.L.: Electrolytes and hypertension; in Paul, O. (eds.) Epidemiology and control of hypertension, pp. 119–130 (Stratton, New York 1975).

30 Levy, R.L.; White, P.D.; Stroud, W.D.; Hillman, C.C.: Overweight: its prognostic

significance in relation to hypertension and cardiovascular renal diseases. J. Am. med. Ass. *131:* 951–953 (1946).

31 Lowe, C. R.: Arterial pressure, physique, and occupation. Br. J. prev. soc. Med. *18:* 115–124 (1964).

32 Luft, F. C.; Rankin, L. I.; Bloch, R.; Weyman, A. E.; Willis, L. R.; Murray, R. H.; Grim, C. E.; Weinberger, M. H.: Cardiovascular and humoral responses to extremes of sodium intake in normal black and white men. Circulation *60:* 697–706 (1979).

33 Magrini, F.: Haemodynamic determinants of the arterial blood pressure rise during growth in conscious puppies. Cardiovasc. Res. *12:* 422–428 (1978).

34 Makela, M.; Barton, S. A.; Schull, W. J.: The multinational Andean genetic and health program. IV. Altitude and the blood pressure of the Aymara. J. chron. Dis. *31:* 587–603 (1978).

35 Meneely, G. R.; Ball, C. O. T.; Youmans, J. B.: Chronic sodium chloride toxicity: the protective effect of added potassium chloride. Ann. intern. Med. *47:* 263–273 (1957).

36 Meneely, G. R.; Lemley-Stone, J.; Darby, W. J.: Changes in blood pressure and body sodium of rats fed sodium and potassium chloride. Am. J. Cardiol. *8:* 527–532 (1961).

37 Meneely, G. R.; Tucker, R. G.; Darby, W. J.; Auerbach, S. N.: Chronic sodium chloride toxicity in the albino rat. II. Occurrence of hypertension and of a syndrome of edema and renal failure. J. exp. Med. *98:* 71 (1953).

38 Miall, W. E.: Follow-up study of arterial pressure in the population of a Welsh mining valley. Br. med. J. *ii:* 1204–1210 (1959).

39 Miller, A. W.; Bohr, D. F.; Schork, A. M.; Terris, J. M.: Hemodynamic responses to DOCA in young pigs. Hypertension *1:* 591–597 (1979).

40 Mujais, S. K.; Fouad, F. M.; Tarazi, R. C.; Bravo, E. L.; Dustan, H. P.: Hemodynamics of hypertension (Ht) in obese (Ob) subjects (Abstract). Circulation *62:* 111–112 (1980).

41 Oberman, A.; Lane, N. E.; Harlan, W. R.; Graybiel, A.; Mitchell, R. E.: Trends in systolic blood pressure in the thousand aviator cohort over a twenty-four-year period. Circulation *36:* 812–822 (1967).

42 O'Connell, M.; Danforth, E., Jr.; Horton, E. S.; Salans, L.; Sims, E. A. H.: Experimental obesity in man. III. Adrenocortical function. J. clin. Endocr. Metab. *6:* 323–329 (1973).

43 Oliver, W. J.; Cohen, E. L.; Neel, J. V.: Blood pressure, sodium intake, and sodium-related hormones in the Yanomamo Indians, a 'no-salt' culture. Circulation *52:* 146–151 (1975).

44 Onesti, G.; Kim, K. E.; Greco, J. A.; DelGuercio, E. T.; Fernandez, M.; Schwartz, C.: Blood pressure regulation in endstage renal disease and anephric man. Circulation Res. *36:* suppl. I, pp. 145–152 (1975).

45 Padmavati, S.; Gupta, S.: Blood pressure studies in rural and urban groups in Delhi. Circulation *19:* 395–405 (1959).

46 Paffenbarger, R. S., Jr.; Thorne, M. C.; Wing, A. L.: Chronic disease in former college students. VIII. Characteristics in youth predisposing to hypertension in later years. Am. J. Epidemiol. *88:* 25–32 (1968).

47 Page, L. B.; Damon, A.; Moellering, R. C., Jr.: Antecedents of cardiovascular disease in six Solomon Island societies. Circulation *49:* 1132–1146 (1974).

48 Page, L.B.; Vandervert, D.; Nader, K.; Lubin, N.; Page, J.R.: Blood pressure of Qash'qai pastoral nomads in Iran in relation to culture, diet, and body form. Am. J. clin. Nutr. *34:* 527–538 (1981).

49 Prineas, R.J.; Gillum, R.F.; Horibe, H.; Hannan, P.J.: The Minneapolis children's blood pressure study. 2. Multiple determinants of children's blood pressure. Hypertension *2:* suppl. I, pp. 24–28 (1980).

50 Prior, I.A.M.; Grimley-Evans, J.; Harvey, H.P.B.; Davidson, F.; Lindsey, M.: Sodium intake and blood pressure in two Polynesian populations. New Engl. J. Med. *279:* 515–520 (1968).

51 Reisin, E.; Abel, R.; Modan, M.: Effect of weight loss without salt restriction on the reduction of blood pressure in overweight hypertensive patients. New Engl. J. Med. *298:* 1–6 (1978).

52 Report of Salt and Water Subgroup of the Hypertension Task Force of the National Heart, Lung, and Blood Institute, NIH Publication No. 79–1630 (1979).

53 Report of the Task Force on Blood Pressure Control in Children. Pediatrics, *59:* 797–820 (1977).

54 Sasaki, N.: High blood pressure and the salt intake of the Japanese. Jap. Heart J. *3:* 313–324 (1962).

55 Schteingart, D.E.; Gregerman, R.I.; Conn, J.W.: A comparison of the characteristics of increased adrenocortical function in obesity and in Cushing's syndrome. Metabolism *12:* 484–497 (1963).

56 Selye, H.: Textbook of endocrinology. Acta Endocrinilogica (University of Montreal, Montreal 1947).

57 Short, J.J.; Johnson, H.J.: An evaluation of the influence of overweight on blood pressures of healthy men. Am. J. med. Sci. *198:* 220–224 (1939).

58 Siervogel, R.M.; Frey, M.A.B.; Kezdi, P.; Roche, A.F.; Stanley, E.L.: Blood pressure, electrolytes, and body size: their relationships in young relatives of men with essential hypertension. Hypertension *2:* suppl. I, 83–92 (1980).

59 Stamler, J.: On the natural history and epidemiology of hypertensive disease; in Cort, Fencl, Hejl, Jirka, The pathogenesis of essential hypertension, pp. 67–107 (State Medical Publishing House, Prague 1960).

60 Tarazi, R.C.; Ibrahim, M.M.; Bravo, E.L.; Dustan, H.P.: Hemodynamic characteristics of primary aldosteronism. New Engl. J. Med. *289:* 1330–1335 (1973).

61 Truswell, A.S.; Kennelly, B.M.; Hansen, J.D.L.; Lee, R.B.: Blood pressures of Kung bushmen in Northern Botswana. Am. Heart J. *84:* 5–12 (1972).

62 Tuck, M.L.; Dornfeld, L.; Kledzik, G.; Maxwell, M.H.: Weight loss lowers renin and aldosterone levels in obese patients. Clin. Res. *27:* 318A (1979).

63 Tyroler, H.A.; Heyden, S.; Hames, C.G.: Weight and hypertension: Evans County studies of blacks and whites; in Paul, O. (ed): Epidemiology and control of hypertension, pp. 177–204 (Stratton, New York 1975).

64 Watkin, D.M.; Froeb, H.F.; Hatch, F.T.; Gutman, A.B.: Effects of diet in essential hypertension. Am. J. Med. *9:* 441–493 (1950).

65 Watson, R.L.; Langford, H.G.; Abernethy, J.; Barnes, T.Y.; Watson, M.J.: Urinary electrolytes, body weight, and blood pressure. Pooled cross-sectional results among four groups of adolescent females. Hypertension *2:* suppl. I, 93–98 (1980).

66 Whyte, H.M.: Body build and blood pressure of men in Australia and New Guinea. Aust. J. exp. Biol. *41:* 395–404 (1963).

67 Whyte, H.: Behind the adipose curtain. Am. J. Cardiol. *15:* 66–80 (1965).
68 Williams, L.T.; Lefkowitz, A.M.; Watanabe, D.R.; Besch, H.R.: Thyroid hormone regulation of beta-adrenergic receptor number: possible biochemical basis for the hyperadrenergic state in hyperthyroidism. Clin. Res. *25:* 458A (1977).
69 Wood, J.E.; Cash, J.R.: Obesity and hypertension: clinical and experimental observations. Ann. intern. Med. *13:* 81–90 (1939).

H.P. Dustan, Director, Cardiovascular Research and Training Center, The University of Alabama in Birmingham, School of Medicine, 1002 Zeigler Building, University Station, Birmingham, AL 35294 (USA)

Experimental Studies on Nutrition, Hypertension, and Cardiovascular Diseases

C. R. Sirtori, M. R. Lovati, G. Gianfranceschi, R. Farina, G. Franceschini

Center E. Grossi Paoletti, Institute of Pharmacology and Pharmacognosy and Chemotherapy Chair, University of Milan, Milan, Italy

Introduction

Animal studies on the effect of nutrients on the development of cardiovascular disease, most notably in the areas of hyperlipidemia and hypertension, have generally followed knowledge deriving from clinical epidemiological data. Such is the case of studies on electrolyte metabolism and the development of arterial changes typical of the hypertensive process.

The dietary induction of hyperlipidemias is, in particular, a well-established procedure, which has recently involved nutrients like ethanol and different types of proteins. In the latter case, the availability of animal models already displaying genetic forms of the disease may provide significant help in understanding the mechanism of dietary hyperlipidemia.

Diet and Hypertension

Dietary Salt. A direct correlation between salt intake and arterial blood pressure was suggested at the beginning of this century [1] and has received support from numerous epidemiological studies in different populations [19, 46]. The restriction of salt intake, as a preventive measure [50] and even as a treatment for nonsevere forms of the disease, has been also advocated with some success.

The hypothesis that a genetic susceptibility, indicated in clinical studies [42, 45], may interact with salt intake in eliciting hypertension has found support in animal models. The Dahl strain of rats only develops hypertension when exposed to a high-salt diet; in contrast to other strains, hypertension is maintained following salt withdrawal [16]. Increased salt intake in strains of spontaneously hypertensive rats (SHR) exacerbates hypertension; conversely, salt restriction almost inevitably fails to correct the abnormality [74]. Indeed, salt loading may reverse experimental malignant hypertension [49].

In the Milan SHR model, renal changes characterized by reduced glomerular filtration (GFR) [5] probably provide a mechanism for the permanent establishment of hypertension. In these rats, in fact, ingested sodium is retained to a significantly greater extent than in normal animals. The severity of the ensuing hypertension, which corrects the reduced GFR is, therefore, also related to the Na^+ intake [4]. Comparative clinical findings, also suggesting that young subjects genetically prone to hypertension have a reduced GFR [6], support the reduction of Na^+ intake as a preventive measure for young subjects at risk. The demonstration of a reduced Na^+ efflux from cells, particularly erythrocytes, in essential hypertensive patients [22], is a further reason for suggesting a reduced dietary salt intake. On the other hand, recent clinical evidence rules against excessive salt reduction, in view of possible severe toxic effects (e.g. renal tubular necrosis) in salt-restricted patients [38].

Proteins. The effect of dietary proteins on the development of hypertension and on the prevalence of arterial disease has never been analyzed in detail. The deficiency of certain essential amino acids, e.g. tryptophan, has been associated with resistance to some hypotensive agents, e.g. α-methyldopa [17]. In this case, in fact, tryptophan deficiency would imbalance central serotonergic mechanisms, thereby reducing drug effectiveness.

On the other hand, more recent studies in SHR, administered diets with different protein contents, have shown that the percentage of energy from protein may markedly affect development of hypertension and survival of the animals. Stroke-prone SHR, fed on a low-protein diet with excess salt, as commonly done in Japan, rapidly become severely hypertensive and die of stroke in a short period. In contrast, when a high-protein diet, still with excess salt, is given, stroke incidence is markedly reduced [80]. With 17–24% of the dietary calories coming from protein, in fact, the development of hypertension is less severe and, at the 10-month follow-up, stroke incidence is 0%, versus 56% with a low-protein diet.

Further studies on this topic have indicated that methionine is the most influential amino acid on blood pressure development, its dietary concentration being inversely related to blood pressure, followed by lysine and proline. Both methionine and taurine, a metabolite of methionine, significantly reduce blood pressure when added to the diet [51]. High-protein diets exert a saluretic effect in these animals and, moreover, a diet enriched in soybean proteins seems to increase the resistance of the arterial wall to stress [80].

Extrapolation of the experimental findings to the clinical situation, i.e. to the different incidence of cerebrovascular disease in the Japanese (low-protein diet) versus the US population (high-protein diet), is obviously difficult but should be considered with interest [57].

Calorie Restriction. Obesity is a major risk factor for hypertension, insofar as the frequency of hypertensive disease in overweight persons is more than double that of normal weight people of the same age [70]. In SHR, the sympathoadrenal system is apparently affected by caloric intake [82]. Sucrose feeding to these animals markedly stimulates the sympathetic nervous system and raises blood pressure [81]. Conversely, caloric restriction is remarkably effective in lowering blood pressure, with a mean fall between 14 and 19% [83]; the same effect is not noted in normotensive rats.

Polyunsaturated Fatty Acids. Decreased blood pressure has been reported in moderately hypertensive patients following diets with increased polyunsaturated (PU)/saturated fatty acid ratios [29]. It has been suggested that increased synthesis of prostaglandin metabolites with vasodilator effects may occur in this condition.

An inhibitor of the renin-renin substrate system has, however, also been described in plasma [14], and hyperreninemia may result from the deficiency of the circulating inhibitor. Although the nature of this is not known (unsaturated lysophosphatidyl derivatives with renin inhibitory activity have been developed) [76], decreased plasma concentration of PU fatty acids has been shown in the plasma of rats with different types of experimental hypertension [47]. In vitro studies have also suggested that the renin reaction may be inhibited by linoleic and arachidonic acid, as well as by short chain fatty acids (capric and lauric), and not by long chain saturated fatty acids [33]. This observation would be consistent both with the observed increased renin activity in hyperlipidemia, as well as with the possible protective effect of unsaturated fatty acids in hypertension.

Diet and Plasma Lipids

Dietary studies in normolipidemic or hyperlipidemic animal models have been numerous. Most nutrients, e.g. lipids, carbohydrates, proteins, ethanol, etc., have been examined. Often, however, the investigated models seem to bear only a distant physiological relationship with the human type of hyperlipidemias. Comparative evaluations between diet-induced animal hyperlipidemic models and the human types of disease were, therefore, carried out in the reported examples. Moreover, new models of spontaneous hyperlipidemias are described.

Diet-Induced Models of Hyperlipidemia

Cholesterol. Dietary induction of hypercholesterolemia goes back to the beginning of this century, when *Anitschkow* [2] demonstrated a dramatic hypercholesterolemic response in rabbits administered crystalline cholesterol.

Experimental cholesterol feeding has been repeated innumerable times during this century. Many authors still remain critical about this mode of dietary induction of hypercholesterolemia, in view of obvious differences between the most commonly used animal models (rabbits and even some primate strains) and the human types of disease. However, recent findings indicate that the model is not really remote from the human spontaneous diseases.

Although a hypercholesterolemic response to dietary cholesterol in humans may be elicited only when cholesterol is given within formula diets and not in solid food [13, 18], some of the changes in lipoprotein distribution and composition found in animals have clear counterparts in man. Rabbits, as well as some primates [58, 65], develop a characteristic cholesterol-enriched very low-density lipoprotein (VLDL) fraction following dietary cholesterol; moreover, a cholesterol-enriched subfraction of high-density lipoproteins (HDL-I or HDL_c) has also been described [12]. In humans, even in the absence of increased total cholesterol in plasma, both of the typical lipoprotein modifications occurring in rabbits are observed [48, 55]. In particular, the latter change, i.e. increased HDL_c, is of interest, since this subfraction is receptorially active, inhibiting cell cholesterol biosynthesis in vitro [40]. The different responsiveness to dietary cholesterol in humans and rabbits could, therefore, be partly ascribed to the relative proportion of the two changes, in VLDL and HDL, the latter being far more evident in humans.

On the other hand, the VLDL changes are probably quite significant in the development of atherosclerosis. In dietary-induced, propylthiouracil-treated dogs, the formation of cholesterol-enriched VLDL is seen only in the animals where atherosclerotic changes are most severe [41]. The cholesterol-rich VLDL show a marked increase of apolipoprotein E (apoE) [65], a component most likely of liver origin that is selectively induced by cholesterol feeding [53]. VLDL, with a high apoE and cholesterol content, are receptorially active (i.e. they inhibit cell cholesterol biosynthesis), and have a high affinity for tissue macrophages [23]. This latter property of inducing both cholesterol ester biosynthesis and accumulation in peripheral tissues (including the arterial walls) is considered a marker of atherogenic lipoproteins [39].

The effects of dietary cholesterol on plasma lipoproteins may be specific for different animals. *Guinea pigs* are particularly responsive to cholesterol feeding. Just after 1 week of diet, plasma, red cell, and liver cholesterol are markedly elevated. After 10–12 weeks, tissue lipids increase several times and a fatal hemolytic anemia usually develops [61]. In these animals, a relatively deficient cholesterol esterification may be responsible for the observed pathological changes [15].

Among *primates,* some strains are almost refractory to cholesterol feeding, whereas others (e.g. *Saimiris sciureus)* exhibit lipoprotein changes similar to those of rabbits [12]. Still, in others *(Cercopithecus),* low-density lipoprotein (LDL) increases are seen, resembling the human type II disease.

Lipoproteins, particularly VLDL, modified by dietary cholesterol, have a markedly reduced turnover rate, with an increased arterial wall uptake [59]. These findings, recently confirmed by studies on apoB turnover [37], suggest a significant similarity between the fate of cholesterol-enriched VLDL and that of VLDL from type III hyperlipidemic patients [11]. The altered VLDL composition and turnover may also have a more general significance. In fact, hypertriglyceridemic patients, with a relative cholesterol enrichment in VLDL, have a higher incidence of coronary disease [8]; they also show an enhanced liver VLDL secretion [10]. These modified VLDL possibly exhibit an increased affinity for tissue macrophages, resulting in an enhanced risk of atherosclerosis.

Fat Intake. Administration of fat-enriched diets, in the absence of dietary cholesterol, has received considerable attention, especially in view of human epidemiological data linking the intake of fat, in particular saturated fat, to hypercholesterolemia and the development of atherosclerosis [69]. So-called *semisynthetic* diets, containing elevated percentages of saturated fat with sucrose and, in some cases, vitamin D, have frequently been tested in different animal species.

In almost all studies, the administration of experimental diets resulted in increased levels of plasma cholesterol. Differently from the administration of a strict cholesterol regimen, semisynthetic diets may induce a selective increase in LDL cholesterol, e.g. in rabbits [60]. On the other hand, the atherogenicity of the two regimens is similar [60].

An additive effect may be observed between dietary fat and carbohydrates. This is typically seen with diets administered to rodents [52]. In primates, however, the increases in plasma cholesterol induced by fat- and carbohydrate-enriched diets are remarkable; in this case the administration of cholesterol leads to further increases in the levels of plasma cholesterol, while triglycerides (TG) decrease [68].

The role of proteins in these experimental dietary regimens has never been examined in detail. It appears, however, as also indicated below, that animal proteins have a considerably more marked hypercholesterolemic activity in all models.

Polyunsaturated Fatty Acids. This widely applied mode of treatment for human hyperlipoproteinemias has been given considerable attention in animal models. Lipoprotein composition is significantly altered by PU fatty acid feeding. By examining the composition of lymph lipoproteins following duodenal instillation to rats of various long chain fatty acids in the form of micelles, *Ockner* et al. [54] showed, both with saturated and PU fatty acids, a significant increase in chylomicron TG. However, whereas in the case of palmitate instillation lymph cholesterol is equally distributed between chylomicrons and VLDL, after linoleate, the largest cholesterol fraction is in chylomicrons, with a lesser percentage in VLDL and d >1.006 lipoproteins. The selective increase of chylomicron cholesterol is associated with a more rapid catabolism in treated animals. More recently, *Sheehe* et al. [62] confirmed a larger size of intestinal absorptive particles in rats fed PU fatty acids; these changes were concomitant with decreased cholesterol esterification.

The more rapid catabolic rate of chylomicrons, as compared to VLDL, may be explained with the rapid transfer of phospholipids and apolipoprotein components from chylomicrons to HDL [71]. In recent human experiments of this type, administration of saturated fat in single or multiple loads during the day resulted in decreased HDL cholesterol levels; these were more significant following single loads [30]. In contrast, apoA-I levels were increased, and these increases were statistically significant following multiple fat ingestions. These observations suggest that both the degree of saturation of dietary fat and the mode of administration may modify the plasma transport of ingested fat. Available experimental data, on the other hand, do not contribute to understanding of the reduced levels of HDL cholesterol and apoA-I in hyperlipoproteinemic patients treated with PU regimens [63, 77]. The different physicochemical properties of HDL may be responsible for these latter observations [73].

Administration of PU fatty acid-enriched diets is the most frequently described method for inducing atherosclerosis regression in animals. So-called *prudent* diets, when given to primates after dietary induction of atherosclerosis, have a remarkable effect on plaque regression [3, 79].

Nondigestible Fibers and Saponins. Administration of undigestible fibers (i.e. plant polysaccharides and lignin resistant to hydrolysis by digestive enzymes) has recently gained considerable reputation, in view of epidemiological evidence linking low fiber intake to atherosclerotic disease and large bowel disorders in the Western world [9]. Animal and in vitro studies have suggested a significant absorption of bile acids to the polysaccharide components of indigestible fibers, resulting in enhanced elimination of neutral and acidic steroids.

Experimental data on some types of vegetable fibers have shown a quite remarkable activity on dietary-induced hyperlipidemias. In particular, alfalfa seeds, when given together with a fat-enriched, animal protein-based diet in rabbits, almost completely prevent the rise in plasma cholesterol [36]. The effect of alfalfa is additional to that of concomitantly administered vegetable proteins. Other types of fibers, e.g. bran and cellulose, are far less active, thus confirming human data [24].

The hypothesis that saponin components may be responsible for the lipid-lowering activity of fibers has been supported by rat experiments showing a selective decrease in cholesterol absorption following digitonin or alfalfa saponin administration [44]. Soybean saponins, on the other hand, have recently shown some activity on fecal cholesterol excretion in swine, with negligible effects on cholesterolemia [75]. Unfortunately, human studies with purified saponins have been hampered by significant side effects (e.g. hemolytic anemia) in some subjects [43].

Vegetable Proteins. An antiatherosclerotic activity of vegetable versus animal proteins was suggested early in this century by *Ignatowski* [28]. These old observations, soon overshadowed by the cholesterol theory of atherosclerosis, have received considerable interest in recent years following experimental and, particularly, clinical studies indicating a significant hypocholesterolemic activity of vegetable proteins. Clinical studies in type II hypercholesterolemic patients have generally substituted textured soybean products for animal proteins within low lipid-low cholesterol regimens. In comparative studies, as well as in the presence of cholesterol and of saturated fatty acids, the 'soybean protein diet' always induced remarkable decreases in plasma cholesterol [66, 67].

Experimental studies on the soybean protein diet have provided limited contributions to understanding the mode of action. In rabbits on a semisynthetic hypercholesterolemic diet, *Fumagalli* et al. [20] reported increased fecal excretion of neutral steroids upon shifting from casein to soybean protein. Similar experiments were recently carried out by *Huff and Carroll* [25] who, in addition, observed increased bile acid excretion and accelerated cholesterol catabolism following the dietary change. In swine, *Kim* et al. [31] confirmed the remarkable hypocholesterolemic activity of the regimen but, in a first study, failed to note any meaningful change in fecal steroid excretion, although describing a significant reduction of carcass cholesterol with the soybean protein diet. More recently, the same authors (fig. 1), by analyzing data from individual animals, have shown increases of both neutral and acidic steroids, changing from casein to soy protein [32].

Human studies on this topic have yielded divergent answers. In newborns, soybean milk causes a greater fecal steroid excretion than cow milk [56]. More recent data by our group in type II patients failed, however, to detect significant changes in either fecal steroid excretion or in the slope of the plasma-specific activity of injected labelled cholesterol [21], in spite of the remarkable drop in plasma cholesterol.

The amino acid composition of soy proteins seems to be a significant factor in eliciting the hypocholesterolemic effect. In earlier studies, administration to rabbits of the amino acid mixture corresponding to soybean proteins only exerted a minimal activity, whereas a soybean trypsin digest had the same activity as the intact product [26]. More recently, *Kritchevsky* et al. [35] suggested that the relative arginine richness of soybean proteins, with a resulting high arginine/lysine ratio, might be responsible for the hypocholesterolemia. The rationale for this hypothesis is that arginase activity is less inhibited in the presence of low dietary lysine [34], and this possibly favors arginine-rich protein (apoE) formation. In preliminary studies, the addition of lysine to the soybean diet reduced the hypocholesterolemic activity.

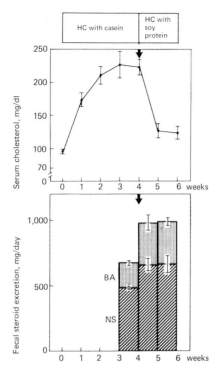

Fig. 1. Changes in serum cholesterol and fecal sterol excretion in swine fed a moderately lipid-enriched diet, first with casein and in the last 2 weeks with soy protein. BA = Bile acids; NS = neutral sterols; HC = hypercholesterolemic [32].

Tissue cholesterol biosynthesis and transport have been studied by us as well as by *Kim* et al. [32]. These last authors showed reduced hydroxymethylglutaryl CoA reductase activity in the swine liver upon addition of soybean proteins to a mash diet. Our group [7] did not demonstrate significant changes of liver 7α-hydroxylase in rats following the change from a hyperlipidemic diet with casein to one with soy protein. Plasma lecithin:cholesterol acyltransferase activity was also unaltered, whereas liver cholesterol esters and acylcholesterol acyltransferase increased.

Spontaneous Models of Hyperlipidemia

As pointed out earlier, most animal models fail to adequately mimic the human type of disorders. For this reason, it has been attempted for a long time, within the different animal species, to select strains more or less

Table I. Lipid and lipoprotein data from a spontaneously hypertriglyceridemic rat (CR–S) as compared to standard Sprague-Dawley animals (CR)

Basal levels	CR		CR-S	
	cholesterol	triglycerides	cholesterol	triglycerides
	mg/dl, $\bar{x} \pm$ SEM, n = 6			
Total	60.7 ± 4.7	69.9 ± 5.9	63.5 ± 5.9	114.5 ± 8.3*
VLDL	6.8 ± 1.5	49.6 ± 14.6	13.7 ± 1.8*	97.1 ± 11.3*
LDL	7.7 ± 2.2	8.8 ± 0.5	4.3 ± 1.4	7.6 ± 1.4
HDL	46.1 ± 1.8	7.9 ± 0.7	43.1 ± 7.3	5.4 ± 1.1

Response to Nath diet**	CR				CR-S			
	cholesterol		triglycerides		cholesterol		triglycerides	
	pre	post	pre	post	pre	post	pre	post
	55.0 ± 4.4	310.0 ± 53.2	64.7 ± 7.9	126.4 ± 15.7	53.8 ± 1.7	222.9 ± 23.0*	145.0 ± 16.3*	185.3 ± 19.5*

* $p < 0.01$ vs corresponding CR; ** 1% cholesterol, 1% cholic acid. VLDL = Very low-density lipoproteins; HDL = high-density lipoproteins; LDL = low-density lipoproteins.

susceptible to dietary induction of hyperlipidemia and/or to development of the human type of atherosclerosis. As pointed out earlier, the response to cholesterol and fat feeding among primates is quite variable among the different strains. *Shore and Shore* [64] also described significant differences in the lipoprotein profiles of Dutch-Belted and New Zealand rabbit strains, the latter being significantly more prone to lipoprotein modifications induced by cholesterol diets, as well as to the development of atherosclerosis.

Although of considerable theoretical and possibly pharmacological interest, these findings still fall short of the typical spontaneous type of human disease. The description of two animal models of spontaneous hyperlipidemia have, therefore, recently received considerable attention. The first case is that of a spontaneous rabbit hypercholesterolemia. Repeated breedings of an original New Zealand strain allowed the development of the so-called Watanabe (WHHL) rabbit which spontaneously exhibits plasma cholesterol levels above 500 mg/dl with TG above 300 mg/dl [78]. These lipid changes result in a dramatic development of atherosclerosis and early death. Although lipoprotein studies have been limited, the hyperlipoproteinemia phenotype would correspond to a human type IIb or III disease. Receptor studies [72] have, however, recently suggested that these animals have a complete absence of peripheral LDL receptors, thus mimicking the human homozygous type IIa disease.

More recently, in our laboratory, we observed a spontaneous hypertriglyceridemia in a rat strain deriving from the original Sprague-Dawley Charles River, and apparently indistinguishable from the standard rat. These animals have a doubling of VLDL cholesterol and TG levels, with a small reduction of HDL cholesterol (table I). The composition of LDL is similar to that of standard Sprague-Dawley rats; intermediate density lipoproteins are essentially absent in both strains. These rats show a higher sensitivity to the dietary induction of hypertriglyceridemia, less so to hypercholesterolemia. The availability of the new animal model may allow a better understanding of the atherogenicity of type IV hyperlipoproteinemia, which is currently debated based on ambiguous clinical evidence [27].

References

1 Ambard, L.; Beaujard, E.: Causes de l'hypértension arterielle. Archs gén. Méd. *1:* 520–533 (1904).
2 Anitschkow, N. N.: Über die Veränderungen der Kaninchen Aorta bei experimenteller Cholesterinsteatose. Beitr. Path. Anat. *56:* 379–383 (1913).
3 Armstrong, M. L.: Evidence of regression of atherosclerosis in primates and man. Postgrad. med. J. *52:* 456–461 (1976).
4 Baer, P. G.; Bianchi, G.; Duzzi, L.: Renal micropuncture study in normotensive and Milan hypertensive rats before and after development of hypertension. Kidney int. *13:* 452–466 (1978).
5 Bianchi, G.; Baer, P. G.; Fox, U.; Duzzi, L.; Pagetti, D.; Giovanetti, A. M.: Changes in renin, water balance, and sodium balance during development of high blood pressure in genetically hypertensive rats. Circulation Res. *36/37:* suppl. I, pp. 53–161 (1975).
6 Bianchi, G.; Cusi, D.; Gatti, M.; et al.: A renal abnormality as a possible cause of 'essential' hypertension. Lancet *i:* 173–177 (1979).
7 Bosisio, E.; Ghiselli, G. C.; Galli-Kienle, M.; Galli, G.; Sirtori, C. R.: Effects of dietary casein and soy protein on liver catabolism and plasma transport of cholesterol in rats. J. Steroid Biochem. *14:* 1201–1207 (1981).

8   Brunzell, J.D.; Schrott, H.G.; Motulsky, A.G.; Bierman, E.L.: Myocardial infarction in the familial forms of hypertriglyceridemia. Metabolism 25: 313–320 (1976).
9   Burkitt, D.P.; Walker, A.R.P.; Painter, N.S.: Dietary fiber and disease. J. Am. med. Ass. 229: 1068–1074 (1972).
10   Chait, A.; Albers, J.J.; Brunzell, J.D.: Very low density lipoprotein overproduction in genetic forms of hypertriglyceridemia. Eur. J. clin. Invest. 10: 17–22 (1980).
11   Chait, A.; Hazzard, W.R.; Albers, J.J.; Kushwaha, R.P.; Brunzell, J.D.: Impaired very low density lipoprotein and triglyceride removal in broad beta disease: comparison with endogenous hypertriglyceridemia. Metabolism 27: 1055–1066 (1978).
12   Clarkson, T.B.; Prichard, R.W.; Bullock, B.C.; et al.: Pathogenesis of atherosclerosis: some advances from using animal models. Exp. molec. Path. 24: 264–286 (1976).
13   Connor, W.E.; Hodges, R.E.; Bleiler, R.E.: The serum lipids in men receiving high cholesterol and cholesterol free diets. J. clin. Invest. 40: 894–901 (1961).
14   Craig, M.; Sullivan, J.M.; Saravis, C.A.; Hickler, R.B.: Studies of the renin-renin substrate reaction in man; kinetic evidence for inhibition by serum. Am. J. clin. Sci. 274: 45–54 (1977).
15   Crocker, P.J.; Fitch, M.; Ostwald, R.: Effects of unsaturation of dietary fat and of arachidonate supplementation on cholesterol pool expansion in the guinea-pig. J. Nutr. 109: 927–938 (1979).
16   Dahl, L.K.; Heine, M.; Tassinari, L.: Effects of chronic excess salt ingestion: evidence that genetic factors play an important role in susceptibility to experimental hypertension. J. exp. Med. 115: 1173–1190 (1962).
17   Fernstrom, J.D.: Effects of the diet on brain transmitters; in Howard and McLean Baird, Recent advances in clinical nutrition, pp. 42–47 (John Libbey, London 1981).
18   Flynn, M.A.; Nolph, G.B.; Flynn, T.C.; Kahrs, R.; Krause, G.: Effect of dietary egg on human serum cholesterol and triglycerides. Am. J. clin. Nutr. 32: 1051–1057 (1979).
19   Freis, E.D.: Salt, volume and prevention of hypertension. Circulation 53: 589–595 (1976).
20   Fumagalli, R.; Paoletti, R.; Howard, A.N.: Hypocholesterolaemic effect of soya. Life Sci. 22: 947–952 (1978).
21   Fumagalli, R.; Soleri, L.; Farina, R.; Musanti, R.; Mantero, O.; Noseda, G.; Gatti, E.; Sirtori, C.R.: Fecal cholesterol excretion studies in type II hypercholesterolemic patients treated with the soybean protein diet. Atherosclerosis 43: 341–353 (1982).
22   Garay, R.P.; Meyer, P.: A new test showing abnormal net Na^+ and K^+ fluxes in erythrocytes of essential hypertensive patients. Lancet ii: 349–353 (1979).
23   Goldstein, J.L.; Ho, Y.K.; Brown, M.S.; Innerarity, T.L.; Mahley, R.W.: Cholesteryl ester accumulation in macrophages resulting from receptor-mediated uptake and degradation of hypercholesterolemic canine B-very low density lipoproteins. J. biol. Chem. 255: 1839–1848 (1980).
24   Heaton, K.W.; Pomare, E.W.: Effect of bran on blood lipids and calcium. Lancet i: 496–497 (1974).
25   Huff, M.W.; Carroll, K.K.: Effects of dietary protein on turnover, oxidation and absorption of cholesterol, and on steroid excretion in rabbits. J. Lipid Res. 21: 546–558 (1980).
26   Huff, M.W.; Hamilton, R.M.G.; Carroll, K.K.: Plasma cholesterol levels in rabbits fed low fat, cholesterol-free semipurified diets: effects of dietary proteins, protein hydrolysates and amino acid mixtures. Atherosclerosis 28: 187–195 (1977).

27 Hulley, S. B.; Rosenman, R. H.; Bawol, R. D.; Brand, R. J.: Epidemiology as a guide to clinical decisions – the association between triglyceride and coronary artery disease. New Engl. J. Med. *302:* 1383–1389 (1980).
28 Ignatowski, A.: Über die Wirkung des tierischen Eiweisses auf die Aorta und die parenchymatösen Organe der Kaninchen. Virchows Arch. *198:* 248–256 (1909).
29 Iacono, J. M.; Marshall, M. W.; Dougherty, R. M.; Wheeler, M. A.; Mackin, J. F.; Canary, J. J.: Reduction in blood pressure associated with high polyunsaturated fat diets that reduce blood cholesterol in man. Prev. Med. *4:* 426–443 (1975).
30 Kay, R. M.; Rao, S.; Arnott, C.; Miller, N. E.; Lewis, B.: Acute effects of the pattern of fat ingestion on plasma high density lipoprotein components in man. Atherosclerosis *36:* 567–572 (1980).
31 Kim, D. N.; Lee, K. T.; Reiner, J. M.; Thomas, W. A.: Effect of a soy protein product on serum and tissue cholesterol concentrations in swine fed high-fat, high-cholesterol diet. Exp. molec. Path. *29:* 385–399 (1978).
32 Kim, D. N.; Lee, K. T.; Reiner, J. M.; Thomas, W. A.: Increased steroid excretion in swine fed high-fat, high-cholesterol diet with soy protein. Exp. molec. Path. *33:* 25–35 (1980).
33 Kotchen, T. A.; Welch, W. J.; Talwaker, R. T.: In vitro and in vivo inhibition of renin by fatty acids. Am. J. Physiol. *234:* E593–E599 (1978).
34 Kritchevsky, D.: Vegetable protein and atherosclerosis. J. Am. Oil Chem. Soc. *56:* 135–140 (1979).
35 Kritchevsky, D.; Tepper, S. A.; Story, J. A.: Influence of soy protein and casein on atherosclerosis in rabbits. Fed. Proc. *37:* 747 (1978).
36 Kritchevsky, D.; Tepper, S. A.; Williams, D. E.; Story, J. A.: Experimental atherosclerosis in rabbits fed cholesterol-free diets. 7. Interaction of animal and vegetable protein with fiber. Atherosclerosis *26:* 397–403 (1978).
37 Kushwaha, R. S.; Hazzard, W. R.: Catabolism of very low density lipoproteins in the rabbit: effect of changing composition and pool size. Biochim. biophys. Acta *528:* 176–189 (1978).
38 Levinsky, N. G.; Alexander, E. A.: Acute renal failure; in Brenner, Rector, The kidney, pp. 806–837 (Saunders, Philadelphia 1976).
39 Mahley, R. W.: Cholesterol feeding: effects on lipoprotein structure and metabolism; in Gotto et al., Atherosclerosis, vol. V, pp. 641–652 (Springer, New York 1980).
40 Mahley, R. W.; Innerarity, T. L.; Bersot, T. P.; Lipson, A.; Margolis, S.: Alterations in human high-density lipoproteins, with or without increased plasma-cholesterol, induced by diets high in cholesterol. Lancet *ii:* 807–809 (1978).
41 Mahley, R. W.; Weisgraber, K. H.; Innerarity, T.: Canine lipoproteins and atherosclerosis. II. Characterization of the plasma lipoproteins associated with atherogenic and non-atherogenic hyperlipidemia. Circulation Res. *35:* 722–733 (1974).
42 Malhotra, S. L.: Dietary factors causing hypertension in India. Am. J. clin. Nutr. *23:* 1353–1363 (1970).
43 Malinow, R.: Reduction in cholesterolemia associated with ingestion of alfalfa seeds. 5th Int. Symp. on Atherosclerosis, Houston 1979, Abstr. 302.
44 Malinow, M. R.; McLaughlin, P.; Stafford, C. V.: Prevention of hypercholesterolemia in monkeys *(Macaca fascicularis)* by digitonin. Am. J. clin. Nutr. *31:* 814–818 (1978).

45 McDonough, J.R.; Garrison, G.E.; Hawes, C.G.: Blood pressure and hypertensive disease among negroes and whites. Ann. intern. Med. *61:* 208–228 (1964).
46 Meneely, G.R.; Battarbee, H.D.V.: High sodium low potassium environment and hypertension. Am. J. Cardiol. *38:* 769–785 (1976).
47 Michailor, M.C.: Modification de la composition en acids gras libres de rats soumis à differents types de hypertension. Biomédicine *21:* 393–397 (1974).
48 Mistry, P.: Nicoll, A.; Niehaus, C.; Christie, I.; Janus, E.; Strain, H.; Lewis, B.: Effects of dietary cholesterol on serum lipoproteins in man. Protides biol. Fluids *25:* 349–352 (1978).
49 Mohring, J.; Petri, M.; Szokol, M.; Haack, D.; Mohring, B.: Effects of saline drinking on malignant course of renal hypertension in rats. Am. J. Physiol. *230:* 849–857 (1976).
50 Morgan T.; Gillies, A.; Morgan, G.; Adam, W.; Wilson, M.; Carney, S.V.: Hypertension treated by salt restriction. Lancet *i:* 227–233 (1978).
51 Nara, Y.; Yamori, Y.; Lovenberg, W.: Effect of dietary taurine on blood pressure in genetically hypertensive rats. Biochem. Pharmacol. *27:* 2689–2692 (1979).
52 Nath, N.; Wiener, R.; Harper, A.E.; Elvehjem, C.A.: Diet and cholesterolemia. 1. Development of a diet for the study of nutritional factors affecting cholesterolemia in rats. J. Nutr. *69:* 289–302 (1959).
53 Noel, S.P.; Wong, L.; Dolphin, P.J.; Dory, L.; Rubinstein, D.: Secretion of cholesterol-rich lipoproteins by perfused livers of hypercholesterolemic rats. J. clin. Invest. *64:* 674–683 (1979).
54 Ockner, R.K.; Hughes, F.B.; Isselbacher, K.J.: Very low density lipoproteins in intestinal lymph: role in triglyceride and cholesterol transport during fat absorption. J. clin. Invest. *48:* 2367–2373 (1969).
55 Pinon, J.C.; Bridoux, A.M.: High density lipoproteins in cholesterol-fed rabbits: progressive enrichment with free cholesterol. Artery *3:59–71* (1977).
56 Potter, J.M.; Nestel, P.J.: Greater bile acid excretion with soybean than with cow milk in infants. Am. J. clin. Nutr. *29:* 546–551 (1976).
57 Robertson, T.L.; Kato, H.; Rhoads, G.G.; et al.: Epidemiological studies of coronary heart diseases and stroke in Japanese men living in Japan, Hawaii and California. Incidence of myocardial infarction and death from coronary heart disease. Am. J. Cardiol. *39:* 239–243 (1977).
58 Rodriguez, J.L.; Catapano, A.; Ghiselli, G.C.; Sirtori, C.R.: Very low density lipoproteins in normal and cholesterol-fed rabbits. I. Chemical composition of very low density lipoproteins in rabbits. Atherosclerosis *23:* 73–83 (1976).
59 Rodriguez, J.L.; Catapano, A.; Ghiselli, G.; Sirtori, C.R.: Very low density lipoproteins in normal and cholesterol-fed rabbits: lipid and protein composition and metabolism. II. Atherosclerosis *23:* 85–96 (1976).
60 Ross, A.C.; Minick, C.R.; Zilversmit, D.B.: Equal atherosclerosis in rabbits fed cholesterol-free, low-fat diet or cholesterol-supplemented diet. Atherosclerosis *29:* 301–315 (1978).
61 Sardet, C.; Hansma, H.; Ostwald, R.: Characterization of guinea pig plasma lipoproteins: the appearance of new lipoproteins in response to dietary cholesterol. J. Lipid Res. *13:* 624–639 (1972).
62 Sheehe, D.M.; Green J.B.; Green, M.V.: Influence of dietary fat saturation on lipid absorption in the rat. Atherosclerosis *37:* 301–310 (1980).

63 Shepherd, J.; Packard, C.J.; Patsch, J.R.; Gotto, A.M.; Taunton, O.D.: Effects of dietary polyunsaturated and saturated fats on the properties of high density lipoproteins and the metabolism of apolipoprotein A-I. J. clin. Invest. *61:* 1582–1592 (1978).
64 Shore, B.; Shore, V.G.: Rabbits as a model for the study of hyperlipoproteinemia and atherosclerosis. Adv. exp. Med. Biol. *67:* 123–141 (1976).
65 Shore, V.G.; Shore, B.; Hart, R.G.: Changes in apolipoproteins and properties of rabbit very low density lipoproteins on induction of cholesterolemia. Biochemistry, N.Y. *13:* 1579–1582 (1974).
66 Sirtori, C.R.; Agradi, E.; Conti, F.; Gatti, E.; Mantero, O.: Soybean protein diet in the treatment of type II hyperlipoproteinemia. Lancet *i:* 275–277 (1977).
67 Sirtori, C.R.; Gatti, E.; Mantero, O.; et al.: Clinical experience with the soybean protein diet in the treatment of hypercholesterolemia. Am. J. clin. Nutr. *32:* 1645–1658 (1979).
68 Srinivasan, S.R.; Radhakrishnamurthy, B.; Webber, L.S.; Dalferes, E.R., Jr.; Kokatnur, M.G.; Berenson, G.S.: Synergistic effects of dietary carbohydrate and cholesterol on serum lipids and lipoproteins in squirrel and spider monkeys. Am. J. clin. Nutr. *31:* 603–613 (1978).
69 Stamler, J.: Epidemiology of coronary heart disease. Med. Clins N. Am. *57:* 5–46 (1973).
70 Stamler, R.; Stamler, J.; Riedlinger, W.F.; Algera, G.; Roberts, R.H.: Weight and blood pressure findings in hypertension screening of 1 million Americans. J. Am. med. Ass. *240:* 1607–1610 (1978).
71 Tall, A.R.; Green, P.H.R.; Glickman, R.M.; Riley, J.W.: Metabolic fate of chylomicron phospholipids and apoproteins in the rat. J. clin. Invest. *64:* 977–989 (1979).
72 Tanzawa, K.; Shimada, Y.; Kuroda, M.; Tsujita, Y.; Arai, M.; Watanabe, M.: WHHL-rabbit: a low density lipoprotein receptor deficient animal model for familial hypercholesterolemia. FEBS Lett. *118:* 81–84 (1980).
73 Taunton, O.D.; Morrissett, J.D.; Segura, R.; Pownall, M.J.; Jackson, R.L.; Gotto, A.M.: Effect of dietary fatty acid composition on lipoprotein structure. Circulation *49/50:* suppl. III, Abstr. 1050 (1974).
74 Thurston, H.; Swales, J.D.: Influence of sodium restriction upon two models of renal hypertension. Clin. Sci. *51:* 275–279 (1976).
75 Topping, D.L.; Storer, G.B.; Calvert, G.D.; Illman, R.J.; Oakenfull, D.G.; Weller, R.A.: Effects of dietary saponins on fecal bile acids and neutral sterols, plasma lipids, and lipoprotein turnover in the pig. Am. J. clin. Nutr. *33:* 783–786 (1980).
76 Turcotte, J.G.; Boyd, R.F.; Quinn, J.G.; Smeby, R.R.: Isolation and renin inhibitory activity of phosphoglyceride from shark kidney. J Med. Chem. *16:* 166–169 (1973).
77 Vessby, B.; Boberg, J.; Gustaffson, I.B.; Karlström, B.; Lithell, H.; Ostlund-Lindqvist, A.M.: Reduction of high density lipoprotein cholesterol and apolipoprotein A-I concentrations by a lipid-lowering diet. Atherosclerosis *35:* 21–27 (1980).
78 Watanabe, Y.: Serial inbreeding of rabbits with hereditary hyperlipidemia (WHHL-rabbit). Atherosclerosis *36:* 261–268 (1980).
79 Wissler, R.W.; Vesselinovitch, D.: Studies of regression of advanced atherosclerosis in experimental animals and man. Ann. N.Y. Acad. Sci. *275:* 363–378 (1976).
80 Yamori, Y.; Horie, R.; Ikeda, K.; Nara, Y.; Lovenberg, W.: Prophylactic effect of

dietary protein on stroke prevention; in Yamori et al., Prophylactic approach to hypertensive diseases, pp. 497–504 (Raven Press, New York 1979).
81 Young, J. B.; Landsberg, L. V.: Stimulation of the sympathetic nervous system during sucrose feeding. Nature, Lond. *269:* 615–617 (1977).
82 Young, J. B.; Landsberg, L. V.: Suppression of sympathetic nervous system during fasting. Science *196:* 1473–1475 (1977).
83 Young, J. B.; Mullen, D.; Landsberg, L. V.: Caloric restriction lowers blood pressure in the spontaneously hypertensive rat. Metabolism *27:* 1711–1714 (1978).

C. R. Sirtori, Center E. Grossi Paoletti, Institute of Pharmacology and Pharmacognosy and Chemotherapy Chair, University of Milan, I-20129 Milan (Italy).

Interrelationship of Sodium, Volume, CNS, and Hypertension

Louis J. Tobian

Department of Medicine, University of Minnesota, Minneapolis, Minn., USA

The influence of the kidney on hypertension is strong indeed, as was first pointed out by the work of *Bright and Goldblatt*. In more recent years additional evidence strongly linking the kidney with hypertension comes by way of kidney transplantation. A person with end-stage renal disease and severe hypertension can have a good kidney transplanted in, and before long the blood pressure has often returned to normal, giving very strong testimony to the anti-hypertensive action of a normal kidney.

There are two types of rat hypertension which show the same effect. In Dahl rats, for instance, one can transplant a kidney from the S rats (susceptible to hypertension) to the R rats (resistant to hypertension), and the blood pressure of the R rat will rise. One can also do the opposite, taking the kidney from the animal resistant to hypertension and transplanting it into the animal susceptible to hypertension, and the blood pressure will go down. The same pattern of change has also been seen during transplantations in the Milano strains of hypertensive rats.

Both these strains seem to demonstrate that the kidney can somehow carry the message of blood pressure and that the kidney can somehow command the level of blood pressure within the body. How might the kidney be able to do this? The kidney could do it by means of its two obvious endocrine cells. One of these is the juxtaglomerular cell which secretes renin [25]. Renin causes the elaboration of angiotensin I which gets converted into angiotensin II, one of the most powerful pressor agents that we know of. The angiotensin II acts to constrict arterioles directly. It causes the adrenergic nerves to release more norepinephrine. It causes the adrenal medulla to put out more epinephrine. It acts on specific receptors in the brain to cause an increase of sympathetic nerve outflow. It acts on the adrenal zona glomerulosa to cause more aldosterone to be secreted,

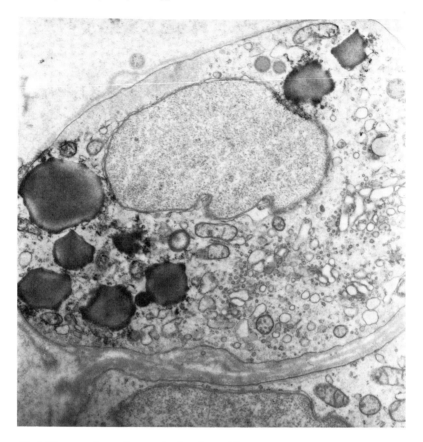

Fig. 1. Electron photomicrograph of an interstitial cell in the rat renal papilla.

and it seems to act on the renal tubules directly, to cause Na retention. Such retention of sodium in the body tends to raise the blood pressure. So juxtaglomerular cells could definitely enable the kidney to control the level of blood pressure in the body.

There is another kidney cell which also has vasoactive properties, the interstitial cell (fig. 1), which is seen in the inner medulla of the kidney of all mammals [12, 26, 27]. It has many cytoplasmic cisterns and contains unusual lipid-filled granules in the cytoplasm. These cells definitely secrete prostaglandins. They secrete a neutral lipid described by *Muirhead*. They have angiotensin II receptors in the cell wall. There is really good evidence that these cells put out vasoactive materials which act on blood pressure. For instance, one can transplant these cells under the skin of a hypertensive rat and the blood pressure will go down. One can remove the transplant and the blood pressure will rise. This could be another cell with which the kidney could control the level of arterial pressure.

The kidney can also control blood pressure with relatively non-hormonal means, that, is, by the ease with which it excretes Na. The most extreme example would be someone with renal parenchymal disease who has a fourth of his nephrons still remaining. Such patients come into sodium balance with a high level of body sodium and are very apt to have hypertension. Contrast that with someone possessing excellent kidney function with no difficulty excreting sodium. Such a person comes into Na balance with a normal level of body sodium and has no stimulus for a rise in blood pressure. Thus, the ease with which the kidney can excrete Na is another means by which the kidney can command the level of blood pressure in the body.

This brings us to the relevance of sodium to hypertension. This realization really started with physicians who treated hypertensive patients with low-salt diets and brought the blood pressure down to normal or near-normal. This is a strong piece of evidence that Na is involved with the mechanism of essential hypertension.

When the diuretic drugs were introduced, it was quickly recognized that they had a very strong influence for lowering blood pressure in hypertensive patients. This influence is mainly sodium-related, since one can nullify the anti-hypertensive effect of the diuretic merely by combining it with a high-salt diet [32]. Thus, a moderate dose of thiazide normally will lower blood pressure; but, if it is combined with a very high salt intake, it often fails to alter the blood pressure.

So it seems that thiazides lower blood pressure mainly by lowering body sodium. If the thiazide is combined with a very high Na intake, the body sodium does not go down and the blood pressure stays elevated. Thus, the thiazide effect is quite similar to the effect of a low-Na diet.

Salt (NaCl) can also raise blood pressure. This was first pointed out by *McQuarrie* et al. [15] of Minnesota. They had a young boy with a peculiar craving for salt. He would eat 60 g daily of NaCl and had hypertension. When he was brought into the hospital and fed a normal NaCl intake of 5 g per day, the blood pressure came down to normal. When he then resumed his 60 g of NaCl a day, his blood pressure would rise again. This was the first clear demonstration that a really high-salt diet would cause hypertension. *McQuarrie* et al. [15] repeated this on 3 or 4 other children and established the fact that a high NaCl diet could indeed raise the blood pressure.

One should also note the interplay between heredity and the ability of a high-NaCl diet to raise the blood pressure. For instance, in a study at the University of Iowa, a moderately high-salt diet, 24 g NaCl a day, was given to some normotensive individuals and the blood pressure did not rise at all. The same diet was given to other people with borderline hypertension and the blood pressure did indeed rise. The same high-salt diet may have a quite different effect on someone with a genetic resistance to hypertension compared to someone else with a genetic susceptibility to hypertension. In this study in Iowa, one sees the remarkable paradox which occurs when one feeds a high-salt diet to genetically resistant people and genetically susceptible people. In the resistant ones, the blood pressure remains the same and the forearm arterioles become vasodilated. On the other hand, the high-salt diet in the borderline hypertensives caused the blood pressure to rise and the forearm arterioles to constrict. In the one case, the high-salt diet produces arteriolar dilation; in the other case, arteriolar constriction.

Falkner et al. [2] did the same sort of study on teenage normotensive children of normotensive parents and on teenage normotensive children of hypertensive parents. They were fed a diet of about 16 g of NaCl a day. The children of normotensive parents had no rise in pressure, whereas the children of hypertensive parents did indeed have a rise of blood pressure, again showing the link between salt-induced hypertension and some hereditary factor [2]. The identity of this hereditary factor is not easy to ascertain, but it could very well be partly related to the ease with which the kidney excretes sodium. A study done at the University of Indiana more or less bore this out. They had two groups of people: normotensive relatives of hypertensives and normotensive relatives of normotensive people. They gave them a salt challenge, 2 liters of isotonic saline intravenously, and followed the sodium excretion over the next 24 h. In the relatives of normotensive people, the sodium was excreted quickly in the urine. When the same salt challenge was given to normotensive relatives of hypertensive people, there was a delay in the excretion of sodium, again suggesting that the hereditary factor for hypertension could be a delay in the renal excretion of sodium in those with the genetic susceptibility to hypertension.

Though there is this genetic susceptibility or resistance, ultimately everyone has his 'salt price', as indicated by a recent study at Indiana. They fed normotensives very large amounts of NaCl: 400, 800, 1,200, 1,500 mEq a day. At 1,200 mEq a day, everyone had a rise of blood pressure. Thus, no matter whether one is genetically resistant or susceptible, if the salt intake goes high enough, everyone will ultimately get some rise of blood pressure [16].

In mineralocorticoid hypertension, the Na intake has a decisive influence. For instance, combining the mineralocorticoid, either aldosterone or deoxycorticosterone, with a high salt intake gives quick rises in blood pressure. If the same mineralocorticoid is combined with a very low salt intake, one sees no rise in blood pressure whatsoever. It is the one certain way to keep the mineralocorticoid from causing a rise of blood pressure.

The sodium story with regard to hypertension also involves people in unacculturated villages all over the world. Such people have no hypertension whatsoever and their blood pressure does not rise with advancing age. The 70-year-olds have the same blood pressure as the 20-year-olds [13, 14, 17, 20–22].

It is not that the people in these remote areas have a genetic resistance to hypertension. When some of them leave the village and migrate to highly acculturated regions, they begin eating generous amounts of salt and often develop blood pressure over 140/90 mm Hg in 50% of such migrants.

What seems to be the common denominator here is the lifelong, low intake of salt, less than about 60 mEq per day. If one has this lifelong low salt intake, even though he has the genetic susceptibility to hypertension, he still will not develop the rise in blood pressure. One tribal group in the Amazon basin under the care of missionaries ate copious amounts of salted canned food and had their share of hypertension. A neighboring tribal group eating native low-Na food had no hypertension.

Shaper described the Samburu people that have virtually no hypertension when on their farms eating a low-salt diet. When they are drafted into the army of Kenya and begin eating 18 g of salt a day in army food for a few months, they begin to have their share of hypertension.

Page described an unacculturated group that lives on an offshore island and cooks their food in sea water and, of course, has a high Na intake. Even though this group has a low state of acculturation and is lean and active, they still have considerable hypertension which is seemingly related to their high Na intake.

Page also described a nomadic group in Iran that are primitive, lean, and work very hard, but they happen to have a high Na intake and develop considerable hypertension.

So, it would seem that the critical factor which determines this absence of hypertension in primitive peoples is the fact that they have this lifelong low intake of Na. With such a diet, it does not matter whether one has a genetic resistance or genetic susceptibility. He still does not get hypertension.

Regarding salt and hypertension, however, many people are able to eat a high-salt diet and still not get hypertension. *Dahl* in Brookhaven was aware of this paradox and created two strains of rats which would imitate what he saw in people. He developed an S strain (susceptible to hypertension) and an R strain (resistant to hypertension). When both strains eat a low-salt diet, neither group has a blood pressure above the normal range through young adulthood. When both groups are fed a high-salt diet, however, the R strain gets virtually no rise of blood pressure at all, whereas the S strain develops quite a rise in blood pressure. He promulgated a working hypothesis concerning essential hypertension. He postulated the existence of two factors in the hypertension of his rats and in the hypertension of men, namely a hereditary and an environmental factor. He believed the environmental factor was salt intake. In order to get hypertension, one had to have both factors operating, a fairly high salt intake as well as the hereditary susceptibility to hypertension [1].

With regard to salt and hypertension, one important consideration is the tendency for sodium to accumulate in the body. And this, of course, results from both the salt in the diet and the ability of a kidney to excrete the dietary salt. With an elevation in body sodium, even though transient, an increase in blood pressure will tend to occur in susceptible people. The susceptibles will have a vasoconstriction of arterioles instead of a vasodilation, following a NaCl challenge.

Such a pattern also occurs in the Dahl rats. We measured cardiac output and calculated peripheral resistance on these rats after they were fed salt for 3 days. In the resistant Dahl rats at the end of the third day, the blood pressure was unchanged and the calculated peripheral resistance had gone down 14%. The high-salt diet had produced vasodilation in the peripheral circulation. With the S rats that are susceptible to hypertension, 3 days of the high-salt diet produced a rise in the blood pressure; and the high salt, instead of causing vasodilation, caused a 9% increase in calculated peripheral resistance. Again, this demonstrates the principle that a high-salt diet in the susceptibles gives constriction and in the resistants gives vasodilation [7].

When we talk about salt in relation to hypertension, one can ask which is really the preferred diet. People in Japan, North America, and Europe have a much higher salt intake than in the remote areas. If one decides things on a basis of 'tradition', tradition is all on the side of the low-salt diet. It is now believed that man has been on earth for 3 million years. And for almost every day of that 3 million years, except maybe the last 5,000 years, every inland man on the face of the earth was eating a low-salt diet. There is really no other way that he could have operated. He was living in a salt-deficient world. Evolution was helping him cope with this salt-poor world, and he had no possible access to a high-salt diet. In fact, he has a taste for sodium in the tongue and a liking for salt just to help him survive in a low-salt world. And then, 'all of a sudden', maybe within the last 500 years, ordinary people have been able to get all of the salt that they desired. This was unprecedented with regard to man's previous history, and it is possible that maybe 1 out of 5 is not able to cope with this excessive amount of salt and develops a rise in blood pressure. In fact, one can speculate that this 1 person out of 5 might be a superb conserver of sodium, which would be a great advantage in salt-poor prehistoric times but a distinct handicap in a salt-surfeited modern world.

What might be the mechanism of NaCl-induced hypertension? I believe that there are at least two links in the chain of causation. The first link would be that one has to get some extra sodium in the body, at least transiently. The second link would be that the extra sodium in the body causes a rise in blood pressure. If one does not have both of these links in operation, one will not get salt-induced hypertension. And accumulating extra sodium in the body, as said before, results from both Na in the diet and the ability of the kidney to excrete that Na.

Let us consider evidence for the second link: that increases in body sodium lead to hypertension, at least in most people. There is good clinical evidence for this. In renal parenchymal disease, for instance, one comes into sodium balance with high levels of body sodium. Such people have a high prevalence of hypertension. In primary hyperaldosteronism, one comes into sodium balance during the 'escape' phase with a high level of body sodium, and this frequently causes hypertension. Dialysis patients provide another example for this. As they proceed from dialysis to dialysis, the body sodium swings up and down and the blood pressure swings up and down in the same pattern.

Onesti et al. [18] did more direct experiments to investigate this. They obtained blood pressure and cardiac output measurements so that peripheral vascular resistance could be calculated. They had a number of patients with end-stage renal disease. In the course of a series of dialyses, they increased body sodium for a number of weeks. Most of these patients had a rise of blood pressure, and the main cause of the rise of blood pressure was an increase of peripheral vascular resistance. In these subjects the increase in the body sodium resulted mainly in vasoconstriction [18].

They had about 20% of people whose blood pressure went up after salt loading solely because of an increase in cardiac output. The peripheral resistance in these people stayed at the baseline level, but one could say that this level of resistance was actually elevated in relation to the raised level of cardiac output. But, in the main, the blood pressure went up because of an increase in peripheral vascular resistance. Recent evidence suggests that this narrowing is partly a passive structural narrowing as well as being due to an active smooth muscle shortening.

They also had about 20% of patients with end-stage renal disease who did not show this pattern at all. In these, an increase of body sodium did not lead to any rise in pressure at all. This group indicates that one must have the 'second link' operating in order to get NaCl hypertension. Thus, in order to get salt-induced hypertension, one must have both links in the chain of causation operating.

What are the mechanisms by which the increases in body sodium lead to rises in blood pressure and vasoconstriction? These answers are still uncertain but there is much recent information concerning what might be involved. Key experiments in this area were done by *Takeshita* and *Mark* [24]. Dahl S rats were fed a high-Na diet and they developed increased peripheral vascular resistance in the hindquarters. They cut the sympathetic nerves to the hindquarters and removed 50% of the NaCl-induced increase in vascular resistance. This is a strong indication that the sympathetic nerves play an important part in salt-induced hypertension [24].

They did another important experiment. They gave 6-hydroxy dopamine peripherally to baby Dahl S rats. This maneuver destroys most peripheral sympathetic nerves. They then fed these S rats a high-salt diet and they could not produce any hypertension at all, again suggesting the importance of sympathetic nerves in the production of salt-induced hypertension.

When they cut the sympathetic nerves, they deleted only half of the increase of peripheral vascular resistance. This prompted us to do studies searching for other causes of the increase of vascular resistance. We fed Dahl S and R rats a high-salt diet for a month, and we arranged to have the blood of these rats perfuse at a constant flow the hindquarters from another rat used as a bioassay organ. After 1 month of high Na feeding, the S rats were hypertensive and the R rats were normotensive. We could detect vasoconstrictor or vasodilator humoral agents in this way. When the hindquarters were perfused with R blood at a constant flow, we had a certain vascular resistance; if they were perfused with S blood, we had a higher vascular resistance, 17% higher, suggesting that there is some type of vasoconstrictor humoral effect in the blood of the S rats that are developing salt-induced hypertension [31]. It is not likely to be angiotensin II since the S rats had a 39% lower level of plasma renin than the R rats. The experiments since then suggest that the rat has to have a high-salt diet producing hypertension in order for the vasoconstrictor material to appear. If the S rat is on the low-salt diet and has a normal blood pressure, the vasoconstrictor agent is absent. If we remove the kidney about an hour before the perfusion in salt-fed S rats, this humoral vasoconstrictor activity is greatly reduced.

Searching further for factors which influence salt-induced hypertension, we investigated the effect of lesions in the central nervous system. Some Dahl S rats were fed a high-NaCl diet and became hypertensive over the course of many weeks. Some Dahl R rats were also fed salt but did not become hypertensive. Other Dahl S rats were injected with 6-hydroxydopamine in the lateral brain ventricle and developed only half of the expected hypertension when these rats were given the high-salt diet. The 6-hydroxydopamine destroys only catecholamine-containing neurons in the brain. This experiment indicates that S rats will not get the full expression of salt-induced hypertension unless catecholamine-containing neurons in the brain are intact [8].

Searching for other central nervous system factors in NaCl hypertension, we investigated the paraventricular nucleus on either side of the third brain ventricle. This nucleus secretes vasopressin. It has neural connections with the vasomotor center. It receives neural signals from the kidney. It seemed a likely nucleus for involvement in salt-induced hypertension. In Dahl S rats that have sham lesions of these nuclei, the feeding of a high-salt diet gradually produces hypertension. Dahl R rats with the sham lesion get no change in blood pressure at all. On the other hand, when S rats had both paraventricular nuclei completely destroyed by anodal lesions, they only had about half of the hypertension on salt feeding that is seen in the rats with sham lesions. The results indicate that one must have two intact paraventricular nuclei in the brain in order to get the full expression of salt-induced hypertension [8].

Similarly, we produced an electrolytic lesion in Dahl S rats in the anterior inferior end of the third brain ventricle, the so-called 'AV3V' area, and then fed them an 8% high-NaCl diet. The S rats with this lesion had a greatly attenuated rise in blood pressure, less than half the rise seen in S rats with sham lesions. This study indicates that the nuclei and nerve tracts in the periventricular region around the anterior inferior end of the third brain ventricle must be intact in order to achieve the full expression of NaCl-induced hypertension. This area includes the OVLT region which has receptors for angiotensin II.

Other lesions in specific areas of the brain can also exacerbate hypertension. *Goto* et al. [8] induced bilateral lesions of the suprachiasmatic nuclei which lie on either side of the third brain ventricle in the ventral part of the anterior hypothalamus. These nuclei secrete vasopressin but the vasopressin does not reach the pituitary. These nuclei have many neural connections with several hypothalamic nuclei. The S rats with sham lesions had the usual rise in blood pressure when fed a high-NaCl diet. However, in the rats with bilateral suprachiasmatic nucleus lesions, the blood pressure went up even higher during the high-NaCl diet ($p<0.001$) [8] and several lesioned rats died prematurely with severe hypertension. Here is a nucleus that tends to keep the blood pressure down. When it is destroyed, a particularly severe salt-induced hypertension appears which accelerates mortality rates in these rats and increases the heart weight: body weight ratio by 15% ($p<0.001$).

We carried out still another study of the central nervous system in relation to a genetic susceptibility to salt-induced hypertension. If one puts 500 ng of angiotensin II into the lateral brain ventricle, one regularly gets a transient increase in blood pressure. *Ikeda* et al. [11] did this in S rats and in R rats on a low-salt diet, with both groups normotensive. They found that a small amount of angiotensin II in the lateral brain ventricle caused a rise of blood pressure which was twice as great in S rats as in R rats. They also produced a transient rise of pressure by injecting a small amount of hypertonic saline into the lateral brain ventricle. And again the rise in the blood pressure was twice as great in the normotensive S rats as it is in the normotensive R rats. These two studies suggest that these pressor responses in S rats are poised to react in a particularly strong fashion whenever the pressor stimuli are introduced. Moreover, *Gotoh* et al. [10] studied hypertensive patients whose blood pressure rose when they were fed salt. In these particular patients they found a 5-mEq/l increase in the sodium concentration of the cerebrospinal fluid as these patients switched from low- to high-NaCl diets [10].

Dahl et al. [1] fed some S rats a high-salt diet and they became hypertensive, while other S rats were fed the same high-salt diet but with equimolar KCl added to it and they had very much smaller rises in pressure. We wondered whether the central nervous system pressor responses could be influenced by adding potassium to the diet. *Goto* et al. [9] placed Dahl S and R rats on a low-salt diet, which will result in a normal blood pressure through young adulthood. They gave as a pressor stimulus a small amount of angiotensin II into the lateral brain ventricle. As before, the S rats had a much greater rise in pressure than the R rats. They not only fed the usual diets but in some groups also added either potassium chloride or potassium citrate to the diet. They found that potassium feeding had a profound effect on the hyperactive pressor responses to angiotensin II in S rats, bringing them down almost to the low level seen in the R rats [9]. Again, using small amounts of hypertonic NaCl in the lateral brain ventricle as a stimulus, *Goto* et al. [9] examined the pressor responses in the S rats, which again were much greater than in R rats. However, when the S rats had been given either potassium chloride or potassium citrate in the diet, the hyperactive pressor response of the S rats was almost brought down to the low level seen in the R rats. Thus, *Goto* et al. [9] found that either of these potassium salts, when added to the diet, will almost normalize the hyperactive pressor responses in S rats. This effect may explain, in part, the prevention of NaCl hypertension by the feeding of potassium salts.

The first link in the chain of causation of salt-induced hypertension is the necessity for at least a transient increase of sodium in a body. This extra body sodium results from both the salt taken in with the food and the ability of the kidney to excrete that salt. In considering what produces the genetic susceptibility to NaCl hypertension, one obvious reason could be that those genetically susceptible have more trouble excreting salt (NaCl) rapidly, even though the kidney may not have obvious pathologic lesions.

In the Indiana study, the relatives of hypertensive people excreted salt slower after a salt challenge, adding support to the concept. This problem can be validly approached in the Dahl rats since they can be studied at a time when both S rats and the R rats have a blood pressure within the normal range. In this study both S and R rats were eating a low-sodium diet

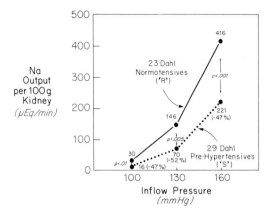

Fig. 2. Sodium excretion of isolated kidneys from S and R rats at varying inflow pressures. The distance between the large black dot (the mean value) and the tip of the arrowhead represents the standard error of the mean.

and were both normotensive. We obtained an isolated kidney from each S and R rat, put it in a chamber without ischemia, and perfused it with blood from two neutral Sprague-Dawley rats. In 1951, *Selkurt* observed that, as pressure in the renal artery rises, the kidney excretes increasing amounts of sodium. Our findings would confirm this. The top line of figure 2 represents isolated kidneys from R rats. As inflow pressure rises, they progressively excrete increasing amounts of sodium. The bottom line of figure 2 represents kidneys from pre-hypertensive S rats. Again, as renal artery pressure increases, progressively more salt is excreted. But, note that there is a distinct shift in the pressure natriuresis curve for the S kidneys. It takes more pressure to excrete a given amount of sodium in the S kidneys than it does in the R kidneys. With this kind of a study, one can compare the sodium excretion at equal levels of inflow pressure. And, at 100, 130, and 160 mm Hg, the S kidneys put out about a half as much Na per minute as the R kidneys (fig. 2) [29]. It is also possible to overcome this defect of natriuresis in S kidneys by applying a greater renal arterial pressure. Thus, if one perfuses the S kidney at 160 mm Hg, that kidney will excrete 50% more sodium than an R kidney perfused at a normal pressure of 130 mm Hg. This study allowed us to extend Dahl's working hypothesis. In this working hypothesis, one could propose that the shift in the pressure natriuresis curve in the S rats can partially account for their great suscep-

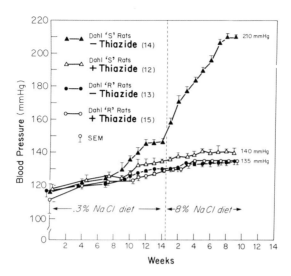

Fig. 3. Blood pressure of Dahl S and R rats as influenced by high and low levels of sodium in the diet, and by treatment with a thiazide diuretic agent.

tibility to NaCl hypertension. However, as long as both S and R rats are on the low-salt diet, this shift in the pressure natriuresis curve would be of little consequence. There is so little salt taken in on this diet that even with the handicap in Na excretion, there would be almost no increase in body sodium and thus very little stimulus for a rise of blood pressure in the S rats.

On the other hand, if both strains begin eating a high-salt diet, this delay in natriuresis in S rats will cause a transient increase in body sodium. The increase in body sodium will trigger a rise in blood pressure. When that rise in pressure occurred, it would accelerate and normalize sodium excretion and make it just as fast in S rats as in the R rats. In fact, as long as the S rats continued to have hypertension, their natriuresis will test out as being perfectly normal. One could give them a sodium load and it would be excreted very rapidly, just as fast or even faster than the rate one finds in R rats. Since this was just a working hypothesis, we sought to get some independent data either for or against it. This prompted another experiment. The working hypothesis here is that the delay in natriuresis is responsible for the salt-induced hypertension. Control S rats were fed a high-salt diet as indicated by the dotted line in figure 3, and they promptly had quite a rise in blood pressure up to very hypertensive levels in 8 weeks. Other S rats were given a thiazide diuretic for a few weeks and were then challenged with the same high-NaCl diet. Instead of the usual large rise of blood pressure, their blood pressure proceeded without any increase and, in fact, was parallel to that of the R rats and only 5 mm Hg above it. One interpretation could be that the S rats have a renal defect in sodium excretion even though they do not have obvious renal lesions. Thus, when they receive a salt challenge, they have a transient increase in body sodium and this produces a higher blood pressure. On the other hand, the S rats receiving thiazide had their sodium excretion facilitated and accelerated by the drug. Hence, when they receive the sodium challenge, they are able to excrete the Na in a normally rapid manner. Thus, there was no accumulation of body sodium and, hence, no stimulus for a rise in pressure. As long as they had the facilitation of the thiazide, their kidneys put out sodium fast enough to prevent any accumulation of body sodium [30].

We then sought to determine why the S kidney had a reduced rate of sodium excretion. We had either an S rat or an R rat perfusing a normal Sprague-Dawley kidney at a constant pressure. We found that when the normal kidney was perfused with R blood, it excreted about 130 µEq of sodium per 100 g per minute. When it was perfused with S blood, it put out about half as much sodium per minute [30]. There seemed to be some humoral effect in S blood which retarded the rate of natriuresis in the neutral 'normal' kidney by about 50%.

Again, searching for reasons why the S kidney is slow in excreting salt, we developed a technique for quick freezing rat kidneys. We plunged them into liquid nitrogen to achieve almost instantaneous freezing. When the kidney was thawed to −5 °C, we dissected out the renal papilla and determined the prostaglandin E_2 concentration in the renal papilla of R rats and S rats. When both strains are on a 0.3% low-NaCl diet, the S rats had a 60% reduction in prostaglandin E_2 concentration in the papilla compared to R rats [28]. If both strains are fed a 4% high-NaCl diet for 4 weeks or 11 weeks, prostaglandin E_2 concentration in the papilla doubles in the R rats and doubles in the S rats, but the concentration is always about 50% lower in the S rats compared with the R rats. For some reason there seems to be an intrinsically low concentration of prostaglandin E_2 in the renal papilla of S rats. It is known that the prostaglandin in the renal medulla has an influence on sodium reabsorption, at least in the ascending limb of the loop of Henle and in the collecting tubule and collecting duct. One piece of evidence for this is that intravenous indomethacin will double the Na concentration in the rat papilla 20 min after its injection [4]. The reason for this would appear to be that the indomethacin had reduced severely the level of prostaglandin and this had increased transport of sodium by the ascending limb and collecting tubule and duct, thereby exaggerating the countercurrent mechanism. Utilizing isolated tubules, *Stokes* has shown that increasing prostaglandin E_2 retards sodium transport in the ascending limb of Henle and in the collecting tubules. *Olson* et al. have shown a similar effect in the collecting duct. Thus, the reduced prostaglandin E_2 in S papillas would tend to enhance Na reabsorption and reduce Na excretion. And, any effect which retards natriuresis would also facilitate the development of NaCl-induced hypertension. Along this same line, *Ferris* has produced a low PGE_2 state in the rat kidney by reducing essential fatty acids in the diet. These rats have a diminished capacity to excrete a Na load and develop hypertension when fed a high-salt diet.

There is an additional likely interrelationship between low papillary prostaglandin E_2 in S kidneys, retarded Na excretion in S kidneys, and the susceptibility to NaCl hypertension in S rats. On either low-NaCl diets or high-NaCl diets, plasma flow to the renal papilla is always about 25–35% lower in S rats than in R rats [5]. This low flow in the papilla could be important with regard to sodium excretion. In any normal rat fed a high-salt diet for about a week, one finds a large 32–45% increase in the flow to the papilla [5]. This increase in papillary plasma flow is part of the physiological response which a rat makes when faced with a high NaCl challenge. The S rat is not able to increase the flow to the papilla to anywhere near the extent which we see in an R rat. This may partially explain why the S rat has a handicap in sodium excretion.

Moreover, during the Na retention phase in caval dogs or salt-depleted dogs, a reduced plasma flow to the renal papilla was the only significantly abnormal alteration in renal hemodynamics or glomerular filtration rate [3, 23]. *Osgood* et al. [19] have also produced evidence that increasing papillary plasma flow during volume expansion favors natriuresis by greatly reducing the Na gradient between the ascending thin limb of deep nephrons and the ascending vasae rectae. When this gradient is abolished, passive Na-reabsorption from the thin ascending limb would also be largely abolished, thereby favoring increased natriuresis by the deep nephrons [19]. In the Na-retaining caval dogs or Na-depleted dogs, the low plasma flow to the papilla presumably was largely responsible for the increased papillary osmolality, which removes more water from the descending limb [19, 23]. In the volume-expanded rats, the high plasma flow to the papilla was presumably largely responsible for the reduced papillary osmolality which removes less water from the descending limb [19]. The physiologic pattern appears to be low papillary plasma flow, high papillary osmolality, enhanced Na retention; or high papillary plasma flow, low papillary osmolality, enhanced Na excretion.

In this connection, it is noteworthy that the Dahl S rats always have greatly reduced papillary plasma flows compared to Dahl R rats, regardless of whether the diet contains low or high amounts of NaCl [5]. This should favor an enhanced passive reabsorption of Na out of the thin ascending limb of the deep nephrons which would significantly diminish natriuresis. Furthermore, the papillary prostaglandins also have a place in this scenario. We have recently found strong evidence that prostaglandins act as vasodilator substances upon the vessels supplying the rat renal papilla [6]. Thus, the low level of PGE_2 in the papillas of S rats would therefore encourage relative vasoconstriction and reduced papillary flow; and, as mentioned above, this would increase the tendency to Na retention which, in turn, would increase susceptibility to NaCl hypertension.

We sought to determine if we could correct the low papillary prostaglandin E_2 level in S rats by feeding large amounts of its precursor, linoleic acid, and whether this might influence the course of NaCl-induced hypertension. We placed two groups of S rats on 20% high-fat diets containing 5% NaCl. One group had 1.5% of linoleic acid in the diet, more than enough to prevent essential fatty acid deficiency. The other group had 16% linoleic in the diet. This study compares a 'normal' linoleic intake with a high linoleic intake. The S rats with the normal linoleic intake promptly developed hypertension when they were fed this high-salt diet. The S rats on the high linoleic intake had a much different course. Their rise of blood pressure was delayed for about 6 weeks and, when it did rise, it only went up to about half the extent of the normal linoleic group [28]. In fact, it rose to just about the level one would find in an S rat on a low-sodium diet that has reached this relatively advanced age (fig. 4).

Moreover, the 16% linoleic diet tripled the prostaglandin E_2 content in the S papilla, bringing it to about normal levels. Seemingly, the increased amount of linoleate in the diet increased the prostaglandin concentration in renal papilla and also greatly retarded the extent of NaCl hypertension.

Now to describe some of the implications of these findings with regard to human essential hypertension. First of all, if there is a delayed excretion of sodium in those with the genetic susceptibility to hypertension, and if

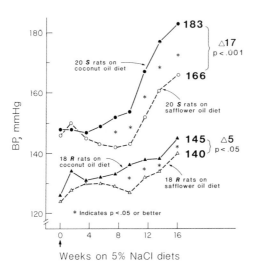

Fig. 4. Blood pressure of Dahl S and R rats as influenced by 5% NaCl diets containing coconut oil (1.5% linoleic acid diet) or safflower oil (16% linoleic acid diet).

one can pinpoint the individual with this genetic susceptibility, then an obvious way to prevent hypertension is to have such individuals on a lifelong low NaCl intake. Such a diet is compatible with excellent health and might literally prevent the hypertension from ever making its appearance.

With regard to the treatment of established essential hypertension, if there is a defect in sodium excretion, the practice of treating hypertension with a combination of a low-NaCl diet and a modest dose of a diuretic agent would constitute good strategy, because one would thereby overcome the tendency toward a defect in natriuresis. One might ask why the diuretic agent should be combined with a low NaCl intake. There are several reasons for this. A modest reduction of salt intake, down only to about 60 mEq per day, is not too difficult to achieve. Such a diet in conjunction with a thiazide diuretic agent will produce less hypokalemia. One also achieves a better control of blood pressure. There is no 'drown out' of the diuretic agent with a high salt intake. Moreover, thiazides are likely to increase the appetite for salt, and salt restriction in the diet will obviate this. Furthermore, one needs much less diuretic agent if it is combined with this modest reduction of NaCl in the diet. A non-specific fatigue often accompanies the use of strong doses of diuretic. A modest dose of a diuretic does this to a much smaller extent. And one can effectively use such a modest dose of a diuretic agent if it is combined with this 60-mEq sodium restriction in the diet. Thus, a reasonable anti-hypertensive strategy could include both a modest dose of diuretic and a modest restriction of NaCl in the diet.

The resulting reduction in body sodium brings about arteriolar vasodilation, which helps bring the blood pressure down. If thiazide diuretics are used in this manner, premature beats often appear because of the slightly reduced plasma potassium levels. But this is very easily managed. One can use a low-salt diet or potassium supplements or spironolactone, triamterine, or amiloride, and the premature beats usually disappear. Thiazide can also raise uric acid, but one can continue treatment by adding probenecid. With this addition, the plasma uric acid level returns to normal, joint symptoms disappear, and the hypertension is still controlled.

Refractory hypertensive patients often have an inordinate rise of aldosterone during treatment. In such patients, the use of spironolactone or converting-enzyme inhibitors can be quite helpful. All the anti-hypertensive agents are more effective when combined with this low-NaCl diet, mild diuretic combination. Even the beta-blockers are more effective when combined with a low-NaCl diet–diuretic combination, often requiring only one fourth of the dose of beta-blocker to achieve the same reduction in blood pressure. Moreover, this is true for all drugs which lower blood pressure by reducing angiotensin II levels, including the converting enzyme inhibitors and beta-blockers. The low-salt diet–diuretic combination makes the blood pressure of almost all individuals more angiotensin-dependent. Therefore, all such angiotensin II-reducing drugs have a much more powerful action when combined with a low-NaCl diet–diuretic combination.

In view of the fact that hypertensives appear to have a defect in natriuresis, a rational part of the treatment would include the low-NaCl diet, modest diuretic combination. This slightly reduces the body sodium level, which causes vasodilation, a decrease of peripheral vascular resistance, and a drop in blood pressure. There are fewer side effects using mild doses of diuretic, but the hypertension will often not be brought within the normal range with just the mild diuretic, low-NaCl diet regimen. This may necessitate the concomitant use of other anti-hypertensive drugs.

If we think of essential hypertension in general, it is a combination of increased arteriolar narrowing in the main, combined with an early tendency to a high cardiac output and a high oxygen consumption. It depends on age. In the teens and in the 20s, one sees relatively more of an elevation of cardiac output with smaller degrees of elevation of peripheral vascular resistance. But, in these youthful subjects, peripheral resistance could be considered high for that level of cardiac output. Thus, even in the young, there are two factors raising blood pressure: an increase in cardiac output and an increase in peripheral resistance. As one goes into the 30s and beyond, the high blood pressure is mainly maintained by the increase in peripheral resistance, with no obvious increase in cardiac output. Hence, what is needed is an attack on one or the other or both of these factors. Since the main problem in hypertension is the vasoconstriction, we certainly want to remedy that, partly by the use of the low-salt diet–diuretic combination. The use of drugs which cause relaxation of vasoconstriction is also warranted. As mentioned before, there is evidence for a hyperactivity of the sympathetic nervous system in hypertension. Thus, the use of drugs such as peripheral α_1-adrenergic blockers (e.g. prazosin) or the use of certain centrally acting α_2-agonist drugs which reduce peripheral sympathetic traffic (e.g. clonidine or methyldopa or, by other actions, reserpine) tend to reduce sympathetic vasoconstriction and thereby decrease blood pressure. Drugs which act directly on arterioles as vasodilators (e.g. hydralazine) also lower blood pressure but concomitantly raise cardiac output. Converting-enzyme inhibitors reduce angiotensin II levels and thus also act as direct vasodilators which raise cardiac output to a lesser degree.

As mentioned above, one can also reduce high blood pressure by slightly reducing cardiac output. Reducing cardiac output by itself would not likely be effective treatment. If cardiac output is reduced on a chronic basis, as in patients with mitral stenosis, the cardiac output is lower but the peripheral resistance is raised and the blood pressure remains the same. This long-term physiological adjustment to low cardiac output can go on for years. For successful treatment of hypertension, one would wish to have the reduction in cardiac output accompanied by some physiologic force which prevented this rise in peripheral vascular resistance. *Lund-Johansen* of Norway studied people who had been treated for 5 years with beta-blockers alone. At that time, all he could find to explain the drop in blood pressure was a decrease in cardiac output. These patients were at the same level of peripheral vascular resistance that existed before they started treatment. Since the peripheral resistance had not gone even higher in the face of reduced cardiac output, something was occurring to keep the peripheral resistance lower than it might have been expected to go. In order for a beta-blocker to lower blood pressure, something has to happen to keep the peripheral resistance from rising. The beta-blocker itself does this to a certain extent by lowering renin secretion and angiotensin II levels. The low-salt diet–diuretic combination also encourages vasodilation, and it is noteworthy that beta-blockers are much more effective in lowering blood pressure when combined with the diuretic–low-NaCl regimen. Other drugs which encourage vasodilation will also help the beta-blockers to lower blood pressure, including arterial vasodilators, sympathetic suppressants, and alpha-blockers.

This general strategy for treating the hypertensive patient attempts to correct the special physiological aberrations associated with essential hypertension and is an effective way to treat it.

References

1 Dahl, L.K.; Heine, M.; Tassinari, L.: Effects of chronic salt ingestion. Evidence that genetic factors play an important role in susceptibility to experimental hypertension. J. exp. Med. *115:* 1173 (1963).
2 Falkner, B.; Onesti, G.; Angelakos, E.: The effect of salt loading on cardiovascular response to mental stress in adolescents with high and low genetic risk for essential hypertension. Proc. 8th Scientific Meet. of the Int. Society of Hypertension, Abstract 123, Milan 1981.
3 Faubert, P.F.; Chou, S.Y.; Porush, J.G.; Belizou, I.J.; Spitalewitz, S.: Papillary plasma flow and tissue osmolality in chronic caval dogs. Am. J. Physiol. (in press, 1982).
4 Ganguli, M.; Tobian, L.; Azar, S.; O'Donnell, M.: Evidence that prostaglandin synthesis inhibitors increase the concentration of sodium and chloride in rat renal medulla. Circulation Res. *40:* suppl. 1, pp. 135–139 (1977).
5 Ganguli, M.; Tobian, L.; Dahl, L.: Low renal papillary plasma flow in both Dahl and Kyoto rats with spontaneous hypertension. Circulation Res. *39:* 337 (1976).
6 Ganguli, M.; Tobian, L.; Ferris, T.; Johnson, M.A.: Prostaglandin inhibition decreases plasma flow to the renal papilla in rats. Fed. Proc. (in press, 1982).
7 Ganguli, M.; Tobian, L.; Iwai, J.: Cardiac output and peripheral resistance in strains of rats sensitive and resistant to NaCl hypertension. Hypertension *1:* 3 (1979).
8 Goto, A.; Ikeda, T.; Tobian, L.; Iwai, J.; Johnson, M.A.: Brain lesions in the paraventricular nuclei and catecholaminergic neurons minimize NaCl hypertension in Dahl salt-sensitive rats. Clin. Sci. *61:* suppl. 7, pp. 53–55 (1981).
9 Goto, A.; Tobian, L.; Iwai, J.: Potassium feeding reduces hyperactive central nervous system pressor responses in Dahl salt-sensitive rats. Hypertension *3:* suppl. 1, p. 3 (1981).
10 Gotoh, E.; Ohnishi, T.; Fujishima, S.; Kaneko, Y.: Relation of sodium concentrations

in cerebrospinal fluid to systemic blood pressure levels in salt-sensitive and nonsalt-sensitive, essential hypertensive patients. Proc. 8th Scientific Meet. of the Int. Society of Hypertension, Abstract 158, Milan 1981.

11 Ikeda, T.; Tobian, L.; Iwai, J.; Goossens, P.: Central nervous system pressor responses in rats susceptible and resistant to NaCl hypertension. Clin. Sci. mol. Med. *55:* 225 (1978).

12 Ishii, M.; Tobian, L.: Interstitial cell granules in renal papilla and the solute composition of renal tissue in rats with Goldblatt hypertension. J. Lab. clin. Med. *74:* 47 (1969).

13 Lowenstein, F.W.: Blood pressure in relation to age and sex in the tropics and subtropics. Lancet *i:* 389 (1961).

14 Maddocks, I.: Blood pressures in Melanesians. Med. J. Aust. *i:* 1123 (1967).

15 McQuarrie, I.; Thompson, W.H.; Anderson, J.A.: Effects of excessive ingestion of sodium and potassium salts on carbohydrate metabolism and blood pressure in diabetic children. J. Nutr. *11:* 77 (1936).

16 Murray, R.G.; Luft, F.C.; Block, R.; Weyman, A.E.: Blood pressure responses to extremes of sodium intake in normal man. Proc. Soc. exp. Biol. Med. *159:* 432 (1978).

17 Oliver, W.J.; Cohen, E.L.; Neel, J.V.: Blood pressure, sodium intake and sodium-related hormones in the Yanomamo Indians, a 'no-salt' culture. Circulation *52:* 146 (1975).

18 Onesti, G.; Kim, K.E.; Greco, J.A.; Del Guercio, E.T.; Fernandes, M.; Swartz, C.: Blood pressure regulation in end-stage renal disease and anephric man. Circulation Res. *36/37:* suppl.1, p.1 (1975).

19 Osgood, R.W.; Reineck, H.J.; Stein, J.H.: Further studies on segmental sodium transport in the rat kidney during expansion of the extracellular fluid volume. J. clin. Invest *62:* 311 (1978).

20 Page, L.B.; Danion, A.; Moellering, R.C., Jr.: Antecedents of cardiovascular disease in six Solomon Islands societies. Circulation *49:* 1132 (1974).

21 Prior, I.A.M.; Evans, J.G.; Harvey, H.P.B.; Davidson, F.; Lindsey, M.: Sodium intake and blood pressure in two Polynesian populations. New Engl. J. Med. *279:* 515 (1968).

22 Sinnet, P.F.; Whyte, H.M.: Epidemiological studies in a total highland population, Tukisenta, New Guinea. Cardiovascular disease and relevant clinical, electrocardiographic, radiological and biochemical finds. J. chron. Dis. *26:* 265 (1973).

23 Spitalewitz, S.; Chou, S.Y.; Faubert, P.F.; Porush, J.G.: Effects of chronic salt depletion on intracortical blood flow distribution and papillary plasma flow in the dog (Abstract). Clin. Res. *28:* 462A (1980).

24 Takeshita, A.; Mark, A.L.: Neurogenic contribution to hindquarters vasoconstriction during high sodium intake in Dahl strain of genetically hypertensive rats. Circulation Res. *43:* suppl.1, p.1 (1978).

25 Tobian, L.: The interrelationship of electrolytes, juxtaglomerular cells and hypertension. Physiol. Rev. *40:* 280 (1960).

26 Tobian, L.; Ishii, M.: Interstitial cell granules and solutes in renal papilla in 'post-Goldblatt' hypertension. Am. J. Physiol. *217:* 1699 (1969).

27 Tobian, L.; Ishii, M.; Duke, M.: Relationship of cytoplasmic granules in renal papillary interstitial cells to 'post-salt' hypertension. J. Lab. clin. Med. *73:* 309 (1969).

28 Tobian, L.; Johnson, M. A.; Ganguli, M.; Goto, A.; Iwai, J.: Prostaglandin E_2 (PGE_2) in renal papilla in NaCl hypertension. Hypertension (in press, 1982).
29 Tobian, L.; Lange, J.; Azar, S.; Iwai, J.; Koop, D.; Coffee, K.; Johnson, M.A.: Reduction of natriuretic capacity and renin release in isolated blood-perfused kidneys of Dahl hypertension-prone rats. Circulation Res. *43:* suppl.1, p.92 (1978).
30 Tobian, L.; Lange, J.; Iwai, J.; Hiller, K.; Johnson, M.A.; Goossens, P.: Prevention with Thiazide of NaCl-induced hypertension in Dahl 'S' rats: evidence for a Na-retaining humoral agent in 'S' rats. Hypertension *1:* 316 (1979).
31 Tobian, L.; Pumper, M.; Johnson, S.; Iwai, J.: A circulating humoral pressor agent in Dahl 'S' rats with salt hypertension. Clin. Sci. *57:* 345s (1979).
32 Winer, B. H.: The antihypertensive actions of benzothiadiazines. Circulation *23:* 211 (1961).

L. J. Tobian, Professor of Internal Medicine, Chief, Hypertension Section, Department of Medicine, Box 285, Mayo Building, University of Minnesota, Minneapolis, MN 55455 (USA)

An Italian Preventive Trial of Coronary Heart Disease: The Rome Project of Coronary Heart Disease Prevention

Alessandro Menotti

Research Group of the Rome Project of Coronary Heart Disease Prevention (PPCC), Laboratory of Epidemiology and Biostatistics, National Institute of Health, Rome, Italy

Introduction

The role of nutrition in hypertension and cardiovascular diseases has been recognized in epidemiological observational studies conducted in many countries all over the world. These studies mainly point to a relationship between saturated fatty acids in the diet, serum cholesterol levels, and incidence of coronary heart disease (CHD) on the one hand, and between salt intake and hypertension on the other. Such relationships have also been supported by a number of studies in basic sciences and disciplines other than epidemiology which have yielded some of the most important advances of the last few decades [8, 9, 13]. Attempts to verify hypotheses concerning the causal relationship between diet and nutrition and cardiovascular diseases and to achieve the prevention of some conditions, mainly CHD, have been conducted through controlled trials which have yielded positive or doubtful results. However, during the last decade, multifactorial preventive trials [4, 12, 14], including the attack on several risk factors and through several means, have prevailed over monofactorial dietary trials [2, 3, 11, 16]. Using this kind of approach, a multifactorial preventive trial of CHD has been under way in Italy since 1973, under the title of Rome Project of Coronary Heart Disease Prevention.

The Rome Project of Coronary Heart Disease Prevention

This Project (PPCC) represents the Italian section of the World Health Organization (WHO) European Multifactor Preventive Trial of CHD,

Table I. Structure of the PPCC trial, Rome

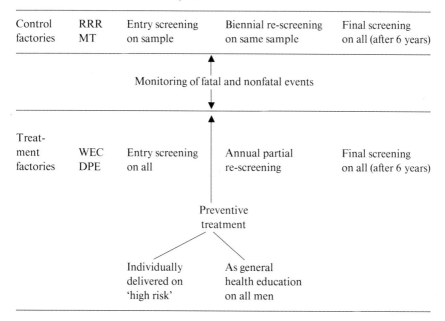

coordinated by the WHO Regional Office for Europe in Copenhagen [15, 17]. The study includes centers in Great Britain, Belgium, Poland, and Spain.

The statistical units are made up of occupational groups, called 'factories', of men aged 40–59 at entry. Members of pairs of 'factories' are randomly allocated to either control or treatment groups. The Rome Project has enrolled only four factories, two under treatment and two as controls, for totals of 3,131 and 2,896 men, respectively, and for an overall total of 6,027 individuals. The overall European pool includes 44 pairs of factories for a total of over 62,000 individuals.

The general structure of the trial as conducted in Rome is presented in table I; table II shows the size of the samples and the participation at entry. An entry screening was offered to all men belonging to the treatment factories, whereas only a randomly chosen subsample was offered the same procedure in the control factories (in Rome about 25%). In this way, baseline comparison between the two halves of the study was possible and further reexamination of the same samples in the controls and of random samples in the treatment groups allowed comparison of the changes of risk

Table II. Men enrolled in treatment (T) and control (C) factories, PPCC, Rome

Factory	Number enrolled	Number invited	Number examined	Participation %
WEC (T)	2,220	2,220	1,936	87
RRR (C)	1,061	114	97	85
DPE (T)	911	911	706	77
MT (C)	1,835	592	494	83
All men	6,027	3,837	3,233	84
All (T)	3,131	3,131	2,642	84
All (C)	2,896	706	591	84

Table III. Measurements at entry screening, PPCC, Rome

Compulsory measurements	Optional measurements
General data	Marital status
Type of work	Physical activity at leisure
Physical activity at work	Car driving habit
Smoking habit	Stress
Height and weight	Family history
Blood pressure	Alcohol, coffee, and tea intake
Serum cholesterol	Some previous disease history
Resting ECG	Use of drugs
Questionnaire on angina pectoris, myocardial infarction, claudicatio intermittens, chronic bronchitis, dyspnea	Dietary habits
	Sitting height, two diameters, Skinfold thickness
	Serum triglycerides
	Lipoprotein electrophoresis
	Basal and afterload blood glucose
	Serum uric acid

factors over time. The list of both compulsory (WHO protocol) and optional (Rome) measurements is given in table III.

High-risk subjects are defined as those who, on the basis of the entry examination, are in the upper 20% of a risk score suggested by the WHO European Collaborative Group [17], which includes the entry levels of age,

Table IV. Additive coronary risk score (from the WHO protocol) [17], PPCC, Rome

Factor	Coronary risk score						
	0	0.5	1.0	1.5	2.0	2.5	3.0
Age, years	50		50				
Physical activity at work	heavy		moderate		sedentary		
Cigarettes/day	5		5–19		20		
Systolic blood pressure, mm Hg	120		120–139		140–159		160
Serum cholesterol, mg/dl	210	210–219	220–229	230–239	240–249	250–259	260

systolic blood pressure, serum cholesterol, physical activity at work, and cigarette consumption (table IV). Within the treatment factories preventive action of an individual type is available to subjects defined as high risk; the remaining individuals are offered only a mass education campaign against risk factors for CHD. No intervention is provided to control factories which are left to the ordinary health services.

In the Rome section, the option has been taken to extend individual treatment on the largest possible scale, actually offering it also to (a) men belonging to the upper 20% of a risk probability rank provided by the multiple logistic equation (MLF) (derived from the Italian data of the Seven Countries Study); (b) men with hypertension (160 mm Hg systolic and/or 95 mm Hg diastolic as the average of two measurements); and (c) men with entry levels of serum cholesterol greater than 249 mg/dl and/or triglyceride greater than 199 mg/dl. Most of these choices overlapped with the previously mentioned WHO score system, and eventually the overall percentage of individually treated subjects rose to about 30%.

In the treatment groups, a re-screening was organized at each anniversary for a 10% randomly chosen sample and for all individually treated subjects. In the control factories the subsample initially examined is reexamined every 2 years. Such examinations allow comparison of the trends of risk factor levels in both control and treatment groups and evaluation, for both the whole population and the high-risk individuals, of the effect of the preventive action on risk factors. Actually, in the treatment groups measurements of risk factor levels (for individually treated subjects) were performed more frequently than once a year (at least twice for serum lipids,

Table V. Guidelines for individual treatment of risk factors, PPCC, Rome

Serum cholesterol	diet: low fat, low cholesterol, low saturated fat, relatively high polyunsaturated fat, reducing if overweight, high residue + drugs in selected cases
Blood pressure (160 and/or 95 mm Hg or more)	drugs (diuretics preferred) + low salt diet
Cigarette smoking	advice for cessation or reduction, or for preference to pipe and cigars
Physical inactivity	advice to increase physical activity at leisure
Overweight (relative weight 115% or more)	reducing diet
Serum triglycerides (170 mg/dl or more)	diet: reducing if overweight, low simple carbohydrate, low fat, low or no alcohol if needed
Impaired glucose tolerance (170 mg/dl or more after glucose load, 50 g-1 h)	diet: reducing if overweight, low simple carbohydrate
Overt diabetes	diet + oral hypoglycemic agents and/or insulin
Serum uric acid (7 mg/dl or more)	diet: low purine + drugs in selected cases

body weight, smoking habit, and level of physical activity, and at least six to eight times for blood pressure, depending on the schedule of the individually planned sessions); for comparison within individuals and between treatment and control subjects, only annual measurements were considered. It should be mentioned in this context that participation of individually treated men at the annual screening has never been particularly high, ranging between 65 and 80%.

The type of intervention aimed at modifying the levels of risk factors is summarized in table V. In high-risk subjects, and generally in individually treated subjects, sessions with doctors, dietitians, and other health workers were arranged 4–10 times the first year and 2–8 times in the following years. The general guidelines for intervention were as follows:

(1) *Serum cholesterol* lowering diet (for any level of cholesterol) included the following suggestions: low fat, low cholesterol, low saturated fatty acid diet with relatively high polyunsaturated fats and high residue, reduction of overweight. This implied avoiding or limiting dairy products,

fat meats, sausages, cheese, and in general, food rich in animal fat while giving preference to vegetable oils and to cooking and dressing with soft margarines instead of butter and hard margarines, and to legumes, vegetables, and fruit in general. In Rome, individuals with serum cholesterol levels of 300 mg/dl or more were immediately given drug treatment, mostly clofibrate.

(2) As suggested by the WHO European Collaborative Group, hypertensive subjects as previously defined were considered for treatment if the *high blood pressure* found at screening was confirmed at the first individual treatment session. First-choice medication was a combination of reserpine, clopamide, and dihydroergochristine, per os, in amounts varying from case to case. Patients already satisfactorily controlled with other types of drugs were allowed to continue as before, except for intensifying treatment if necessary. Further, appropriate dietary recommendations were offered: in addition to limiting sodium intake, overweight subjects with hypertension were advised to follow a low-calorie regimen.

(3) *Antismoking advice* was given in individual sessions trying to provide positive rather than negative messages. Smokers were invited, by several means of persuasion, to stop smoking altogether or, if this proved impossible, to decrease the number of cigarettes smoked per day, to avoid inhalation, to use low tar and low nicotine brands, or to shift to a pipe or cigars.

(4) Efforts to increase *physical activity* in sedentary people were made by suggesting a brisk 20- or 30-min walk every day and, in some cases, to engage in some noncompetitive type of exercise.

(5) Following the recommendations of the International Group, even in the absence of high serum cholesterol or blood pressure values, *overweight* subjects (115% or more relative body weight as compared to local standards) were encouraged to reduce by observing a low-calorie, low-fat, low-carbohydrate diet, and by limiting their alcohol intake if this was excessive, as well as by increasing the amount of physical activity.

(6) Intervention on high levels of *serum triglycerides,* on *diabetes or glucose intolerance,* and on high *serum uric acid* represented an option of the Rome group and has been reported in detail elsewhere [15].

The *mass education programs* consisted of a variety of means and procedures which are summarized in table VI. The general principles were those described for the individual approach; the tendency was to provide positive rather than negative messages and to promote both information and behavioral changes. Moreover, a questionnaire on attitudes toward

Table VI. Mass education procedures against CHD risk factors, PPCC, Rome

Booklet on CHD prevention
Booklet on cholesterol-lowering diet
Booklet on hypertension and its treatment
Booklet on cigarette smoking
Booklet on reducing diet
Booklet on physical activity
Several series of posters on the same items as above
Printed guide to the choice of food on the basis of saturated fat and cholesterol content
Cooking recipes
Other printed material on the same principles and items
Presentations of films on risk factors
Group discussions on diet
Group discussions on cigarette smoking
Group discussions on risk factors
Group discussions on results obtained in the trial
Circular letters on prevention, healthy habits and results obtained in the trial
Invitation to personal doctor to cooperate with the study staff
Special medical examination for those aiming at reassuming noncompetitive sport activities.

prevention and of knowledge in this field has been administered to subsamples of men at different stages of the intervention.

Monitoring of incidence included collection of fatal and nonfatal events for both control and treatment groups. Mortality and causes of death in all men have been obtained through notifications from the factory management and from systematic check of the files of local registry offices (mainly for men who had retired during the course of the study), whereas nonfatal events were registered only in men still at work when absence for illness of 20 consecutive days or more had been ascertained. Fixed criteria were used for validating both fatal and nonfatal cardiovascular events (fatal and nonfatal myocardial infarction, other fatal and nonfatal CHD, fatal and nonfatal strokes, all causes of death). Men retired during the trial were not considered exposed to the risk of nonfatal events.

Entry Examination Data of the PPCC

Entry levels of main risk factors [5, 10] as measured in control and treatment individuals are summarized in table VII. Sampling and ran-

Table VII. Mean entry levels of main risk factors, PPCC, Rome

Risk factors	Treatment (n=2,642)	Control (n=591)
Serum cholesterol, mg/dl	219.91 ± 43.58	211.48 ± 39.74*
Systolic blood pressure, mm Hg	133.62 ± 19.82	130.90 ± 20.92*
Cigarettes per day, number	11.98 ± 11.58	9.73 ± 10.91*
Body weight, kg	78.43 ± 11.17	75.75 ± 9.61*
Sedentary people, %	51.7	58.5*
MLF estimate (5 years) with age, serum cholesterol, systolic blood pressure, body mass index (probabilities of CHD)	0.0169 ± 0.0169	0.0175 ± 0.0151

* Significant difference: $p < 0.05$.

domization were not particularly lucky and some differences have been found between treatment and control groups, the former having higher mean levels than the latter for serum cholesterol, blood pressure, cigarette consumption, and body weight, whereas the proportion of sedentary men was higher in the control groups. The number of physically active people was negligible in both treatment and control factories. Since the mean age of control groups was slightly higher, these differences were re-balanced when the entry data were introduced into the MLF (one of the European solutions of the Seven Countries Study) with an estimated risk very similar in the two halves of the study (0.0169 in treatment groups, 0.0175 in controls) and not significantly different.

Changes of Risk Factors in the PPCC

Some results of the first 4-year follow-up concerning changes of some risk factors are reported in table VIII [7]. Changes are given separately for high-risk men (WHO definition, see above) and for samples of all men. In the treatment groups, subjects examined at the fourth anniversary represented 90% of a random sample of about 10%, and 71% of the men orig-

Table VIII. Changes in some risk factors in 4 years in %, PPCC, Rome

Groups	Serum cholesterol	Systolic blood pressure	Cigarette consumption	Body weight	MLF risk
High risk					
Treatment (n=375)	−9.97*	−5.92*	−23.18*	−2.96*	−31.43*
Control (n=72)	−1.36	+0.92	−16.59*	−0.81	+ 7.54
Difference (T–C)	−8.61*	−6.84*	− 6.59*	−2.15*	−38.97*
Sample of all men					
Treatment (n=238)	+0.88	−1.74*	−23.27*	−1.75*	−10.98*
Control (n=384)	+5.28*	+4.81*	−18.79*	+0.48	+20.16*
Difference (T–C)	−4.40*	−6.55*	− 4.46	−2.23*	−31.14*

* Significant change: $p < 0.05$. += Increase; −= Decrease.

inally defined as high-risk subjects; in the control groups, those examined after 4 years represented 65% of the original random sample examined at entry. Changes are computed in percentages, starting from an entry zero level in both treatment and control men, in spite of the above-mentioned differences of levels of treated as compared to controls.

Changes are expressed as $\frac{y_A}{x_A} - \frac{y_B}{x_B} \cdot 100$,

where y_A and y_B are the final levels for treatment and control groups, and x_A and x_B are the entry levels again for the treatment and control groups, respectively. The focus was on describing trends rather than absolute values since the correction for age differences was eventually provided by the MLF estimates.

For all factors and for both high-risk and sample of all men the overall trend was in favor of treated as compared to control groups. In 4 years, the algebraic differences between treated and controls ranged from 2 to 8%, depending on different factors and on high-risk or samples of all men. For the main purpose of the study and of the international analysis, the main changes concerned a 4.4% difference for serum cholesterol, a 6.5% difference for systolic blood pressure, a 2.2% difference for body weight, and a 4.5% difference in cigarette consumption, all in favor of the treated groups.

When four factors were introduced into the MLF equation (plus age as a correcting factor) the risk estimated at entry, substantially identical in control and treated men, changed markedly with a net 39% difference in high-risk men and a 31% difference in samples of all men, all differences being in favor of the treated groups. Some of the peaks observed along the trends of the curves (not reported here) are likely to be due to sampling variations since (especially in the treated random sample) a new sample was drawn each year. The overall picture showed that the controls were not stable in time, following in some circumstances some age or 'natural' secular trends.

It became clear that in this kind of intervention what really counts is a diverging trend when comparing control with treatment groups. If individually considered, all factors in the treatment groups decreased in a statistically significant measure (except serum cholesterol in random samples), whereas in the control groups only cigarette consumption decreased significantly (in both high-risk and random sample), and blood pressure and serum cholesterol increased in a significant measure in the random sample. The MLF estimated risk of CHD diverged in both high-risk and random sample, all changes being individually significant except the increase for high-risk controls. The overall picture of the 5-year data is not substantially different for treatment groups, indicating that the results achieved are stable, while the corresponding data for the controls are not available.

Since the measurements of risk factors in controls have always been done in the same samples, it is expected that a favorable screening effect might have been induced and therefore the comparisons with the treated groups should be more favorable to the latter if considering that the majority of the controls was not screened at entry. This might also counterbalance the possible source of error due to sampling variation as a result of the relatively low participation, the explanation for which has been given above. The main lesson from these results is that putting a brake on the natural increase of some risk factors in middle-aged men might be as important as a decrease, when controls maintain an increasing age trend. The differences observed in risk changes as estimated by the MLF are satisfactory, and they might correspond to important changes in incidence

Table IX. Mean composition of dietary fats in subsamples of high-risk men randomly selected from an intervention factory: evaluation at entry and after 4 years, PPCC, Rome

	Saturated g	Mono-unsaturated g	Poly-unsaturated g	P/S ratio	Fat energy kcal	Cholesterol mg
Entry	29.4	48.0	10.4	0.35	785	301
(n=50)	± 6.7	± 6.8	± 5.1	±0.26	±215	±109
4 years	21.1	26.8	9.4	0.48	517	182
(n=40)	± 4.2	± 5.7	± 5.9	±0.39	±185	± 64

if the hypothesis of the reversibility of risk is true and if it works with short-term latency.

Dietary Behavior in the PPCC

The results presented above suggest that, altogether, the contribution to the reduction of the estimated risk depended, although to different extents, upon changes induced in all the factors considered. At this stage of the analysis, it is almost impossible to detect the contribution of the changes in dietary habits, also because they may have influenced at the same time serum cholesterol, body weight, and blood pressure.

Self-reported changes (not described here in detail) in fat consumption indicated differences in the right direction when comparing treated subjects and controls [1]. In small samples of men belonging to the treatment factories (40–50 men), an estimation of fat energy intake and of the various types of fats has been made at entry and after 4 years, on the basis of a 7-day meal diary. Fat energy dropped from 785 to 517 kcal, saturated fat from 29 to 21 g, monounsaturated fat from 48 to 27 g, and polyunsaturated fat from 10.4 to 9.4 g. Cholesterol intake decreased from 301 to 182 mg, whereas the P/S ratio increased slightly from 0.35 to 0.48 (table IX).

However, a more objective measure of such changes was desirable. At the fourth anniversary examination the measurement of fatty acids in red blood cells by gas chromatography was possible in subsamples of 140 control men and 121 treated men. Data on the linoleate/oleate (L/O) ratio, linoleate/palmitate (L/P) ratio, and the P/S ratio have been reported separately in subgroups of controls (as a whole), in men treated only with mass

Table X. Mean red blood cell fatty acid composition in 140 control and 121 treated subjects, PPCC, Rome

	P/S	L/O	L/P
Controls (n=140)	0.83 ±0.78	0.48 ±0.10	0.32 ±0.08
Collective treatment (n=61)	1.03* ±0.26	0.50 ±0.10	0.33 ±0.07
Individual treatment (n=60)	1.16** ±0.64	0.51* ±0.13	0.36*** ±0.10

Treatment vs controls: * $p < 0.05$; ** $p < 0.01$; *** $p < 0.001$.
P/S = Polyunsaturated/saturated ratio; L/O = linoleate/oleate ratio; L/P = linoleate/palmitate ratio.

Table XI. Prevalence of consumers of different types of cooking and seasoning fats at the fourth anniversary (samples), in %, PPCC, Rome

	Controls (n=140)	Individual treatment (n=60)	Collective treatment (n=61)
Butter	58.0	44.9	30.2
Olive oil	98.5	89.8	86.8
Seed oil	48.8	90.9	91.4
Margarine	22.3	34.5	43.1

education (61 men), and in individually treated high-risk subjects (60 men) (table X). The trends were in the direction expected on the basis of the intensity of treatment, but unfortunately the data are of limited value due to the absence of a baseline for comparison. Questionnaires administered to the same individuals confirmed, in terms of fat consumption from shortening and seasoning, the tendency towards a better compliance with advice in individually treated men, as compared to those treated by mass education, both behaving quite differently from control men (table XI). These data also reflect the type of changes obtained when comparing entry data with data collected at the fourth anniversary, mentioned above.

Table XII. 4-year follow-up of hypertensive subjects, PPCC, Rome

	Control	Treatment
Exposed to risk	106	497
Mean change in systolic blood pressure, %	− 1.7	− 8.4*
Mean change in diastolic blood pressure, %	− 3.2	−10.8*
Mean change in serum cholesterol, %	+ 3.5	− 4.1*
Mean change in daily cigarette consumption, %	−19.7**	−17.2*
Mean change in body mass index, %	+ 0.2	− 2.7*
Cumulative incidence of		
fatal and nonfatal CHD		
fatal and nonfatal strokes, %	13.2	5.8*, **

* Significant change: $p < 0.05$; ** significant difference controls vs treated: $p < 0.001$.

Incidence Data on Hypertensive Subjects in the PPCC

Incidence and mortality data are already available for the fifth anniversary, but they cannot be released, as agreed upon by the WHO Collaborative Group, until the end of the total 6-year follow-up. The Rome group, however, has obtained permission to publish partial data, concerning the first 4-year follow-up *limited* to men with hypertension belonging to both the intervention group and the subsample of the control group examined at entry [6].

Table XII summarizes the changes observed in mean levels of some risk factors and the incidence of lumped fatal and nonfatal cases of CHD (myocardial infarction, sudden death, other coronary deaths) and stroke (fatal and nonfatal) in the age-adjusted subgroups. Changes of risk factor levels and incidence of cardiovascular events run parallel and the reduced incidence (−56%) seems well-justified by the trends in risk factors, mainly in blood pressure. This preliminary result is encouraging in view of the final data analysis coming at the end of the 6-year follow-up.

Comments

The overall problems, data, and results presented in this paper belong strictly to the Rome Project and have only limited significance since they are simply part of a much wider enterprise. The initial statistical assumption was that none of the single national studies could ever reach statistical significance in demonstrating a measurable reduction of CHD incidence, insofar as the final data analysis will take into account the statistical problems of comparing pairs of factories instead of lumping together, as was done in a preliminary way here, all control men on the one hand and all treated men on the other.

In spite of these problems, the preliminary trends seem to indicate that: (a) mixed intervention is feasible in occupational groups where intensive individual treatment is offered to high-risk subjects and more general collective mass education is provided for the others; (b) perhaps, within single studies or within special groups (such as the hypertensive subjects), differences of cardiovascular incidence between control and treatment groups can be elicited. The end of the trial is planned for spring, 1981, and final data on comparative incidence should become available by 1982.

References

1 Angelico F.; Amodeo P.; Menotti, A.; Ricci G.; Urbinati G. C.: The Rome Project of Coronary Heart Disease Prevention: evaluation of dietary changes after a four-year intervention, abstract. 5th Int. Symp. on Atherosclerosis, Houston 1979.
2 Archer, M.: The diet and coronary heart disease study of the New York City Department of Health. Cardiovasc. Epidem. Newsletter, Council of Epidemiology (Am. Heart Ass., New York 1972).
3 Dayton S.; Pearce M. L.; Hashimoto S.; Dixon W. J.; Tomiyasy U.: A controlled clinical trial of a diet high in unsaturated fat in preventing complications of atherosclerosis. Circulation *40:* suppl. II, pp. 1–63 (1969).
4 Farquhar, J. W.; Wood, P. D.; Breitrose, H.; Haskell, W. L.; Meyer, A. J.; Maccody, N.; Alexander, J. K.; Brown, B. W.; McAlister, A. L.; Nash, J. D.; Stern, M. P.: Community education for cardiovascular health. Lancet *i:* 1192–1195 (1973).
5 Gruppo di Ricerca del Progetto Romano di Prevenzione della Cardiopatia Coronarica: Distribuzione di alcuni fattori di rischio coronarico in popolazioni lavorative di Roma: Il Progetto Romano di Prevenzione della Cardiopatia Coronarica. G. Arterioscler. (NS) *1:* 77–89 (1976).
6 Gruppo di Ricerca del Progetto Romano di Prevenzione della Cardiopatia Coronarica: La prevenzione di alcune complicanze cardiovascolari negli ipertesi con trattamento multifattoriale. G. ital. Cardiol. *11:* 164–169 (1981).
7 Gruppo di Ricerca del Progetto Romano di Prevenzione della Cardiopatia Coronarica: Variazioni dei fattori di rischio in quattro anni nel Progetto Romano di Prevenzione della Cardiopatia Coronarica (PPCC). G. ital. Cardiol. *10:* 204–215 (1980).

8 Inter-Society Commission for Heart Disease Resources: Primary prevention of the atherosclerotic disease. Circulation 52: A55–A95 (1970).
9 Levy, R.I.; Rifkind, B.M.; Dennis, B.H.; Ernst, N.D. (eds): Nutrition, lipids and coronary heart disease. A global view (Raven Press, New York 1979).
10 Menotti, A.; Ricci, G.; Urbinati, G.C. (eds): Presentazione e stato attuale del Progetto Romano di Prevenzione della Cardiopatia Coronarica (PPCC). Quad. di 'Cuore e Vasi' No.1, pp.1–77 (Edizioni Internazionali Gruppo Editoriale Medico, Roma 1979).
11 Miettinen, M.; Turpeinen, O.; Karvonen, M.J.; Elosuo, R.; Paavilainen, E.: Effect of cholesterol-lowering diet on mortality from coronary-heart disease and other causes. A twelve-year clinical trial in men and women. Lancet ii: 835–838 (1972).
12 Multiple Risk Factor Intervention Trial (MRFIT) Research Group: A national study of primary prevention of coronary heart disease. J. Am. med. Ass. 235: 825–827 (1976).
13 NHLI: Report by the Task Force on Arteriosclerosis. NIH, vol. I and II. DHEW Publ. No. 72-219 (DHEW, Washington 1971).
14 Puska, P.: North Karelia Project. Design and experiences with a programme for community control of cardiovascular diseases. Archs Mal. Cœur. 68: no. spécial, p.133 (1975).
15 Research Group of the Rome Project of Coronary Heart Disease Prevention: The Rome Project of Coronary Heart Disease Prevention. Ann. Ist. Sup. Sanit. 12: 316–330 (1976).
16 USPHS: The national diet-heart study final report. Circulation 37: suppl. I, pp.1–428 (1968).
17 WHO European Collaborative Group: An international controlled trial in the multifactorial prevention of coronary heart disease. Int. J. Epidemiol. 3: 219–224 (1974).

Alessandro Menotti, MD, Research Group of the Rome Project of Coronary Heart Disease Prevention (PPCC), Laboratory of Epidemiology and Biostatistics, Istituto Superiore di Sanità, Viale Regina Elena 299, I-00161 Rome (Italy)

Nutrition-Related Risk Factors for the Atherosclerotic Diseases – Present Status

Jeremiah Stamler

Department of Community Health and Preventive Medicine, Northwestern University Medical School, Chicago, Ill., USA

Introduction and Background

Since the theme of the US-Italy Symposium is Nutrition and Cardiovascular Diseases, this presentation focuses on nutrition-influenced risk factors for the atherosclerotic diseases. The focus is particularly on coronary heart disease (CHD), since the data are most extensive for this end point. The available evidence indicates that the established major risk factors for CHD are also significantly related to the atherosclerotic diseases of other sites (cerebral, trunk, peripheral).

As to present status of knowledge in this area, it is first appropriate to emphasize that the extensive studies of recent years – epidemiologic, animal-experimental, clinical, pathologic – have yielded large bodies of data confirming fundamental concepts and conclusions formulated in the late 1950s and 1960s, based on the first great post-World War II wave of research on this problem. For perspective in this regard, it is relevant to cite the basic theoretical generalizations formulated in 1958, in the monograph *Nutrition and Atherosclerosis,* which the writer co-authored with *Katz* and *Pick:*

'Atherosclerosis is a distinct entity... it is not inevitable, nor is it irreversible. Rather it is preventable and (at least up to a point) curable.

'Further, atherosclerosis is a metabolic disease, in which altered cholesterol-lipid-lipoprotein metabolism plays a critical and decisive (but not exclusive) rule. Elevated levels of circulating plasma cholesterol-lipids-lipoproteins are cardinal signs of this metabolic abnormality and constitute the metabolic prerequisites for atherogenesis in most persons afflicted during middle-age.

'Finally – and this is the decisive point in terms of the problem of the relationship between diet and atherosclerosis – the atherogenic alterations in cholesterol-lipid-lipoprotein metabolism are frequently brought about by the life-span pattern of diet. They are byproducts of an habitually unbalanced diet excessive in total calories, empty calories, total fats, saturated fats, cholesterol, refined carbohydrates, salt...' [32].

Almost 10 years later, after a detailed review in four chapters of his monograph, *Lectures on Preventive Cardiology,* the writer set down the following Recapitulation and Summary:

'Prior to the discussion of prevention, it may be worthwhile to summarize the material reviewed in chapters 4–7 on the etiology and pathogenesis of atherosclerotic coronary heart disease, by listing the key facts unequivocally demonstrated to be valid and proved:

(1) Severe atherosclerosis is the underlying pathologic process in most cases of clinical coronary disease.

(2) A several-fold increase in cholesterol – particularly esterified cholesterol – is the biochemical hallmark of the atherosclerotic plaque.

(3) The excess cholesterol in the plaque is derived from the cholesterol-bearing lipoproteins of the circulating plasma.

(4) Sustained hypercholesterolemic hyperlipidemia – be it due to diabetes mellitus, the nephrotic syndrome, hypothyroidism, endogenous metabolic derangement of unknown cause (essential familial hypercholesterolemia or hypertriglyceridemia), or (as is most frequently the case in our country) habitual dietary habits – is associated with frequent, premature, severe atherosclerotic CHD.

(5) In groups of middle-aged patients with clinical CHD, higher mean serum cholesterol-lipid-β-lipoprotein levels are found than in matched control groups.

(6) Sustained ingestion of diets containing increased quantities of cholesterol and fat is a virtual prerequisite for the production of significant atherosclerosis in a wide range of experimental animals.

(7) In animals ingesting high-cholesterol, high-fat diets, other factors (e.g. hypertension, hypothyroidism, renal damage) will act synergistically to intensify atherogenesis, or to prevent or reverse the pathologic process (e.g. estrogens).

(8) Marked contrasts exist among populations in different parts of the world – particularly between the economically developed versus the

underdeveloped countries – in habitual diets, levels of serum cholesterol-lipid-β-lipoprotein and occurrence rates of premature atherosclerotic coronary disease. These three variables are generally correlated.

(9) From data on social class differences, on the influences of migration, and on the effects of two world wars, it is evident that the marked international differences in occurrence rates of premature CHD are due to socioeconomic factors (i.e. differences in living habits, principally dietary habits), and not to racial, ethnic, climatic or geographical factors.

(10) Where the mean serum cholesterol levels of populations are low (e.g. 140–175 mg/100 ml), clinical CHD and severe coronary atherosclerosis at postmortem are rare, particularly in middle age.

(11) High mean serum cholesterol levels in populations (e.g. 220 mg/100 ml or greater), and high rates of middle-age clinical coronary heart disease, occur only where the habitual diets are high in calories, total fat, saturated fat, and cholesterol.

(12) In populations studied prospectively, risk of premature atherosclerotic disease is increased in the presence of hypercholesterolemia-hyperlipidemia.

(13) In populations with the nutritional-metabolic prerequisites for severe premature atherosclerotic disease, risk is also increased by hypertension, diabetes, overweight, cigarette smoking, a positive family history of premature vascular disease. Physical inactivity (sedentary living habit) and psychologic stress are in all likelihood additional significant risk factors.

(14) Atherosclerosis is, at least in part, a reversible disease.

(15) Serum cholesterol-lipid-β-lipoprotein levels are amenable to dietary influence, being responsive particularly to intake of cholesterol and saturated fat (less so to intake of calories, unsaturated and polyunsaturated fats, type of carbohydrate, meal-eating pattern, etc.).

(16) The other major coronary risk factors – hypertension, diabetes, overweight, cigarette smoking, physical inactivity – are amenable to control and correction by nutritional-hygienic-pharmacologic means.

The foregoing phenomena have been unequivocally proved and firmly established. They have been massively documented – to an extent far exceeding most phenomena in medicine. They compel the conclusion that diet is of key importance among the multiple factors involved in the etiology and pathogenesis of atherosclerotic disease. They are the solid foundation stones of the nutritional-metabolic theory of atherogenesis. They also are the firm scientific bases for practical approaches to the primary and secondary prevention of premature atherosclerotic CHD [69].'

These conclusions have stood the test of time and further research over the last 15 years, in this writer's judgment. Yet – after all these years of extensive multifaceted investigation – one hears a great deal about the 'controversial' nature of the diet-heart relationship (a useful short-hand term). The main purpose of this paper is to examine several of the specific assertions that have been put forward to support the notion that the field is enmeshed in uncertainty and unclarity. It is the writer's assessment that, in fact, these assertions are not sound, the conclusions set down in the two above-cited quotations remain valid, and the diet-heart relationship has been illuminated in depth and breadth.

By way of further background, prior to examination of the assertions under question, it may be of value briefly to expand on the data base demonstrating certain key relationships in definitive and unequivocal fashion, beyond debate or challenge.

The Established Relationship in Cross-Population Studies among Dietary Lipid, Serum Cholesterol, and CHD

Several different types of data are available in quantity in this area. For example, at least a score of studies have been reported over the last 25 years using data from the Food and Agriculture Organization (FAO) and the World Health Organization (WHO) to relate nutrition and CHD for sets of countries [74, 93]. Consistently, significant associations have been demonstrated in univariate analyses between total fat, saturated fat, dietary cholesterol (also total calories, total protein, animal protein), and CHD mortality; also between animal products (dairy products, meats, poultry, eggs), processed sugars, and CHD mortality; also an inverse relationship between vegetable products (fruits, grains, legumes, vegetables) and CHD mortality.

For economic reasons, many of these variables are highly correlated, making bivariate regression or partial correlation analysis treacherous as a means to dissect out independent effects. However, with use of 2 × 2 cross-classification and analysis of variance, plus combination of the three key dietary lipids (saturated fats, polyunsaturated fats, and dietary cholesterol) into a single score (with the equations of *Keys* et al. [38] or *Hegsted* et al. [22]), it is possible to go a bit further [46]. The data show that the significant relations between dietary lipid and CHD persist with control for vegetable products or refined sugars. The latter two, on the other hand, do not relate significantly to CHD mortality when dietary lipid is simultaneously considered.

The data from international autopsy studies, epitomized by the International Atherosclerosis Project, also show a relationship across populations between dietary fat and severe coronary atherosclerosis, and in addition between dietary fat and serum cholesterol, and between serum cholesterol and severe coronary atherosclerosis [52].

In addition, of course, there are similar comparisons among living population cohorts, demonstrating similar positive associations among dietary lipids, serum cholesterol, and CHD rates. These investigations are epitomized by the Seven Countries Study and the studies of migrants, e.g. the Ni-Hon-San study involving men in Japan, Japanese-American men in Hawaii, and Japanese-American men in the San Francisco Bay area [28, 30, 33, 34, 36, 50, 62, 63, 74, 85, 92, 96].

The massive epidemiologic data on the 3-way relationship among dietary lipid, serum cholesterol, and CHD risk of populations are consistent with findings from animal-experimental and clinical research, including metabolic-ward-type and other controlled experiments, supporting the conclusion that the associations are etiologic in nature.

The Established Relationships within Populations between the Serum Cholesterol of Individuals and Their CHD Risk

The data are massive – more than 30 prospective studies in many countries – on what are known as the major risk factors, serum total cholesterol, blood pressure, and cigarette smoking, and their independent relationship to risk of premature CHD for individuals [74, 78, 93]. At least 10 clinical investigations using angiography have yielded like findings, and the animal experimental data are confirmatory, all indicative of a cause-and-effect relationship [78, 93].

The Established Relationship between Caloric Imbalance and Two Established Major CHD Risk Factors, Hypertension and Hypercholesterolemia

Relative weight and serum cholesterol of individuals are significantly correlated, and weight gain or loss are associated with increase or decrease respectively in serum total cholesterol, low-density lipoprotein (LDL) cholesterol, very low-density lipoprotein (VLDL) cholesterol, and triglycerides. Further, weight gain is associated with a fall in high-density lipoprotein (HDL) cholesterol; weight loss, with an increase in HDL cholesterol [14, 69, 73]. Relative weight also has a significant direct relationship to blood pressure. Moreover, marked overweight is the single best-defined correlate of non-insulin-dependent diabetes (NIDD), the mass form of this disease also implicated as a CHD risk factor [9, 93].

The Diet-Heart Relationship Goes beyond the Relationship of Dietary Lipid to Serum Cholesterol-Lipoproteins

The significance of the foregoing facts on the relationship of relative weight to serum cholesterol-lipids-lipoproteins, blood pressure, diabetes needs to be understood clearly and explicitly: contrary to a frequently held impression, the diet-heart relationship is not limited to the influence of dietary lipid composition on serum cholesterol-lipids-lipoproteins, as important as that is. Thus, recent data indicate a relationship of dietary lipid to CHD independent of serum cholesterol, blood pressure, cigarettes, relative weight (see below) [66]. Further, calorie imbalance with consequent obesity – and its prevention and correction – are of major consequence, especially when the habitual diet is atherogenic in its lipid composition, i. e. high in cholesterol and saturated fat.

New evidence has been added to old indicating that other nutritional considerations are relevant for the effort to prevent or correct elevated serum total cholesterol and its atherogenic components. In particular, in addition to dietary lipid composition and calorie balance, the pectins and gums in fruits, legumes, and other vegetable products play a role, possibly other fibers and vegetable proteins as well [42].

In regard to the prevention and correction of hypertension, in addition to calorie imbalance, there are the matters of dietary electrolytes (especially sodium) and alcohol. As to sodium, hypertension has been induced in experimental animals by salt feeding; dietary Na is a prerequisite for the production of experimental hypertension by other means; and Na is viewed as pivotal in the pathophysiology of hypertension [1, 82]. Depletion of extracellular Na, by oral diuretics or (in the pre-diuretic era) by low Na diet, is a cornerstone of the treatment of hypertension. Cross-population studies indicate that the higher the habitual Na intake, the higher the prevalence rate of hypertension. Preliterate isolated populations who have not developed the custom – common to the rest of the human race since the development of agriculture and animal husbandry [1] – of adding Na to food have little or no hypertension, little or no rise in blood pressure with age.

As to alcohol, cross-sectional data are available from several studies in the USA and other countries showing that heavy drinkers – i.e. daily consumers of 4–5+ drinks per day – have sizable and significantly higher mean blood pressures and prevalence rates of hypertension than others in the population [17, 18]. (A drink is defined as 12 ounces or about 360 ml of beer, 4⅓ ounces or about 130 ml of table wine, 3 ounces or about 90 ml of aperitif wine, 1.0–1.5 ounces of distilled spirits, each containing about 13 g alcohol.) At least two sets of prospective data are available confirming this relationship [18], as well as results of an intervention study [65]. All this is in addition to and independent of the contribution of heavy alcohol consumption to calorie excess, to hypertriglyceridemic hyperlipidemia (at least in some people metabolically sensitive to this agent), and to heart disease via a direct toxic effect on heart muscle.

Clearly, then, multiple aspects of nutrition – the eating and drinking habits of tens of millions in modern industrialized society – play an important role in the etiology and pathogenesis of CHD and the other atherosclerotic diseases. In fact – as elaborated below – the role of habitual diet in producing the modern epidemic of premature CHD is almost certainly primary and essential, i.e. of critical importance.

With these data sets as background, then, what about the assertions that have been advanced to challenge the relationship between diet and CHD? What follows is an examination of the most common among these.

Assertion No. 1: No Relation Can Be Shown between the Habitual Dietary Lipid Intake of Individuals within a Population and the Serum Cholesterol and CHD Risk of Individuals

Some studies – e.g. the Framingham [21] and Tecumseh [56] community studies, also the Israeli [29] civil servant study – have reported no association or only a low-order non-significant association between the dietary lipid intake of individuals and their serum cholesterol level. This is clearly an anomalous finding given the massive evidence demonstrating that dietary lipids influence serum cholesterol. In contrast to these negative results, several studies in diverse populations around the world have recorded significant positive relationships of the type anticipated from all the other data [66]. The Tarahumara Indian Study found a highly significant relationship, with the highest correlation coefficient (about 0.9) between dietary cholesterol and plasma cholesterol recorded by any investigation (fig. 1) [15].

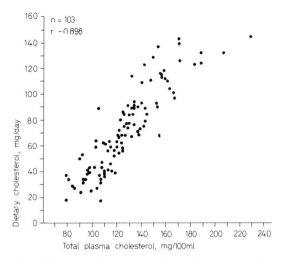

Fig. 1. The correlation between the total plasma cholesterol and dietary cholesterol intake per day (r = 0.898, p < 0.01) in a subsample of the Tarahumara study [15].

Why these paradoxes and contradictions? A key to understanding them is the fact that, in contrast to the other populations, the Tarahumara Indians eat almost identically day after day. There is little intraindividual variation in dietary cholesterol. This contrast merits attention and elaboration. It relates to a key concept that *Ancel Keys* put his finger on years ago. To make the point crystal clear, it is useful to consider traits like serum cholesterol and blood pressure. Despite sizable intraindividual variation in these traits, one measurement is enough to characterize individuals within a population, as the prospective epidemiological studies have repeatedly shown. This is because the intraindividual variation, while sizable, is much less than the interindividual variation: the ratio of the variances is about one to three. This is *not* the case in most populations for the dietary lipids, as *Ancel Keys* noted qualitatively years ago. Recently, *Kiang Liu* of our research group worked out the quantitative aspects of this matter. For such variables as dietary cholesterol and saturated fat, in contrast to serum cholesterol and blood pressure, the intraindividual variance for free-living Americans is two or more times larger than the interindividual variance (table I) [43, 45]. That is, individuals vary from themselves day-to-day much more than they vary from each other. Therefore, any single measurement – e.g. the 24-hour recall used in the Tecumseh

Table I. Ratio of intraindividual to interindividual variances for several diet-related biomedical variables measured for middle-aged American men in prospective studies on the cardiovascular diseases

Variable	Ratio of intra- to interindividual variances
Weight, lb	0.04
Systolic pressure, mm Hg	0.34
Diastolic pressure, mm Hg	0.47
Heart rate, beats/min	0.47
Serum cholesterol, mg/dl	0.38
Serum uric acid, mg/dl	0.38
1-Hour post-load plasma glucose, mg/dl	0.94
Dietary total calories/day	1.47
Dietary cholesterol, mg/day	3.00
Saturated fat, % cal	2.11
Polyunsaturated fat, % cal	4.40
$1.35 (2S-P) + 1.5Z$[1]	2.08
24-Hour sodium excretion	3.20

[1] S is percent of total calories from saturated fatty acids; P is percent of calories from polyunsaturated fatty acids; Z is the square root of dietary cholesterol in mg per 1,000 cal.

study – is entirely inadequate for validly characterizing individuals within a population, however good it may be to describe the group. *Liu* worked out the statistics for estimating degree of error resulting from this phenomenon, e.g. misclassification of individuals arrayed into quantiles or diminution of the true correlation coefficient between dietary lipid and serum cholesterol. Given the observed intra- and interindividual variances for dietary cholesterol or saturated fat within our population, the diminution in the true correlation coefficient between dietary lipid and serum cholesterol with one measurement (such as one 24-hour dietary recall) is about 50%. Thus, if the true correlation is in the order of 0.4, the observed will be 0.2. In populations eating a diet high in saturated fat and cholesterol, there are interindividual differences in genetically influenced endogenous mechanisms that contribute sizably to the differences in serum cholesterol among individuals. The specific biochemistry of these mechanisms is unknown; hence, no measurements can be made to control for these in analyses of the relationship of dietary lipid intake of individuals to

Table II. Relationship between entry dietary lipid intake and serum cholesterol of individuals, linear regression analysis, 1,900 men aged 40–55 at entry in 1958, Western Electric Co. Study

Regressor–Entry dietary lipid variable, 1958	Serum cholesterol	
	coeff.	p
1958 Entry examination trivariate analysis		
Saturated fatty acids, % cal	1.10	0.014
Polyunsaturated fatty acids, % cal	−2.10	0.064
Dietary cholesterol, mg/1,000 cal	0.03	0.044
Constant	229.41	
r	0.08	0.006
Univariate analysis		
Dietary lipid score	0.51	<0.001
Body mass index, kg/m^2	–	–
Age, years	–	–
Constant	216.89	
r	0.08	<0.001
Trivariate analysis		
Dietary lipid score	0.54	<0.001
Body mass index, kg/m^2	1.15	0.003
Age, years	0.38	0.171
Constant	167.90	
r	0.11	<0.001

Dietary lipid score = $1.26(2S-P) + 1.5\sqrt{C}$.

their serum cholesterol level. Hence, the true correlation coefficient in this case, for a population like ours with a high intake of saturated fat and cholesterol (in contrast to the Tarahumara Indians), must be much below 1.0, and if the method of measurement does not do well in classifying individuals, a false negative set of data results.

When the method of measurement is better, the outcome is different, as recent analyses from the Western Electric Study show. To the credit primarily of *Oglesby Paul* who organized that study, and *Anne MacMillan*

Shryock who led its dietary work, the Western Electric Study did not limit itself to either a short dietary history of what the men usually ate or to a 24-hour recall [66]. Rather, for 2 years in a row, at baseline and at the first follow-up year, the habitual recent diet of the 1,900 men was characterized in depth by the Burke method, requiring about an hour per person [12]. With that method, dietary cholesterol and saturated fatty acids were both significantly and independently related to serum cholesterol, dietary polyunsaturates nearly so (table II) [66]. When body mass index was also considered, itself significantly related to serum cholesterol, saturates remained significant. The combined lipid score of *Keys* et al. [38] also showed a highly significant relationship between baseline dietary lipids and baseline serum cholesterol of individuals ($p < 0.001$).

The availability in this study of dietary and serum cholesterol data 2 years in a row taught another important lesson: The baseline year was 1958; at that time (in contrast to nowadays), few men had had a previous measure of serum cholesterol. When they were informed of their serum cholesterol levels at baseline, a sizable proportion of those with levels of 300 mg/dl or higher changed their diets, i.e. reduced their intake of saturates and cholesterol (fig. 2) [67]. Consequently, there was no longer a relationship between dietary lipids and serum cholesterol in the first follow-up year – clearly an artefact produced by the study. Since some other studies (e.g. Framingham) collected dietary data years after the study started, this same artefact could be operating, as noted by *Liu* et al. [43] and *Shekelle* et al. [67], and could account at least in part for their failure to find a significant relationship between dietary lipid and serum cholesterol of individuals. In fact, nowadays, when wide segments of the US population with high serum cholesterol levels have made diet changes, the true relationships between dietary lipids and serum cholesterol of individuals cannot only be obscured, but can actually be inverted, so that those with higher serum cholesterol levels seem to be those with lower habitual saturated fat and cholesterol intakes. In fact, such data were recorded in the baseline findings of the Multiple Risk Factor Intervention Trial [14].

Table III from the Western Electric Study takes the matter a step further, showing that change from baseline to year 1 in dietary saturates and cholesterol, and in combined dietary lipid score (also in body mass index) of the individuals was all significantly related to change in serum cholesterol from baseline to year 1 [66]. (Changes in polyunsaturates were negligible from 1958 to 1959, the probable reason for absence of a significant finding for this variable.) *Jacobs* et al. [25] have emphasized that,

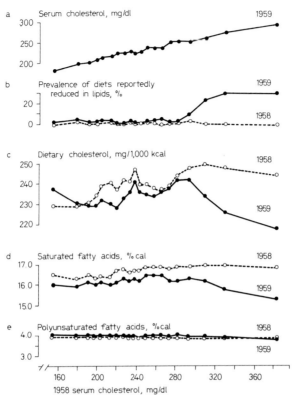

Fig. 2. Western Electric Study: Mean value for serum cholesterol at the second (1959) examination, percentage prevalence of diets reportedly reduced in fat and/or cholesterol at the first (1958) and second (1959) examinations, and mean values of dietary cholesterol, saturated fatty acids, and polyunsaturated fatty acids at both examinations, according to level of serum cholesterol at the first examination. For this purpose, the total group of 1,900 men was divided into 20 groups as equivalent in size as possible based upon the distribution of serum cholesterol concentration at the initial examination. The number of persons in each of the 20 groups, going from lowest to highest, was as follows: 96, 90, 100, 88, 101, 94, 91, 110, 99, 48, 126, 95, 100, 93, 91, 95, 97, 97, 93 and 96. The graphs for dietary cholesterol, saturated fatty acids, and polyunsaturated fatty acids were smoothed using a three-group moving average in order to compensate for excessive sampling variation from group to group [67].

when multiple factors relate to a variable, including intrinsic factors (under genetic influence) for which one can make no measurement, so that in cross-sectional static analyses one has a non-measurable dilution of a true relationship, analysis of effects of change is a much better way to examine the question of a relationship between two variables. In contrast to the

Table III. Relationship between change in dietary lipid intake and change in serum cholesterol of individuals. Entry to first annual examination, 1,900 men aged 40–55 at entry, linear regression analysis, Western Electric Co. Study 1958–1959

Variable: change, 1958–1959	Change in serum cholesterol, 1958–1959			
	3-variable analysis		4-variable analysis	
	coeff.	p	coeff.	p
Change in dietary sat. fatty acids, % cal	1.68	<0.001	1.58	<0.001
Change in dietary poly. fatty acids, % cal	−0.02	0.492	−0.38	−0.334
Change in dietary cholesterol, mg/1,000 cal	0.03	0.041	0.03	0.029
Change in body mass index, kg/m²	–	–	5.47	<0.001
Constant	−10.31		−11.37	
r	0.13	<0.001	0.18	<0.001

Regressor	Univariate analysis		Trivariate analysis	
	coeff.	p	coeff.	p
Change in dietary lipid score	0.64	<0.001	0.62	<0.001
Change in body mass index, kg/m²	–	–	5.53	<0.001
Age, years	–	–	0.10	0.627
Constant	−10.31		−16.37	
r	0.12	<0.001	0.18	<0.001

static look at a single set of data at a given point in time, the analysis of change over time eliminates – or at least markedly reduces – the interfering effect of genetically controlled interindividual differences in endogenous biochemical regulation.

Table IV, Parts A and B, show the relationship of the early examination findings in the Western Electric Study to 19-year risk of CHD and

Table IV. Relationship between dietary lipid intake and 19-year mortality from CHD and all causes[a], 1,884 men aged 40–55 at entry, Chicago Western Electric Co. Study

Baseline variable	Coronary heart disease			All causes		
	multiple logistic coefficient	z^b	p^c	multiple logistic coefficient	z^b	p^c
Part A						
Saturated fatty acids, % cal[d]	0.0320	0.846	0.398	0.0154	0.540	0.589
Polyunsat. fatty acids, % cal[d]	−0.2352	−2.236	0.025	0.0993	−1.332	0.182
Dietary cholesterol, mg/1,000 cal[d]	0.0032	2.436	0.015	0.0023	2.239	0.026
Serum cholesterol, mg/dl[d]	0.0056	3.508	<0.001	0.0031	2.456	0.014
Systolic pressure, mm Hg[d]	0.0234	5.285	<0.001	0.0212	6.204	<0.001
Cigarettes, number/day[e]	0.0384	5.653	<0.001	0.0382	7.323	<0.001
Age, years[e]	0.0945	5.211	<0.001	0.1057	7.750	<0.001
Body mass index, kg/m²[d]	0.0251	1.017	0.309	0.0136	0.727	0.468
Alcohol, drinks/month[e]	−0.0021	−1.321	0.187	0.0003	0.310	0.756
Constant	−12.3815			−11.0762		
Part B						
Dietary lipid score[d]	0.0264	2.427	0.015	0.0161	1.969	0.048
Serum cholesterol, mg/dl[d]	0.0056	3.501	<0.001	0.0030	2.377	0.018
Systolic pressure, mm Hg[d]	0.0242	5.468	<0.001	0.0220	6.415	<0.001
Cigarettes, number/day[e]	0.0379	5.590	<0.001	0.0376	7.200	<0.001
Age, years[e]	0.1026	5.674	<0.001	0.1104	8.111	<0.001
Body mass index, kg/m²[d]	0.0316	1.290	0.197	0.0192	1.033	0.302
Alcohol, drinks/month[e]	−0.0015	−0.996	0.319	0.0006	0.535	0.592
West.-North. European	−0.2136	−1.067	0.286	−0.1144	−0.769	−0.442
Middle European	−0.2192	−1.191	0.234	−0.2715	−1.943	0.052
Other ethnic backgrounds	−0.5226	−1.295	0.195	−0.5454	−1.799	0.072
Constant	−14.1058			−11.9370		

[a] 213 CHD deaths, and 448 deaths from all causes.
[b] Ratio of regression coefficient to its standard error.
[c] Two-tailed tests.
[d] Mean of entry and 1st annual examination for each man, 1958 and 1959.
[e] Entry examination, 1958. Dietary lipid score = $1.26(2S-P) + 1.5\sqrt{C}$.

all-causes mortality [66]. In both univariate and multivariate analyses, dietary cholesterol and polyunsaturates, and the combined dietary lipid score were related to 19-year risk of CHD death. While the relationship of dietary saturates to CHD mortality was not significant in the multivariate analyses, this is probably a reflection of the fact that over the range of observed saturated fat intake the P/S ratio changed little, so that – once polyunsaturates were considered in the analysis – saturates added little prognostic information. In any case, it is noteworthy that over the range of polyunsaturates recorded in 1958–1959 for the Western Electric men – $3.9 \pm 0.9\%$ of total calories, *not* a high polyunsaturated fat intake for any of the men – there was an inverse relationship between this variable and CHD mortality, i.e. higher poly fat intake was independently related to lower 19-year CHD mortality.

The findings of the Western Electric Study of dietary cholesterol merit emphasis: significant positive relationship to serum cholesterol at baseline, to change in serum cholesterol from baseline to first reexamination, to 19-year CHD mortality, to 19-year all-causes mortality (tables II-IV-A) [66] (see below).

In the multivariate analyses of the Western Electric data, the positive relationships of dietary lipid to CHD risk were independent of serum cholesterol, systolic pressure, cigarette smoking, age – all related highly significantly to CHD risk (table IV-A and B) [66]. This is a particular intriguing finding, because it has been tacitly assumed that dietary lipid influences long-term CHD risk solely via its effect on serum total cholesterol. This finding says that it does so over and above its effect on serum total cholesterol. How? Possibly by influencing LDL, HDL, and their subfractions [7, 14, 47–49]; possibly by affecting cell receptors for lipoproteins, with consequent changes in their internalization, metabolism, degradation, and role in atherogenesis [11, 20]; possibly in other ways as well, e.g. via effects on thrombogenesis? Clearly this merits further work.

Conclusion. With nutritional survey methods adequate validly to distinguish one person from another within a population, dietary lipid intake of individuals is significantly and independently related to serum cholesterol of individuals and to long-term CHD risk. Thus, these data from within-population epidemiologic research are complementary and consistent with data from cross-population studies, as well as animal-experimental findings on the important role of dietary lipid in the etiology of epidemic atherosclerotic disease.

Table V. Serum cholesterol and lipoprotein levels in 'macrobiotic' vegetarians and controls in Boston

Group	Serum cholesterol, mg/dl				weight, kg
	total	LDL	VLDL	HDL	
115 controls	184 ± 37	118 ± 34	17.2 ± 11.0	49 ± 12	73 ± 15
115 vegetarians	126 ± 30	73 ± 24	11.8 ± 7.0	43 ± 11	58 ± 9
Mean difference	58 ± 48***	45 ± 44***	5.4 ± 13.3***	6 ± 15***	15 ± 16***
% difference	−31.5	−38.1	−31.4	−12.2	−20.5
Mean difference¹	55 ± 53***	39 ± 46***	4.6 ± 14.4*	9 ± 17**	0 ± 9

* p ≤ 0.05; ** p ≤ 0.01; *** p ≤ 0.001.
¹ Weight-matched pairs (n=42).

Assertion No. 2: Recent Findings on the Association between the Lipoprotein Fractions of Serum Total Cholesterol – Particularly HDL – and CHD Risk Bring into Question Conclusions about the Diet-Heart Relationship that Were Reached Based on Serum Total Cholesterol and the Effects of Diet on It

Some who make this assertion question the propriety of nutritional recommendations to lower serum total cholesterol: Might not adverse effects supervene in levels of lipoprotein fractions, particularly HDL, with reduction in saturated fats and cholesterol, and (for overweight persons) reduction in calories, with partial replacement of saturated fats by complex carbohydrate and polyunsaturates, as repeatedly recommended to lower serum total cholesterol? Before dealing with this question explicitly, let it first be noted that about 70% of the plasma total cholesterol circulates bound to LDL, and LDL is the key atherogenic lipoprotein fraction. A diet that effects and sustains reduction in plasma total cholesterol can thus be expected to lower LDL cholesterol, and the research findings repeatedly have shown this to be the case. The dramatic impact of a Japanese-like diet, low in animal products, i.e. low in saturated fat and cholesterol, and at a caloric level avoiding overweight, in effecting low plasma total and LDL cholesterol, and low VLDL cholesterol as well, without significantly lower

Table VI. Diet composition and effects on serum lipids-lipoproteins. Controlled experiment with 12 healthy men aged 24–60

Variable	Reference diet A	Fat-modified diet B	High-fiber, fat-modified diet C	High-fat, fat-modified, high-fiber diet D
Protein, % cal	14	14	14	14
Vegetable protein, % cal	34	34	52	49
Fat, % cal	40	27	27	40
Linoleic acid, % cal	4.6	8.1	8.4	12.4
Polyunsaturates, % cal	5.2	8.5	8.7	12.8
Saturates, % cal	19.3	8.4	8.7	12.7
P/S ratio	0.27	1.01	1.00	1.01
Cholesterol, mg/2500 cal[1]	617	245	252	245
Available carbohydrates, cal[2]	46	59	59	47
Dietary fiber, g/2,500 cal[1]	19	20	55	43
Pectin, g/2,500 cal[1] as polygalacturonate	1.2	1.8	6.3	6.5
Change, serum lipids-lipoproteins				
Total cholesterol, %	–	– 21.6	– 29.2	– 24.6
LDL-cholesterol, %	–	– 26.5	– 34.5	– 31.5
HDL-cholesterol, %	–	– 12.0	– 10.6	– 5.5
HDL_2-cholesterol, %	–	– 34.6	– 11.5	– 23.1
Total triglycerides, %	–	0.0	– 20.8	– 26.4
VLDL triglycerides, %	–	+ 4.8	– 19.0	– 38.0
Ratios				
Total chol./HDL-chol.	4.49	4.02	3.56	3.56
LDL-chol./HDL_2-chol.	17.7	19.9	13.1	15.8

Alcohol intake was negligible.
[1] Energy intakes varied from 1,550 to 4,250 cal/day.
[2] Monosaccharides and disaccharides contributed 18–20% of calories in all four diets.

HDL cholesterol, and hence with a significantly higher ratio of HDL cholesterol to LDL cholesterol, is well illustrated in the report from a Boston commune (table V) [64].

A recent controlled isocaloric experiment with mixed ordinary foods yielded similar results (table VI) [42]. It is particularly valuable in demonstrating the multiple favorable changes, and their large order, with a diet moderately reduced in total fat, markedly reduced in saturated fat and cholesterol, moderately increased in polyunsaturated fats, complex carbohydrates, and fiber (diet C, table VI) – a marked fall in serum total cholesterol (–29.2%), LDL cholesterol (–34.5%), total triglycerides (–20.8%), VLDL triglycerides (–19.0%), with only a slight fall in HDL cholesterol (–10.6%) and HDL_2 cholesterol (–11.5%), therefore with consequent marked falls in the ratio of total cholesterol to HDL cholesterol (–20.7%) and in the ratio of LDL cholesterol to HDL_2 cholesterol (–26.0%). Note the similarity of the data in tables V and VI – from a free-living group spontaneously electing to eat a diet of this type and from a controlled experiment, respectively. Note also the degree of fall in total cholesterol, LDL cholesterol, and VLDL in both groups. As to the slight fall in HDL cholesterol, note the favorable change in the ratios – of total cholesterol to HDL cholesterol, LDL cholesterol to HDL cholesterol, and LDL cholesterol to HDL_2 cholesterol. It is likely that the small decrease in HDL cholesterol and HDL_2 cholesterol is simply a reflection of the lesser amounts of cholesterol that the putatively protective lipoprotein molecule – i.e. the HDL, or more precisely HDL_2, per se – has to carry with the more favorable diet. In fact, it is reasonable to speculate (pending data) that no change occurs in the level of HDL and especially HDL_2, i.e. of lipoprotein (as distinct from its more easily measured cholesterol, used to approximate the lipoprotein).

The experiences on a massive scale, over the course of 4 years, of the Multiple Risk Factor Intervention Trial (MRFIT) are consistent with the foregoing observations and, in addition, illuminate a key facet, i.e. the critical importance of weight reduction when people are obese. For the obese high-risk men in the Special Intervention Group of MRFIT, loss of 10 or more pounds on its fat-modified diet (reduced in cholesterol and saturated fat) resulted in a significant fall in plasma total cholesterol, LDL cholesterol, triglycerides (VLDL), and a significant increase in HDL cholesterol (table VII) [14]. The fall in serum total cholesterol was substantially greater than expected based on change in dietary lipid composition alone, i.e. moderate weight loss has a sizable enhancing effect.

The importance of prevention and correction (at least partial) of overweight, as well as of lipid composition of the diet, merits emphasis since it is often neglected. The Western Electric Study data cited above, as well as the findings of the National Diet-Heart Study (ND-HS) (tables VIII and IX), all indicate this [14]. Analyses of both the ND-HS and MRFIT data show that, on a fat-modified diet, reduced in saturated fat and cho-

Table VII. Serum-plasma lipids-lipoproteins, special intervention group, Multiple Risk Factor Intervention Trial, baseline and means of years 1–4 or 2, 4, by weight change

Variable	Subgroup				
	weight loss $\geq$ 10 lbs[1]	weight loss 5–9 lbs	little weight change	weight gain $\geq$ 5 lbs	all
Serum total cholesterol					
Number of men	1,489	982	1,906	1,064	5,467
S1, mg/dl	258.1	255.4	253.5	248.3	254.1
x̄ 1–4, mg/dl	233.4	236.6	238.7	239.1	237.0
Change, mg/dl	–24.7	–18.8	–14.8	–9.2	–17.1
% change	–9.6	–7.4	–5.8	–3.7	–6.7
Plasma total cholesterol					
Number of men	1,532	1,005	1,941	1,099	5,607
S2, mg/dl	243.9	240.9	239.9	235.1	240.3
x̄2, 4, mg/dl	224.3	227.8	230.4	231.1	228.4
Change, mg/dl	–19.7	–13.1	–9.5	–4.0	–11.8
% change	–8.1	–5.4	–4.0	–1.7	–4.9
Plasma triglycerides					
Number of men	1,532	1,005	1,941	1,099	5,577
S2, mg/dl	203.9	206,5	188.9	179.1	194.3
x̄2, 4, mg/dl	164.2	185.0	188.4	213.4	186.1
Change, mg/dl	–39.7	–21.5	–0.5[2]	+34.3	–8.2
% change	–19.5	–10.4	–0.3	+19.2	–4.2
Plasma LDL cholesterol					
Number of men	1,512	982	1,918	1,087	5,528
S2, mg/dl	162.1	159.5	160.4	158.0	160.2
x̄2, 4, mg/dl	147.9	148.7	151.1	149.3	149.4
Change, mg/dl	–14.2	–10.8	–9.3	–8.7	–10.8
% change	–8.8	–6.8	–5.8	–5.5	–6.7
Plasma HDL cholesterol					
Number of men	1,512	982	1,918	1,087	5,528
S2, mg/dl	42.3	41.9	42.4	42.0	42.2
x̄2, 4, mg/dl	44.4	43.4	42.4	41.0	42.8
Change, mg/dl	+2.1	+1.5	0.0[2]	–1.0	+0.6
% change	+5.0	+3.6	0.0	–2.4	+1.4

[1] In this table on weight change, the values of weight were: for baseline, weight at the second screening (S2) visit; for the intervention period, weight at the fourth annual visit.
[2] Not statistically significant (i.e. p > 0.05).

Table VIII. Weight loss and serum cholesterol fall, diet B, National Diet-Heart Study

Loss of weight, %	Serum cholesterol fall, %	
	Open Centers	Faribault
> 3.7	13.7	19.5
1.2–3.7	11.6	19.3
< 1.2 or gain	6.9	14.1
Ratio	13.7/6.9 = 1.99	19.5/14.1 = 1.38

Table IX. Change in nutrient intake, weight, and serum cholesterol (%), multiple linear regression analysis, National Diet-Heart Study, Open Centers

Independent variable	Regression coefficient	
	average of diets B, C, D	pool of diets B, C, D
△ SF, % cal	0.36*	0.33
△ PF, % cal	–0.41*	–0.77*
△ $\overline{V}$ · mg chol./1,000 cal	0.25	0.24*
% △, wt, remote	0.61*	0.59*
% △, wt, recent	0.79*	0.70*
Entry serum cholesterol, mg/dl	–0.036*	–0.040*

* $p \leq 0.05$.
△ is change in.

lesterol, modest weight loss by obese individuals enhances reduction of serum cholesterol – over and above the effect of change in dietary lipid composition – by 30–40% (cf. the experience of the closed population in the ND-HS, Faribault, table VIII) [55]. Moreover, as recent experience shows, weight loss with a fat-modified diet contributes to favorable responses of all components of the serum lipids-lipoproteins, including HDL [14].

Conclusion. The recent resurgence of research on fractions of serum total cholesterol, i.e. on serum lipoproteins, has yielded data complementing and refining – not contradicting – the extensive data on the interrelationship between diet and serum total cholesterol. Long-standing dietary recommendations for prevention of CHD remain valid. Based on the extensive positive experiences of Mediterranean and Far Eastern countries, as well as on controlled experiments in man and animals, these recommendations emphasize moderate reduction in total fat, marked reduction in saturated fat and cholesterol, moderate (not marked) increase of polyunsaturated fat, moderate increase in complex carbohydrates and fiber (from vegetables, fruits, whole grain products, legumes), foods of low caloric density to assist in avoiding or correcting obesity while assuring a high intake of all essential macro- and micronutrients, and avoidance of high intake of sodium and alcohol. When these recommendations are followed, the entire lipid-lipoprotein spectrum is influenced favorably.

Assertion No. 3: Diet Recommendations to Influence Serum Cholesterol and CHD Risk Are Able To Effectuate only Negligible and Insignificant Changes

In regard to ability to effectuate marked reductions in serum total cholesterol and sizable changes in all lipids-lipoproteins in a favorable direction in terms of CHD risk, the data cited in tables V–VIII demonstrate unequivocally that the foregoing assertion is invalid. It is also contrary to the extensive evidence from cross-population studies in man [28, 30, 32–36, 52, 69, 74, 93], e.g. the large diet-related differences in mean serum cholesterol recorded among the cohorts of the Seven Countries Study and the Ni-Hon-San Study and their relation to CHD risk (fig. 3; table X) [28, 30, 33–36]. This assertion is also contrary to the results of all the well-controlled experiments in man, including not only those of *Lewis* and colleagues, summarized in table VI, but also those of *Ahrens* and colleagues, *Connor* and colleagues, *Hegsted* and colleagues, *Keys* and colleagues, *Mattson* and colleagues, to cite the main US research groups [22, 32, 38, 51, 69, 74, 83, 93]. Table XI gives three linear regression equations summarizing experience of three investigative teams [22, 38, 51]. Perusal of these equations and accomplishment of a few simple calculations based on them make it immediately and abundantly clear that even moderate changes in composition of dietary lipid result in sizable changes in serum cholesterol (see below). Moreover, as already noted above, and as the equations predict, with consistent and sizable approaches to enjoying the pleasures of eating Mediterranean and Far Eastern style, the changes are large indeed – even more so (by 30–40% for serum total cholesterol, by

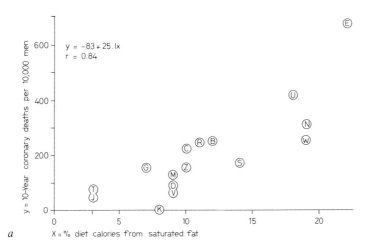
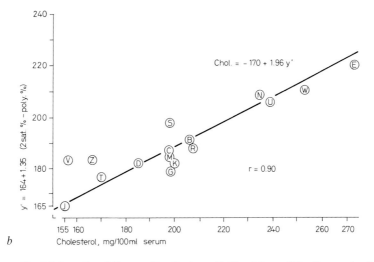

Fig. 3. International Cooperative Study on Epidemiology of Cardiovascular Disease: 10-year coronary death rates of the cohorts plotted against the percentage of dietary calories supplied by saturated fatty acids *(a);* relation of mean serum cholesterol concentration of the cohorts at entry to fat composition of the diet expressed in the multiple regression equation derived from controlled dietary experiments in Minnesota *(b);* CHD age-standardized 10-year death rates of the cohorts versus the median serum cholesterol levels (mg/dl) of the cohorts. All men were judged free of CHD at entry. The coefficient of correlation is r = 0.82 *(c).* The cohorts are: B = Belgrade; C = Crevalcore; D = Dalmatia; E = east Finland; G = Corfu; J = Ushibuka; K = Crete; M = Montegiorgio; N = Zutphen; R = Rome railroad; S = Slavonia; T = Tanushimaru; U = American railroad; V = Velika Krsna; W = west Finland; Z = Zrenjanin [33, 36].

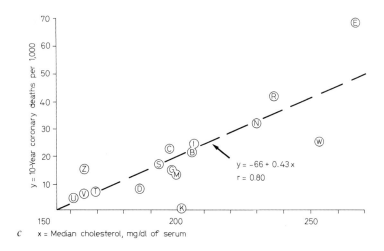

even a higher percentage for serum triglycerides and VLDL) when some loss of excess weight is part of the process of acquiring better eating habits.

As to how much reduction in CHD risk can be anticipated from improved nutrition leading to lower levels of serum cholesterol in the population, both the cross-population and the within-population studies give information permitting estimates of this. Data from comparisons of the experience of middle-aged men in different countries are given in figure 3 and tables X and XII (28, 30, 34, 36). In the early 1960s, estimates were made based on the follow-up experience of the Framingham study (table XIII) [16, 55]. These data indicate that a 5% reduction in serum cholesterol could accomplish a 12.8% fall in CHD incidence; a 10% decrease, a 24.4% fall. Additional data, from five major US prospective studies, combined in the analyses of the national cooperative Pooling Project, yield a more conservative, but still sizable, estimate of the preventive potential from this approach (table XIV) [57, 75]. They indicate that about one quarter of first major coronary events in middle-aged white American men are attributable to elevated levels of serum cholesterol, defined as levels of about 220 mg/dl and above. By inference, shifting the population distribution of serum cholesterol downwards, so that all or almost all persons would have levels under 220 mg/dl, would of itself yield a 25% decrease in incidence of premature CHD. This would mean a shift from the mean levels in the order of 235 ± 45 mg/dl, prevailing for middle-aged American men in the late 1940s, 1950s, and early 1960s, to

Table X. Findings in the Ni-Hon-San Study of Japanese men in Japan, Japanese-American men in Hawaii, Japanese-American men in California

Variable	Japan	Hawaii	California
Dietary acculturation score[a] (% of population)			
0–29	0.2	13.5	24.1
30–69	14.4	75.8	69.7
70+	85.4	10.8	6.2
Age, years	56.9[b] ± 5.8	54.4 ± 5.6	52.4 ± 6.1
Height, cm	160.7 ± 6.0	162.8 ± 6.0	164.0 ± 5.2
Weight, kg	55.2 ± 9.0	63.4 ± 9.6	65.9 ± 9.2
Relative weight[c]	109.1 ± 15.8	121.9 ± 16.1	126.0 ± 14.6
Back skinfold	10.4 ± 4.9	16.5 ± 6.8	16.3 ± 8.0
Total calories	2,164 ± 619	2,275 ± 736	2,262 ± 695
Calories/weight	39.8 ± 11.7	36.5 ± 12.4	35.0 ± 11.7
Total protein, g	76.8 ± 27.3	94.0 ± 35.4	88.5 ± 26.5
Animal, g	39.8 ± 22.8	70.5 ± 32.7	66.0 ± 24.4
Vegetable, g	37.1 ± 13.2	23.5 ± 10.1	22.5 ± 8.6
Total fat, g	36.6 ± 20.4	85.1 ± 38.9	94.8 ± 36.4
Saturated[d], g	16.0 ± 13.3	59.1 ± 32.7	66.3 ± 30.5
Unsaturated, g	20.6 ± 13.7	26.0 ± 19.8	28.5 ± 20.4
Cholesterol, mg	464.1 ± 324.4	545.1 ± 316.4	533.2 ± 297.8
Total carbohydrate, g	339.2 ± 109.8	260.2 ± 96.2	251.3 ± 99.8
Simple, g	61.1 ± 37.4	91.6 ± 54.7	96.4 ± 53.9
Complex, g	278.2 ± 104.4	168.7 ± 73.7	154.9 ± 66.1
Alcohol, g	28.9 ± 39.8	13.3 ± 30.6	8.8 ± 24.2
Protein, % cal	14.3 ± 3.4	16.7 ± 4.0	16.2 ± 4.1
Animal, % cal	7.4	12.4	11.7
Vegetable, % cal	6.9	4.1	4.0
Fat, % cal	15.1 ± 6.9	33.3 ± 9.4	37.5 ± 8.1
Saturated[d], % cal	6.7	23.4	26.4
Unsaturated, % cal	8.6	10.3	11.3
Carbohydrate, % cal	63.0 ± 11.2	46.4 ± 11.0	44.2 ± 9.4
Simple, % cal	11.3	16.1	17.0
Complex, % cal	51.4	29.7	27.4
Alcohol, % cal	8.9 ± 11.5	3.7 ± 7.8	2.5 + 6.0
Serum cholesterol, mg/dl	181.1 ± 38.5	218.3 ± 38.2	228.2 ± 42.2
Serum cholesterol ≥ 260 mg/dl, per 1,000	31.6	124.0	162.5
Serum triglycerides, mg/dl	133.8 ± 87.1	239.9 ± 161.1	233.7 ± 144.4
Age-adjusted CHD prevalence rates (per 1,000)			
Definite CHD by EKG	5.3	5.2	10.8

Table X. Continued

Variable	Japan	Hawaii	California
Definite + possible CHD by ECG	25.4	34.7	44.6
Angina pectoris by questionnaire	11.2	14.3	25.3
Possible infarction by questionnaire	7.3	13.2	31.4
Average annual mortality rate from CHD, 4 to 5-year follow-up, men aged 50–64, per 1,000[e]	1.3	2.2	3.7
Age-adjusted incidence rates of non fatal myocardial infarction and CHD death, men aged 45–64, per 1,000 person-years Japan-Hawaii comparison	1.4	3.0	–
Hawaii-California comparison	–	2.8	4.3
CHD incidence and serum cholesterol			
Cases (mean), mg/dl	209.5	232.9	–
Non-cases (mean), mg/dl	181.0	218.0	–
Standardized multivariate logistic regression coefficients (CHD incidence)			
Serum cholesterol, mg/dl	0.410*	0.318**	–
Systolic pressure, mm Hg	0.465*	0.432**	–
Cigarette smoking per day	0.009	0.626**	–
Relative weight	0.270	0.109*	–
Age, years	0.547*	0.340**	–

[a] A high score indicates a high proportion of the total diet coming from typically Japanese foods.
[b] Mean and standard deviation; where standard deviations are not given, the means were calculated from the published data.
[c] Ratio of observed weight to ideal weight for height and sex, × 100.
[d] Principally saturated fats; saturated fatty acids as such reported in the range 10.8–12.8% of calories for Hawaii men of varying ages, 12.4–15.2% for California men; polyunsaturated fatty acids, 5.3–6.3% and 5.5%–6.3%, respectively; no such data given for Japanese men in Japan.
[e] * $p < 0.05$; ** $p < 0.01$. Average of rates for ages 50–54, 55–59, 60–64.

Table XI. Equations to predict diet-induced serum cholesterol change

$\triangle$ Cholesterol = 1.35 (2 $\triangle$ S- $\triangle$ P) + 1.5 $\triangle$ Z* *(Keys)*
$\triangle$ Cholesterol = 2.16 $\triangle$ S–1.65 $\triangle$ P + 6.77 $\triangle$ C**–0.5 *(Hegsted)*
$\triangle$ Cholesterol = 1.60 + 0.118 $\triangle$ C*** *(Mattson)*

*Z = Square root of diet cholesterol in mg per 1,000 cal
**C = Diet cholesterol in dg per day
***C = Diet cholesterol in mg per 1,000 cal

The Mattson equation is univariate since it was derived from a metabolic ward study with total fat, saturated fat, and polyunsaturated fat held constant at 40, 16 and 5% of calories, respectively (see table XX).

Table XII. Dietary fat, serum cholesterol and 10-year CHD mortality rate, men originally aged 40–59, Seven Countries Study

Cohort	Total fat % cal	Sat. fat % cal	Poly. fat % cal	2 S-P	Serum cholesterol		10-year mortality	
					median	90th %	CHD	all
Greece	36	7	3	11	201	258	9	57
Yugoslavia	31	5	5	5	171[1]	219[1]	12[1]	105[1]
Italy	26	9	3	15	198	253	21	126
Rome RR	–	–	–	–	207	260	22	75
US RR	40	18	5	31	236	294	57	115
Finland	37	20	3	37	259	323	65	167
Netherlands	40	18	5	31	230	291	44	125

[1] Excludes Belgrade and Slavonia.
2 S-P: S is percentage of calories from dietary saturated fatty acids; P: the percentage of calories from polyunsaturated fatty acids; this expression is, therefore, a measure of the combined impact of these two components on serum cholesterol of the group. RR = Railroad.

170 ± 25 mg/dl, i.e. a decrease overall of about 28%. A fall of one third or one half as much would yield smaller, but still substantial, decreases in CHD rates. These estimates are consistent with the data in tables II–IV-B. They are also in keeping with results of the few small-scale randomized controlled trials that have been done (see below).

Table XIII. Estimated relative reduction in CHD incidence (v) associated with relative reduction in serum cholesterol (u)

Relative decrease in serum cholesterol (u)	Relative decrease in coronary heart disease incidence (v)	Relative decrease in serum cholesterol (u)	Relative decrease in coronary heart disease incidence (v)
0.01	0.026	0.26	0.551
0.02	0.052	0.27	0.567
0.03	0.078	0.28	0.583
0.04	0.103	0.29	0.598
0.05	0.128	0.30	0.613
0.06	0.152	0.31	0.627
0.07	0.176	0.32	0.642
0.08	0.199	0.33	0.655
0.09	0.222	0.34	0.669
0.10	0.244	0.35	0.682
0.11	0.267	0.36	0.695
0.12	0.288	0.37	0.707
0.13	0.310	0.38	0.720
0.14	0.330	0.39	0.731
0.15	0.351	0.40	0.743
0.16	0.371	0.41	0.754
0.17	0.391	0.42	0.765
0.18	0.410	0.43	0.776
0.19	0.429	0.44	0.786
0.20	0.448	0.45	0.796
0.21	0.466	0.46	0.806
0.22	0.484	0.47	0.815
0.23	0.501	0.48	0.824
0.24	0.518	0.49	0.833
0.25	0.535	0.50	0.842

Values of v were obtained from the relationship, $v = 1 - (1-u)^{2.66}$.

As to the feasibility of achieving such reductions in serum cholesterol by dietary means, see tables V–VII or, more importantly, in terms of whole populations, and the cardinal task of controlling the epidemic of premature CHD, note the experience in the United States over the last 15 years. This above all, and decisively, refutes the assertion that it is not possible to accomplish much by dietary means. It is now more than 20 years since the first public appeals were made in the USA urging modification of life-

Table XIV. Serum cholesterol and risk of a first major coronary event between ages 40–64, 8,274 white men, pool 5, Pooling Project, final report

Quintile of level and level, mg/dl		Number of events	Risk of an event per 1,000	Relative risk	Absolute excess risk per 1,000	Percent of all excess
I + II	≤218	166	162.7	1.00	–	–
I	≤194	86	172.4	–	–	–
II	194–218	80	153.0	–	–	–
III	218–240	104	186.8	1.15	24.1	8.3
IV	240–268	167	266.3	1.64	103.6	35.8
V	>268	210	324.1	1.99	161.4	55.8
All		647	222.4	–	–	–

$$\frac{\text{QIII–QV excess events, 3,000 men}}{\text{All events, 5,000 men}} = \frac{289.1}{1,112.0} = 25.6\%$$

of all events are excess events, attributable to hypercholesterolemia.

styles in order to control the major risk factors for premature CHD [2, 87]. Over the years, there have been repeated statements of expert groups to the public of this type – on lipid composition of the diet, weight control, avoidance and cessation of smoking, improvement of cardiopulmonary fitness, and control of high blood pressure. These have come from many sources, professional, voluntary, and public (e.g. the American College of Cardiology, the American Heart Association, the Inter-Society Commission for Heart Disease Resources, the White House Conference on Nutrition, the Advisory Committee to the Surgeon General on Smoking and Health, the Senate Select Committee on Nutrition and Human Needs, the National Heart, Lung, and Blood Institute, and recently in regard to nutrition jointly from the US Department of Agriculture and the Department of Health and Human Services) [2, 3, 24, 72, 78, 79, 86, 87, 93].

Extensive evidence is now available indicating that these repeated urgings, widely disseminated although weakly financed as public health efforts, have been heard and listened to by large segments of the American people. Data are now available from several sources indicating that 'critical mass' has been reached, i.e. changes have been widespread and extensive enough to be registered in overall data collected on the whole population and on representative samples. For example, two recent US

Table XV. Information sources influencing household dietary change

	Influential sources[1]	Most influential source, %[2]
Doctor/dentist/nurse	56	38
Magazine	32	7
Television	29	5
Newspaper	27	4
Friend/relative	26	7
Diet/health book	23	4
Food label	21	2
Dietician/nutritionist	13	3
Diet group	10	4
Radio	10	1
Health food store	9	1
Government publication/pamphlet	8	1
Extension worker/public health educator	4	1
Other influences	7	3
No influence/I figured it out myself	9	9
Don't know/no answer	7	10

[1] Percentages add to more than 100 because respondents were allowed multiple answers.
[2] Percentages are based on responses from 862 households (64% of the total sample) which made dietary changes for reasons of health or nutrition in the 3 years prior to the survey.

Department of Agriculture (USDA) surveys indicate that by 1977 fully one half and by 1980 fully two thirds of the population had made nutritional changes for health reasons, first and foremost related to major CHD risk factors [26, 27, 58]. Data from the second of these surveys indicate that health professionals and the mass media play important roles in stimulating people to improve their eating habits (table XV) [58].

These survey findings are in accord with trends in annual food disappearance data from the USDA, at least in regard to dairy and egg fat, and the highly saturated visible fats (butter, lard) [71]. Over the last 25 years, per capita availability of all of these has declined sizably. Correspondingly, the periodic Household Food Surveys of the USDA also indicate substantial decreases in consumption of the main sources of saturated fat and cholesterol – meat fat, egg fat, dairy fat, the visible fats, and commercial baked goods (table XVI) [61]. Studies of local samples of the population

Table XVI. Intake of foods and nutrients per person by age and sex, 1965 and 1977, and percent change, 1977 compared to 1965, US nationwide food consumption surveys

Food or nutrient	age 25–34							age 35–50							age 51–64						
	men			women				men			women				men			women			
	1965	1977	%Δ	1965	1977	%Δ		1965	1977	%Δ	1965	1977	%Δ		1965	1977	%Δ	1965	1977	%Δ	
Milk and milk drinks	318	243	−23.6	204	182	−10.8		236	203	−14.0	152	130	−14.5		203	180	−11.3	151	139	− 7.9	
Cheese	11	21	+90.9	10	19	+90.0		13	18	+38.5	13	18	+38.5		14	17	+21.4	14	19	+35.7	
Eggs	55	38	−30.9	27	26	− 3.7		51	41	−19.6	31	23	−25.8		51	36	−29.4	33	24	−27.3	
Beef	110	89	−19.1	64	50	−21.9		102	79	−22.5	57	51	−10.5		81	74	− 8.6	54	54	0.0	
Pork	98	55	−43.9	54	32	−40.7		82	52	−36.6	50	32	−36.0		91	57	−37.4	49	30	−38.8	
Poultry	32	31	− 3.1	21	25	+19.0		33	32	− 3.0	25	25	0.0		28	33	+17.9	25	27	+ 8.0	
Fish	14	14	0.0	9	10	+11.1		13	17	+30.8	13	14	+ 7.7		18	22	+22.2	9	12	+33.3	
Beef, pork, poultry, fish	254	189	−25.6	148	117	−20.9		230	180	−21.7	145	122	−15.9		218	186	−14.7	137	123	−10.2	
Fats, oils	42	18	−57.1	23	15	−34.8		39	19	−51.3	23	14	−39.1		35	18	−48.6	24	15	−37.5	
Calories	2,917	2,449	−16.0	1,803	1,616	−10.4		2,632	2,314	−12.1	1,652	1,514	− 8.4		2,422	2,148	−11.3	1,619	1,522	− 6.0	
Protein	118.6	98.1	−17.3	72.3	65.9	− 8.9		106.2	95.6	−10.0	68.3	63.9	− 6.4		98.0	90.1	− 8.1	67.4	65.2	− 3.3	
Fat	146.1	114.8	−21.4	86.5	73.7	−14.8		132.4	109.3	−17.4	80.2	70.8	−11.7		121.3	101.6	−16.2	79.9	71.2	−10.9	

%Δ is percent change: (1965−1977)/1965 .

Table XVII. Baseline nutrient intake, Framingham, ND-HS and MRFIT men

Nutrient	MRFIT 12,847 men	Framingham 864 men	ND-HS 1,196 men
Year(s)	1974–76	1967–1970	1963
Calories, per day	2,488	2,608	2,565
Protein, % cal	16.2	15.8	15.6
Carbohydrate, % cal	37.9	38.6	40.7
Alcohol, % cal	7.4	6.7	4.1
Total fat, % cal	38.3	39.0	40.4
Saturated fatty acids, % cal	14.0	15.0	15.6
Monounsaturated fatty acids, % cal	14.8	15.5	17.6
Polyunsaturated fatty acids, % cal	6.4	5.4	3.9
Cholesterol, mg/day	451	530	533

indicate similar trends (e.g. the results of screening in the 1970s for the MRFIT, compared with those of the 1960s for the Framingham and National Diet-Heart Study populations) (table XVII) [14, 21, 55].

As expected from the apparent reduction in intake of saturated fat and cholesterol and the modest increase in polyunsaturated fat, population surveys also show with a high degree of consistency that the serum cholesterol levels were lower in the 1970s compared to the preceding decades (table XVIII; fig. 4) [8, 13, 40, 74]. Evidence is available indicating that these changes have been made to a greater degree by the more highly educated strata of the adult population compared to the less highly educated. A similar situation exists in regard to cigarette smoking, i.e. the proportion of adults smoking cigarettes has declined steadily since the *Report to the Surgeon General on Smoking and Health* in 1964, for both men and women, more so for men, and more so for the more highly educated of both sexes compared to the less highly educated. The same is true for the mass adoption of rhythmic (isotonic) exercise. As to the control of high blood pressure, remarkable progress has been made across the board (i.e. regardless of socioeconomic and ethnic situation), with the proportion of hypertensives detected, treated, and controlled changing from a low 10–15% in the late 1960s and early 1970s to 45–70% by the last years of the 1970s.

Table XVIII. Trend of serum cholesterol: middle-aged US men

Study	1950s to early 1960s		Late 1960s to 1970s	
	number	mean	number	mean
Albany	1,712	231.4	455	215.2
Chicago	3,242	243.4	12,337	211.4
Framingham	1,344	226.6	506	219.4
Minneapolis-St. Paul	630	232.9	628	222.6
NW Railroad	2,424	230.5	–	–
Tecumseh	685	231.0	595	211.0
NHES-HANES	1,634	230.0	2,027	226.3
LRC	–	–	11,708	209.3
MRFIT	–	–	370,599	219.1
All	11,671	233.4	398,855	216.8 △

△ 7.1% Lower.

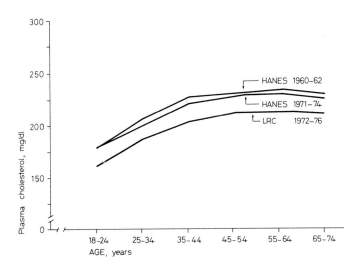

Fig. 4. Mean cholesterol levels by age. Comparison of Health and Nutrition Survey (HANES) and Lipid Research Clinics (LRC) survey [40].

Table XIX. Decline in mortality rates: United States, 1968–1978, persons aged 35–74

Statistic	Coronary heart disease	Cerebrovasc. disease	Major cardiov. dis.	All causes
1968 Rate	598.4	148.1	850.8	1,624.9
1978 Rate	434.5	88.7	615.4	1,323.0
Difference	−163.9	−59.4	−235.4	−301.9
% difference	−27.4	−40.1	−27.7	−18.6
Lives saved [1] 1978	113,908	40,248	163,520	216,206
Lives saved 1969–1978	568,050	185,523	804,359	1,092,638

ICD rubrics 410–413, 430–438, 390–448, 8th revision.
Rates, per 100,000 population, are averages of the rates for persons aged 35–44, 45–54, 55–64, 65–74.
[1] Based on expected deaths if the 1968 age-specific mortality rates had continued to prevail.

These several favorable trends in life-styles and life-style-related major CHD risk factors, sizable enough to be detectable for the whole US population, were accompanied over the years 1968–1978 by a steady and substantial decline in mortality from CHD, encompassing all age groups that have been experiencing the coronary epidemic, young adult, middle-aged, and elderly. For the age group 35–74, this had reached almost 30% by the end of the decade (table XIX) [79, 90, 92]. Concomitantly, an even greater decline in mortality from the cerebrovascular diseases was recorded and, given that these two entities make up most of the major cardiovascular disease rubric, there has also been a marked fall in mortality from this broad set of diseases, attributable chiefly to the impact of hypertension and severe atherosclerosis. As a consequence, all-causes mortality, the ultimate index, has also fallen substantially, by about 20%. The slopes of the declines in the mortality rates were significantly steeper for the years 1973–1978 than for 1968–1973, i.e. the decrease in mortality accelerated significantly in the latter years of the 1968–1978 period (fig. 5) [92]. Altogether, over the decade 1969–1978, about 1,100,000 lives were

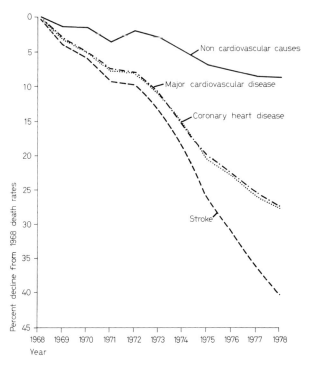

Fig. 5. Trends in cardiovascular disease and non-cardiovascular causes of mortality: decline by age-adjusted death rates, ages 35–74, 1968–1978, United States [92].

saved in the age range 35–74, about 800,000 of them due to the declines in the mortality rates from the major cardiovascular diseases.

The decline in mortality from premature CHD is especially noteworthy since over the prior decades CHD mortality rates rose steadily, particularly for men [92]. It is further remarkable given the additional fact that for many other countries that have been experiencing significant mortality rates from premature CHD, including countries of eastern, northern, and western Europe, the rates continued to rise during the 1970s [90, 92]. Furthermore, among those countries also experiencing a decline (e.g. Australia, Canada), the United States has set the pace, with the largest absolute and relative falls.

Are the declines in mortality due, at least in part, to the improvements in life-style and life-style-related major risk factors? That is not a question amenable to unequivocal answer, given the limitations in available data.

Table XX. Expected decrease in CHD and all-causes mortality based on changes in risk factors, 1950s–1970s, US white men, x̄ age 50 years

Variable	Pooling Project 1950s	Population studies 1970s	% Change
x̄ serum cholesterol	235	210	
x̄ diastolic pressure[1]	86.0	84.3	
% smoking cigarettes	55	40	
Expected mortality rate per 1,000 in 8.6 yrs			
CHD	50.4	39.3	−22.0%
All causes	120.4	103.8	−13.8%

[1] 20% ≥ 95 mm Hg, with ≤ 10% of these controlled in 1950s, 50% in 1970s; x̄ level of controlled hypertensives 85 mm Hg.

Thus, only limited evidence is available as to whether the decline in mortality reflects a decline in disease incidence, or in the case-fatality rate, or in both, although available data suggest that both occurrence of severe atherosclerosis and incidence of acute myocardial infarction plus sudden death have lessened [19, 79]. Simple calculations indicate that the changes in mortality rates from CHD and all causes predicted from findings of prospective epidemiological studies, given the population changes in the major risk factors, correspond closely to the actual observed changes. Thus, the predicted declines for middle-aged white men are 22.0 and 13.8% for CHD and all causes, respectively (table XX) [13]; the observed changes are 25.6 and 18.0%, respectively [92]. Further calculations yield the estimates that the population fall in serum cholesterol (table I) can account for about 40% of the total decline in CHD mortality [13].

There are other sets of data indicating concordance between expected and observed findings. To cite two more: First, given that – in regard to diet, smoking, exercise – the more educated have changed their life-styles more than the less educated, one would expect this former stratum to have experienced a larger decline, at least for the white sector of the population (the situation for blacks is more complex, given their particular problem of hypertension). The one set of available data is in accord with this expectation, i.e. the more highly educated Metropolitan Life Insurance middle-aged white policy holders have indeed experienced a greater decline in

CHD mortality than the general population of age-sex-matched US whites [5].

Second, on the international level, for 20 economically developed countries, statistically significant associations have been demonstrated between trends of life-styles (e.g. intake of animal products and of dietary cholesterol, per capita consumption of cigarettes) and trends of CHD mortality [13].

Of course, these facts do not by themselves constitute definitive proof that the decline in premature mortality from CHD and the major cardiovascular diseases is due primarily to changes in life-styles and life-style-related major risk factors. As already noted, solid proof of this relationship is very hard to come by, in fact probably impossible with the limited available trend data. Nonetheless, the above-cited consistencies are noteworthy. It is the judgment of many experts that it is reasonable to infer a cause-and-effect relationship, accounting at least in part for the observed improvement in life expectancy. It seems that in at least one major area of health policy and action the approaches in the United States have been sound and merit continuation and expansion.

Conclusion. The assertion that diet recommendations to influence serum cholesterol and CHD risk are ineffectual is not in keeping with findings from cross-population and within-population epidemiologic studies, controlled experiments in man, and mass public health experience (national and international) during the last 10–15 years.

Assertion No. 4: Dietary Cholesterol Is of Little or No Consequence in Regard to Serum Cholesterol and CHD Risk in Man

The research of recent years has yielded important new data reinforcing previous conclusions [68–70] that this assertion is completely invalid. Reference has already been made here to some of the key data sets. They include the data of the Western Electric Study showing that dietary cholesterol at baseline was significantly related to 19-year risk of mortality from both CHD and all causes, independent of dietary saturated and polyunsaturated fatty acids, serum cholesterol, blood pressure, cigarette use, body mass index, age, and alcohol use (table IV-A) [66]. They include the Western Electric Study data on the significant independent relationship between dietary cholesterol and serum cholesterol at entry, and between change in dietary cholesterol and change in serum cholesterol from entry to year 1 (tables II, III) [66]. These data showing a significant effect of dietary cholesterol on serum cholesterol, independent of dietary saturated and polyunsaturated fatty acids, are consistent with findings from other studies on man, including controlled experiments both in free-living and institutionalized groups (tables IX, XI) [22, 38, 51, 54, 55]. Scores of painstakingly executed controlled experiments underlie the equations of *Keys* et al. [38] and *Hegsted* et al. [22] (table XI). Table XXI gives the data from the experiment of *Mattson* et al. [51] (cf. table XI). Similar effects of dietary cholesterol on serum cholesterol have been recorded in controlled experiments using natural foods rather than formula diets [83]. The available data do not fully clarify whether the relationship between dietary cholesterol and serum total cholesterol is linear or curvilinear (table XI): whether there is a ceiling somewhere between 700 and 1,000 mg/day, beyond which little or no additional effect occurs, at least in studies lasting only a few weeks (cf. the 'experiment of nature' involving daily ingestion of several eggs for long periods of time, with consequent severe hypercholesterolemia and xanthomatosis) [59]. The extant data also do not fully resolve the issue as to whether or not there is interaction between dietary cholesterol and neutral fat in influencing serum cholesterol, i.e. the possibility that the effect of dietary cholesterol is greater with a high saturated diet than with a high polyunsaturated fat diet. The findings of *Keys* et al. [38], *Hegsted* et al. [22] and *Connor* et al. [15] do not indicate interaction (table XI) [83], but data from the closed population in the National Diet-Heart Study and from at least one other experiment suggest interaction [10, 55]. Also, the available data account only to a limited degree for the observed interindividual differences in response to dietary cholesterol [38]. Intriguing as these incompletely settled matters may be for researchers in this field, they in no way negate or bring into question the mainstream findings. That is, the central tendency (to use the language of statistics) for man as a species is to manifest a rise in serum cholesterol in response to increased dietary cholesterol (within the range of diets eaten by well-nourished man in different contemporary cultures). The scope of the effect, e.g. 6 mg/dl for every 100 mg change in dietary cholesterol (table XX), is one capable of exerting a sizable influence on CHD risk (cf. fig. 3; tables IV-A, IX, XII–XIV).

These findings in man on the relationship of dietary cholesterol to serum cholesterol and CHD risk are consistent with a vast array of animal-experimental data accumulated over the last 70 years, including recent findings in non-human primates showing the key role of dietary cholesterol in atherogenesis [4, 6, 7, 24, 31, 32, 47, 49, 53, 69, 74, 79, 82, 83, 88, 89, 91, 97]. These animal-experimental findings include studies demon-

Table XXI. Controlled experiment on the effect of dietary cholesterol on serum cholesterol in man

Group	Dietary composition				Mean serum cholesterol	
	sat. fat, %	poly-fat, %	total fat, %	cholesterol mg/day	mg/dl	% Change
1	16[1]	5	40	0	160	–
2	16	5	40	245	173	+ 8
3	16	5	40	550	184	+15
4	16	5	40	730	201	+26

Formula diets-dietary cholesterol from eggs. 48 healthy men, average age 26 years, 42 days on diet.
[1] % of total calories.

Table XXII. Plasma lipoprotein cholesterol in rhesus monkeys fed small amounts of cholesterol

Group	Period	Diet chol. mg/1,000 cal	Plasma cholesterol, mg/dl			
			LDL	HDL	total	LDL/HDL
1	baseline	0	90	59	149	1.52
	18 months	0	88	58	146	1.51
2	baseline	0	59	56	115	1.05
	18 months	43	82	48	130	1.70
3	baseline	0	62	55	117	1.12
	18 months	129	125	44	168	2.84

Semisynthetic diet: protein, 20% of cal; carbohydrate, 41% (3:1 starch: sucrose); fat, 39% (30% saturated, 43% monounsaturated, 27% polyunsaturated) (no effect on serum cholesterol); 600 cal/day to maintain positive calorie balance.

strating atherogenesis with feeding small amounts of cholesterol, such that little or no rise in serum cholesterol occurred. Research of this kind – done years ago in rabbits and chickens, with measurement only of serum total cholesterol [4, 37] –, was recently extended to non-human primates, with measurement of both serum total cholesterol and lipoprotein fraction cholesterol [6, 7]. A mixture of neutral fats was fed that produced no effect on serum total cholesterol, and the amounts of dietary cholesterol were small – 43 mg and 129 mg per 1,000 cal, respectively (less than in the diets of a majority of Americans, even nowadays, cf. table XVII). Little or no change in serum total cholesterol occurred, but LDL cholesterol rose substantially, HDL cholesterol fell (table XXII), and atherosclerosis developed [6, 7]. These findings are particularly meaningful when considered together with recent results on the effects of cholesterol feeding on the cholesterol content of chylomicrons and of possibly atherogenic chylomicron remnants, on the induction by cholesterol feeding in animals and man of an abnormal HDL subfraction that behaves like LDL in tissue culture (i.e. it attaches to cell receptors, is internalized, and degraded), and on the tendency of cholesterol feeding to disrupt physiological pathways for LDL disposal [11, 20].

One of the implications of the assertion that dietary cholesterol is inconsequential in man is that the human species is very different from those used in the animal laboratory, including the several species of nonhuman primates for whom development of significant atherosclerosis virtually requires cholesterol feeding. For this implication to be valid, one or both of two conditions must be met: Either one rejects the foundation of experimental medicine based in animal research, as universally accepted by biomedical science since *Darwin* and *Bernard,* or one presents detailed data showing man is exceptional. As to the latter, there are no such data; as to the former, clearly this is not a scientific alternative. Further, one special aspect of this assertion is also clearly invalid in the light of the data cited above, i.e. feeding of massive amounts of cholesterol is necessary to produce atherosclerosis in animals, hence the animal work has little or no relevance for man [4, 6, 7, 31].

Conclusion. Dietary cholesterol in the amounts consumed by hundreds of millions of people in the industrialized countries significantly influences serum total cholesterol, cholesterol-bearing lipoproteins, and CHD risk. The assertion that people in general, including apparently healthy people, need not be concerned about dietary cholesterol and foods high in cholesterol (e.g. eggs) is at variance with a vast array of scientific data on man and animals.

Assertion No. 5: Control of Overweight Is the Decisive Nutrition-Related Aspect of Life-Style for the Prevention of CHD

This assertion is put forward particularly in contradistinction to an emphasis on nutrient composition, particularly dietary lipid composition.

Table XXIII-A. Mean serum cholesterol and prevalence of hypercholesterolemia at different ages, in relation to relative weight[1], white men, Chicago Heart Association Detection Project in Industry, 1967–1973

Age	Relative weight				
	≤104	105–114	115–124	≥125	all
Mean serum cholesterol, mg/dl					
18–24	535[2]	500	388	397	1,820
	161.7	169.5	179.1	185.1	172.7
25–34	935	1,433	1,493	1,758	5,619
	175.9	187.5	192.8	197.4	190.1
35–44	507	989	1,347	1,917	4,760
	195.8	202.7	207.7	212.4	207.3
45–54	435	840	1,243	2,056	4,574
	201.5	209.2	215.6	216.9	213.7
55–64	295	560	765	1,343	2,963
	206.8	209.8	216.3	214.1	213.2
Prevalence of hypercholesterolemia, ≥220 mg/dl, per 1,000					
18–24	33.6	68.0	126.3	141.1	86.3
25–34	90.9	158.4	215.0	242.3	188.5
35–44	230.8	289.2	354.1	395.4	344.1
45–54	285.1	377.4	435.2	449.9	416.9
55–64	359.3	387.5	437.9	431.9	417.8
Prevalence of hypercholesterolemia, ≥240 mg/dl, per 1,000					
18–24	7.5	24.0	25.8	60.5	22.9
25–34	28.9	69.8	89.8	116.6	82.9
35–44	94.7	141.6	173.0	212.8	174.2
45–54	128.7	200.0	244.6	256.3	230.7
55–64	193.2	185.7	227.5	224.1	214.6

[1] Relative weight is the ratio of observed weight to desirable weight for height and sex, from actuarial tables, × 100 [69].
[2] Number of men.

As previous sections of this paper have made clear, overweight and weight gain are indeed a part of the problem. From childhood and youth on, they contribute to the development of key risk factors for CHD, i.e. hypertension, diabetes, elevated levels of serum total cholesterol, LDL cholesterol, VLDL cholesterol, and triglycerides, as well as low levels of HDL. Moreover, at least in some prospective studies, overweight, particularly marked

Table XXIII-B. Mean serum cholesterol and prevalence of hypercholesterolemia at different ages, in relation to relative weight, white women, Chicago Heart Association Detection Project in Industry, 1967–1973

Age	Relative weight				
	<104	105–114	115–124	≥125	all
Mean serum cholesterol, mg/dl					
18–24	1,407[1]	573	326	327	2,633
	169.1	173.7	174.6	176.9	171.8
25–34	986	517	297	368	2,168
	179.6	185.6	188.6	191.9	184.4
35–44	657	646	460	665	2,428
	194.0	197.7	202.3	201.1	198.7
45–54	760	924	906	1,474	4,064
	214.1	218.1	221.8	223.5	220.1
55–64	428	501	538	1,152	2,619
	227.3	234.8	235.0	233.5	233.1
Prevalence of hypercholesterolemia, ≥220 mg/dl, per 1,000					
18–24	65.4	89.0	73.6	94.3	75.2
25–34	102.4	154.7	188.6	192.9	142.1
35–44	185.7	235.3	267.8	278.2	245.5
45–54	411.8	465.4	484.5	528.5	482.5
55–64	542.1	666.7	665.4	619.8	625.4
Prevalence of hypercholesterolemia, ≥240 mg/dl, per 1,000					
18–24	24.2	43.6	15.3	36.7	28.9
25–34	37.5	77.4	84.2	87.0	61.8
35–44	88.3	113.0	143.5	118.8	113.7
45–54	239.9	277.1	305.7	312.8	288.1
55–64	352.8	419.2	427.5	413.2	407.4

[1] Number of women.

overweight, as such is a risk factor for CHD. Thus, in the national cooperative Pooling Project, relative weight (ratio of observed weight to desirable weight for height and sex) was significantly related to risk of a first major coronary event for men age 40–49 at entry (but not for those age 50–59) [57]. This relationship for the 40–49-year-old men held with control for serum cholesterol, blood pressure, and cigarette use, i.e. overweight had an independent and additive effect over and above its unfavorable influence on blood pressure and serum cholesterol. However, the independent effect was a relatively weak one, not nearly as strong as the impact of hypercholesterolemia, hypertension, and cigarette smoking on CHD risk. And in other studies, multivariate analyses show no independent effect of overweight for middle-aged men [37, 84].

It remains true, however, that even moderate weight reduction on a fat-modified diet contributes sizably and significantly to an enhanced improvement of the entire lipid-lipoprotein profile of obese persons, over and above the effect of change in dietary lipid composition (tables III, VII–IX). It also can have a significant impact on blood pressure [69, 73, 81], and it is the main non-pharmacologic means known for the control of NIDD.

Having reiterated all these points to underscore the importance of the prevention and control of overweight, this writer must also emphasize that any tendency to confine nutritional recommendations solely to control of overweight is invalid scientifically and unsound practically. As to the scientific aspects, however much overweight contributes to the prevalence and incidence of unfavorable cholesterol-lipid-lipoprotein patterns in the population, these patterns are common – given the composition, particularly the lipid composition, of the US diet – in those manifesting little or no overweight. This is clearly evident from the data in tables XXIII-A and XXIII-B. Even for those with relative weight less than 1.05 or in the range 1.05–1.14 (i.e. at or near desirable weight), mean serum cholesterol level and prevalence rate of hypercholesterolemia are progressively higher for middle-aged compared to young adults. Even with the improvements in diet composition in the USA in the last 10–15 years and the associated lower mean serum cholesterol levels, compare mean levels for non-obese Chicagoans in young adulthood and middle age with men in Kyushu and Tokyo (table XXIII-A; fig. 6) [35]. These data strongly support the concept that composition of the habitual diet, particularly lipid composition, importantly influences serum cholesterol, and calorie balance is far from being the whole story.

Note in tables XXIII-A and XXIII-B that by age 45–54 over 40% of US men and women at desirable weight in the 1967–1973 period had serum cholesterol levels of 220 mg/dl or greater; 13% of men and 24% of women had levels of 240 mg/dl or greater. By middle-age, millions of such Americans have high cholesterol levels. For those at desirable weight, advice dealing only with weight control is meaningless and useless. Moreover, as data from many cross-population studies indicate (including those

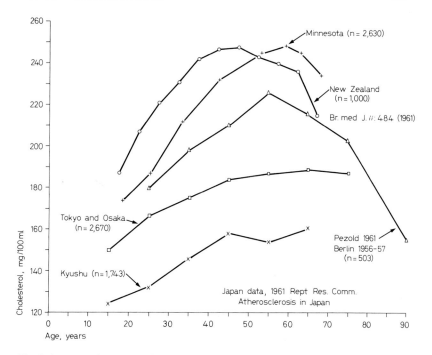

Fig. 6. Averages for serum cholesterol concentration in large samples of clinically healthy men [35].

on southern vs northern Europeans), and as repeated findings of controlled experiments in both free-living and institutionalized groups demonstrate, improved *composition* of the diet – i.e. marked reduction in saturated fat and cholesterol intake, moderate decrease in total fat intake, moderate increase in intake of polyunsaturated fat and fiber-rich (pectin- and gum-rich) complex carbohydrates from fruits, vegetables, whole grain cereal products, and legumes – makes a marked contribution to serum cholesterol reduction and improvement of the entire serum lipid-lipoprotein profile, with or without weight loss (figs. 3, 4; tables II, III, V–XII, XVII, XVIII, XXI).

For the tens of millions who are overweight and all too often hypercholesterolemic-hyperlipidemic as well, concern for the *composition* of their diet is not only relevant for achieving improvement of serum lipids-lipoproteins, i.e. by taking advantage of the synergistic effects of weight loss and improved composition. It is also essential practically, since it is almost impossible without attention to dietary composition to achieve long-term healthful change in eating habits – the real goal, not just 'going on a diet' – with achievement and *maintenance* of weight loss on a diet high in all essential nutrients. Moreover, with no guidance on composition forthcoming from nutrition science, millions in the community who are seeking help will inevitably turn – as has been happening for years – to unbalanced diets promoted by food faddists and self-serving 'experts', including the low-carbohydrate high-animal-fat diet, with its known capacity to raise serum cholesterol [39, 60].

Conclusion. Both massive sets of research data and realistic considerations of practical nutritional guidance for individuals, families, and populations compel the conclusion that attention to nutrient and foodstuff composition of the habitual diet – including its lipid composition – is a cornerstone of the effort to achieve improved eating habits, and thereby prevention and control of overweight and of hypercholesterolemia.

Assertion No. 6: All That is Needed To Improve Serum Cholesterol-Lipids-Lipoproteins Is To Reduce All Dietary Fats Proportionately and Consume Moderate or Low Fat, rather than High Fat

This assertion, a relatively new one on the diet-heart issue, in amplified form explicitly states that with this approach there is no need to be concerned about the composition of the fats consumed, e.g. butter and 'hard' margarine vs 'soft' margarine, a cheese-based salad dressing vs one from olive or corn or safflower oil, baking with lard vs corn oil as shortening. According to the reasoning underlying this assertion, the changes in composition of the visible fats that have occurred in the USA over the years – the marked decline in lard and butter and the marked increase in salad and cooking oils (fig. 7) [79] – are inconsequential.

To this writer's knowledge, those putting forward this assertion have not – in any publication, at least – defined quantitatively what they mean by a moderate or low-fat diet, or what predicted effects on serum cholesterol can be expected from their approach. These are crucial matters in regard to any scientific assessment of this assertion, as will become clear shortly.

First, however, a few words are necessary on the matter of the level of total fat in the diet. As is well known, all fats are concentrated calories (9 cal/g), irrespective of composition, and much more caloric than carbohydrates and proteins (4 cal/g) or even alcohol (7 cal/g). Therefore, one

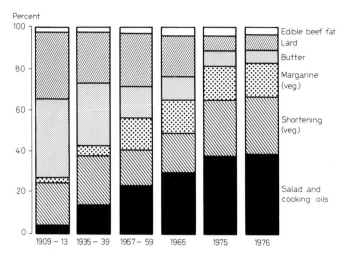

Fig. 7. Nutrient fat from fats and oils (per capita civilian consumption). US Department of Agriculture data [79]. Preliminary figures only for 1976.

way to control calories, for prevention or correction of obesity, a mass problem in the 'western' industrialized countries, is to reduce fat intake. Emphasis on eating less fats, visible and invisible, is an important aspect of the encouragement of an habitual eating pattern made up of foods of low-calorie and high-nutrient density, to assure intake of all essential nutrients at desirable levels, with moderation in total calories. Therefore, for Americans and other populations on *high* total fat diets (35–40+% of calories), this writer has for years supported and disseminated recommendations emphasizing a *moderate* total fat intake, i.e. 20–30% of total calories. (A low-fat diet is one at about the 10–15% level.) In addition to its possible value for calorie control with good nutrient intake levels, this approach may also be useful for reducing risk of thrombogenesis.

Given agreement on the desirability of eating less fat, and a quantitative definition of habitual diets high, moderate, and low in fat, what predictions are possible of the decreases in serum cholesterol with varying degrees of reduction of total fat – without attention to composition vs with attention to composition? The latter encompasses emphasis on decrease in intake of foods high in saturated fatty acids and cholesterol, with partial substitution of unsaturates and polyunsaturates for saturates (as recommended by the American Heart Association, the Inter-Society Commission for Heart Disease Resources, the United States Department of Agri-

Table XXIV-A. Predicted effects on serum cholesterol of various isocaloric approaches to dietary lipid modifications[1]

Total fat	Saturated fatty acids	Polyunsaturated fatty acids	Cholesterol		Predicted fall in serum cholesterol[1]	
			mg/day	mg/1,000		
% cal	% cal	% cal		% cal[2]	mg/dl	%
Baseline						
40.0	17.0	4.0	700	280	–	–
A. Proportionate decrease in dietary lipids						
A. 1. Total fat → 35% of calories						
35.0	14.9	3.5	613	245	6.1	2.5
A. 2. Total fat → 30% of calories						
30.0	12.8	3.0	525	210	12.2	5.1
A. 3. Total fat → 25% of calories						
25.0	10.6	2.5	438	175	18.8	7.8
A. 4. Total fat → 20% of calories						
20.0	8.5	2.0	350	140	25.2	10.5
A. 5. Total fat → 15% of calories						
15.0	6.4	1.5	263	105	31.7	13.2
A. 6. Total fat → 10% of calories						
10.0	4.3	1.0	175	70	38.6	16.1
B. 1. AHA, ICHD, USDA-DHEW type changes[3]						
30.0	8.0	9.0	250	100	37.8	15.8
C. 1. Progressive eating pattern-type changes[3]						
20.0	4.0	10.0	100	40	53.6	22.3
D. 1. Changes à la Greece, Southern Italy						
35.0	7.0	4.0	250	100	33.7	14.0
E. 1. Changes à la Japan						
15.0	4.0	4.0	250	100	42.8	17.8

[1] From a population group mean for middle-aged men of 240 mg/dl; prediction from the equation of *Keys* et al. [38]: Chol = 1.35 (2 $\triangle$ S– $\triangle$ P) + 1.5 $\triangle$ Z, where S and P are percent of calories from saturated and polyunsaturated fatty acids, respectively, and Z is the square root of dietary cholesterol in mg per 1,000 cal.

[2] Mean caloric intake 2,500 per day.

[3] AHA is American Heart Association [2, 3]; ICHD is the Inter-Society Commission for Heart Disease Resources [24]; USDA-DHEW is the United States Department of Agriculture-Department of Health, Education, and Welfare [86]; Progressive Eating Pattern as recommended in the Multiple Risk Factor Intervention Trial [14].

Table XXIV-B. Predicted effects on serum cholesterol of various isocaloric approaches to dietary lipid modifications[1]

Total fat	Saturated fatty acids	Polyunsaturated fatty acids	Cholesterol	Predicted fall in serum cholesterol[1]	
% cal	% cal	% cal	dg/day	mg/dl	%

Baseline					
40.0	17.0	4.0	7.00	–	–
A. Proportionate decrease in dietary lipids					
A. 1. Total fat → 35% of calories					
35.0	14.9	3.5	6.13	9.1	3.7
A. 2. Total fat → 30% of calories					
30.0	12.8	3.0	5.25	18.8	7.8
A. 3. Total fat → 25% of calories					
25.0	10.6	2.5	4.38	28.6	11.9
A. 4. Total fat → 20% of calories					
20.0	8.5	2.0	3.50	38.2	15.9
A. 5. Total fat → 15% of calories					
15.0	6.4	1.5	2.63	47.9	20.0
A. 6. Total fat → 10% of calories					
10.0	4.3	1.0	1.75	57.5	24.0
B. 1. AHA, ICHD, USDA, DHEW type changes[2]					
30.0	8.0	9.0	2.50	57.7	24.0
C. 1. Progressive eating pattern-type changes[2]					
20.0	4.0	10.0	1.00	78.1	32.5
D. 1. Changes à la Greece, Southern Italy					
35.0	7.0	4.0	2.50	51.6	21.5
E. 1. Changes à la Japan					
15.0	4.0	4.0	2.50	58.0	24.2

[1] From a population group mean for middle-aged men of 240 mg/dl; prediction from the equation of *Hegsted* et al. [22]: $\Delta \text{Chol} = 2.16 \Delta S - 1.65 \Delta P + 6.77 \Delta C - 0.5$ where S and P are percent of calories from saturated and polyunsaturated fatty acids, respectively, and C is dietary cholesterol in dg per day.

[2] For other abbreviations, see previous table.

culture, and Department of Health, Education, and Welfare, and others). Quantitative exploration of this question tests the soundness of the assertion about the value of reducing all fats without attention to type (composition), i.e. proportionate reduction. Based on the regression equations from dozens of controlled isocaloric experiments on man (table XI), predicted changes in serum cholesterol with both approaches can be readily calculated (tables XXIV-A, XXIV-B). If the starting point be a total fat intake of 40%, with saturates 17%, polyunsaturates 4%, and dietary cholesterol 700 mg/day (i.e. a diet about like that of the USA circa 1960 and the UK circa 1980), then a sizable balanced reduction of total fat to 30% of total calories (a 25% decrease), with no attention to lipid composition, will yield the diet changes shown in row A.2 of tables XXIV-A and XXIV-B, with predicted falls in serum cholesterol of 5.1 and 7.8%, respectively (mean: 6.4%). With a like total fat intake of 30% of calories, but attention to composition – i.e. emphasis on reduction in saturates (from 17 to 9% of calories), and on reduction in dietary cholesterol (from 700 to 250 mg/day), plus partial replacement of saturates by polyunsaturates, with a consequent moderate increase (from 4 to 9% of calories, *not* a high polyunsaturate diet, see below), as set forth in row B.1 of tables XXIV-A and XXIV-B, the predicted falls in serum cholesterol are 15.8–24.0% (mean: 19.9%) – *three times greater* than predicted with reduction of total fat without attention to lipid composition.

This writer, in discussing improved eating habits for individuals, families, groups, and populations, has repeatedly advised learning to enjoy and enhance the pleasures of eating by turning to the delectable cuisines of Mediterranean and Far Eastern countries. In regard to Mediterranean fare, Greek and southern Italy style [33, 34, 74], note row D.1 in tables XXIV-A and XXIV-B, vs row A.1, both at 35% total fat. Note that the 'classical' Greek-Italian dietary pattern, with olive oil (high in mono-unsaturated oleic acid) as the main visible fat, shows the same level of polyunsaturates as at baseline (4.0%) and is in this regard about the same as with balanced reduction (3.5%). But the saturates and dietary cholesterol are down much more with the Greek-southern Italian pattern and, consequently, the serum cholesterol is predicted to be down 14.0–21.5% (mean: 17.7%) vs 2.5–3.7% (mean: 3.1%), i.e. *a 5-times greater fall.* Only with a truly low-fat diet – 10–15% of calories, for most populations in 'western' industrialized countries a marked change in eating pattern – does the approach of decreasing all fats without attention to lipid composition yield predicted falls in serum cholesterol approaching those with recommendations focusing on lipid composition (e.g. rows A.5 and A.6 vs B.1, D.1, E.1, tables XXIV-A, XXIV-B).

Further, even more can be achieved – i.e. falls in serum cholesterol of 30–35%, and corresponding marked favorable changes in lipoprotein fractions (tables V, VI) – with reduced total fat intake and attention to both lipid and carbohydrate composition, with emphasis also on replacing fats with foods high in complex carbohydrates rich in fibers which lower serum cholesterol.

Finally, in practical terms, extensive experience shows that without detailed recommendations to people on shopping, cooking, and eating, significant changes in eating habits – including meaningful reduction in total fat – are not likely to occur.

Conclusion. Reduction of all fats – without attention to composition of the diet – is not a scientifically sound or practically realistic approach to achieving sizable improvement in serum cholesterol-lipid-lipoprotein levels for populations of 'western' industrialized countries. To achieve such changes, attention must be given to lipid composition of the diet, particularly to sizable reduction in saturated fat and cholesterol intake, plus attention to type of carbohydrate used in place of fat.

Assertion No. 7: The Proposed Nutritional Recommendation for Prevention of CHD Is a High-Polyunsaturated Fat Diet and There Is No Solid Scientific Support for This

This assertion is invalid, since in fact – as spelled out in earlier sections of this paper (e.g. see tables V-XII, XXIV-A, XXIV-B) – the eating style repeatedly recommended for improvement of serum cholesterol-lipoprotein patterns and prevention of CHD is *not* high in polyunsaturated fatty acids. Once again, the years-long dietary lipid recommendations, based on the excellent experience of Mediterranean and Far Eastern populations (see below) and their delectable cuisine, have been: low intake of saturated fatty acids (<10% of calories), low intake of cholesterol (<300 mg/day), moderate (*not* high) intake of polyunsaturated fatty acids (*up to* 10% of calories), and moderate intake of total fat (e.g. 30% of calories). This approach entails *partial* replacement of saturates with unsaturates and polyunsaturates, and partial replacement of saturates with complex carbohydrates from vegetables, whole grain cereals, legumes, and fruits, especially those high in the fibers (pectins, gums) that lower serum cholesterol. It yields a change from high total fat intake (40–45% of calories) to moderate fat intake (20–30% of calories). It does *not* involve high polyunsaturated fat intake (12–20% of calories). Its other central components, to complete the picture, are moderate (not high) alcohol ingestion (e.g. 2–3 drinks per day, if desired), moderate (not high) sodium intake (e.g. <75 mEq/day or <1,760 mg Na or <4,400 mg NaCl), emphasis on foods of high-nutrient and low-caloric density (hence moderate, *not* high intake of refined and processed sugars and of visible fats of desired composition,

Table XXV-A. Dietary goals for the United States

1 To avoid overweight, consume only as much energy (calories) as is expended; if overweight, decrease energy intake and increase energy expenditure
2 Increase the consumption of complex carbohydrates and 'naturally occurring' sugars from about 28% of energy intake to about 48% of energy intake
3 Reduce the consumption of refined and processed sugars by about 45% to account for about 10% of total energy intake
4 Reduce overall fat consumption from approximately 40% to about 30% of energy intake
5 Reduce saturated fat consumption to account for about 10% of total energy intake; and balance that with polyunsaturated and monounsaturated fats, which should account for about 10% of energy intake each
6 Reduce cholesterol consumption to about 300 mg a day
7 Limit the intake of sodium by reducing the intake of salt to about 5 g a day

Table XXV-B. Changes in food selection and preparation to achieve the US dietary goals

1 Increase consumption of fruits and vegetables and whole grains
2 Decrease consumption of refined and other processed sugars and foods high in such sugars
3 Decrease consumption of foods high in total fat, and partially replace saturated fats, whether obtained from animal or vegetable sources, with polyunsaturated fats
4 Decrease consumption of animal fat, and choose meats, poultry and fish which will reduce saturated fat intake
5 Except for young children, substitute low-fat and non-fat milk for whole milk, and low-fat dairy products for high-fat dairy products
6 Decrease consumption of butterfat, eggs and other high cholesterol sources. Some consideration should be given to easing the cholesterol goal for premenopausal women, young children and the elderly in order to obtain the nutritional benefits of eggs in the diet
7 Decrease consumption of salt and foods high in salt content

e.g. corn oil, safflower oil, sunflower oil, olive oil, soft margarine), moderate (*not* high) intake of total calories (tables XXV-A, XXV-B) [79]. This approach offers eminently practical, varied, and enjoyable patterns of eating, international gourmet style if so desired, as millions of American families have learned in recent years as they have moved progressively in this direction. It is completely feasible in terms of the varied fare currently available in US supermarkets and better restaurants. In terms of experience, both among sizable strata of the US population and populations worldwide, it is not a 'strange' or 'radical' or 'draconian' departure. It has nothing in common with 'crash' diets, rather it is an eating pattern for all seasons and all years. It has nothing in common with food faddism, since it is solidly based on a foundation of scientific knowledge and human experience.

The misconception that high polyunsaturate intake is a key component or the key component of the diet pattern recommended for serum cholesterol-lipid-lipoprotein improvement and CHD prevention stems from an historical development in research in this field. In the latter 1950s, the important discovery was made that dietary fats differ in their effects on serum cholesterol level, i.e. saturated fatty acids raise the level, but polyunsaturates have an opposite effect (table XI). The focus of some investigators in those years on this previously unsuspected and intriguing phenomenon led to the use of a verbal shorthand, i.e. the dubbing of diets for serum cholesterol reduction as 'high-polyunsaturated fat diets'. As this writer and his colleagues noted at the time [32], this was at best one-sided terminology, since in fact the diets being used in the human controlled experiments being described were almost always low in saturates and cholesterol, as well as high in polyunsaturates. The tendency of some to use the designation 'high-polyunsaturated fat' diet was perhaps reinforced by the fact that three of the early trials of CHD prevention by diet used a nutritional pattern low in saturates and cholesterol, and high in polyunsaturates, i.e. the Finnish mental hospital study, the Los Angeles Veterans Administration domiciliary facility study, and the New York Anti-Coronary Club [69, 74, 78, 93]. This was also the case for some groups in the National Diet-Heart Study in the early 1960s. However, for one of the experimental groups in that feasibility trial (group B), the diet pattern was low-saturates, low-cholesterol with moderate – *not* high – polyunsaturates. And in all its work, from 1957 through 1973, the Chicago Coronary Prevention Evaluation Program recommended a diet pattern of this latter type, i.e. *not* a diet high in polyunsaturates. In terms of advice to the general public, the statements in 1961 and 1978 by the American Heart Association, in 1970 by the Inter-Society Commission for Heart Disease Resources, in 1977 by the US Senate Select Committee on Nutrition and Human Needs, in 1980 by the US Department of Agriculture and US Department of Health, Education, and Welfare, in 1981 by the Bethesda Conference on Prevention of Coronary Heart Disease, and by the World Health Organization Expert Committee on Prevention of Coronary Heart Disease all emphasize reduced saturated fat and cholesterol intake, and

Table XXVI. Foods and nutrients available for consumption per person per day: USA, Italy, Japan (1954–1965)

Variable	USA		Italy		Japan	
Total calories	3,127	100.0[1]	2,697	100.0	2,226	100.0
Dairy, eggs, meats, poultry, cal	1,129	36.1	384	14.2	82	3.7
Fish, shellfish, cal	23	0.7	23	0.8	81	3.6
Fruits, vegetables, grains, legumes, cal	980	31.3	1,677	62.2	1,782	80.1
Oils, nuts, cal	168	5.4	320	11.9	97	4.4
Sugar, syrup, cal	505	16.1	222	8.2	160	7.2
Total protein, g	94	12.0[1]	79	11.7	68	12.2
Animal protein, g	68	8.7	27	4.0	16	2.9
Total carbohydrate, g	372	47.6	423	62.7	425	76.4
Total fat, g	135	38.9	77	25.7	25	10.1
Sat. fat, g	49	14.1	21	7.0	5	2.0
Poly. fat, g	16	4.6	13	4.3	7	2.8
Cholesterol, mg	586		246		129	
Calcium, mg	998		573		377	
Iron, mg	17		18		19	
Niacin, mg	21		20		20	
Riboflavin, mg	2.7		2.0		0.8	
Thiamine, mg	1.9		2.2		2.1	
Ascorbic acid, mg	125		127		100	

[1] Percentage of total calories.

moderate (*not* high) ingestion of polyunsaturates. Thus, there is a clear record over 20 years in regard to advice to the public.

In 1966, after a series of calculations similar to those in tables XXIV-A and XXIV-B, this writer observed, '... the contribution of polyunsaturated fats to serum cholesterol reduction is a minor one. The decisive changes remain reduction in ingestion of saturated fats and dietary cholesterol. It is a serious disservice to obscure this theoretically and practically important fact by overemphasizing the role of polyunsaturated fats, e.g. by giving the diets that lower serum cholesterol the misnomer of polyunsaturated fat diets. They are correctly characterized first and foremost as low-saturated fat, low-cholesterol diets' [68].

Over the years, this writer has concomitantly emphasized the wisdom in making nutritional recommendations to the general public of basing them on the years – and decades – long experience of populations with lower levels of serum cholesterol, lower rates of hypercholesterolemia, and lower CHD rates than those prevailing in the USA, particularly the experience of Mediterranean and Far Eastern populations (table XXVI) [74, 76]. None of these peoples – in fact, no population on earth, insofar as the writer can determine – has habitually consumed a high-polyunsaturate fare, i.e. 12% or more of calories from polyunsaturates [74]. The decisive characteristics of Mediterranean and Far Eastern fare in terms of lipid composition and serum cholesterol control are the lower levels of saturated fat and cholesterol compared to American fare, plus the higher intake of vegetable products and fiber, and the better calorie balance. These, not polyunsaturates, are the appropriate emphases, repeatedly made over the years.

Conclusion. The decisive lipid nutritional recommendations for the general population for CHD prevention are low saturated fat and low cholesterol intake. High intake of polyunsaturated fat is not one of the recommendations, and has not been for years. The 'issue' of high polyunsaturates is in essence a non-issue.

Assertion No. 8: The Safety of the Diet for Serum Cholesterol Reduction and CHD Prevention Is Open to Serious Question; Therefore, It Should Not Be Recommended

Over the years, several variants of this assertion have been presented. One has been that the recommended diet is high in polyunsaturates, no population has ever eaten such a diet, its long-term effects are unknown, and it may increase risk of cancer and of other diseases (e.g. cholelithiasis), it may influence the aging process unfavorably. As noted in the previous section of this paper, the recommendation is *not* for a high, but rather for a moderate polyunsaturate intake, i.e. up to 10% of total calories. Populations have ingested diets with levels of polyunsaturates of this order (e.g. Switzerland 6.8%, Sweden 7.6%, Netherlands 9.5%, Norway 10.9%) [74]. Despite the sizable toll exacted by CHD and the other atherosclerotic diseases among the middle-aged and elderly of these countries, their age-specific life expectancies are relatively high at all ages for both men and women (see below). During the last 10–15 years, mean level of polyunsaturates for Americans has increased from about 4 to about 7% of total calories [13, 75, 79]. For men and women, black and white, age 35–74 and all ages, these have been years of declining mortality rates from CHD, stroke, all cardiovascular diseases, and all causes (as already noted).

Imputing cause-and-effect in these relationships is very difficult, if not impossible. But the available data give no reason for concern for safety with polyunsaturates at levels up to 10% of calories. And there are reasonable bases for regarding levels in the 6–9% of calories range as desirable (vs 2–5%). In addition to serum cholesterol reduction (modest) and practical considerations of enjoyable eating, there are other possibilities of benefit, e.g. possible favorable influences on thrombogenesis and on blood pressure. (See also the Western Electric data, table IV-A.) In any case, the recommendation for the population is *not* for a high-polyunsaturate diet; hence, any 'issue' about the safety of such a diet is not relevant.

A second assertion about possible risks from the dietary recommendation raises the question of where the serum cholesterol goes when there is a fall induced by shift to a diet low in saturated fat and cholesterol. Specifically, could it be going into arteries, producing atherosclerosis? To this writer's knowledge, there is no evidence in man or animals in favor of this 'possibility'; it is strictly an hypothesis, unsubstantiated by fact. Moreover, considerable evidence against this 'possibility' is available in animals and man.

Another assertion – one of the newer ones – about safety is that cholesterol is an essential constituent of vital cell membranes, and lowering serum cholesterol by the recommended dietary means may do harm to cell membranes. Again, to this writer's knowledge, there is no evidence to support this 'possibility', and there is evidence against it. It too is pure hypothesis, conjecture.

In regard to all the assertions about safety, data on life expectancy would seem to be relevant, especially since they are solid facts pertaining to problems about which it is very difficult to collect truly critical evidence. Such data are especially meaningful, since it has been said in scientific meetings and in the medical literature that 'we should not meddle' with the diet of Americans, especially American children, since on this diet the US population has achieved great progress in life expectancy at birth in the last 30–50 years. The word 'meddle' is of course pejorative. Without any diversion to demonstrate its impropriety as applied to the dietary recommendations under discussion, it is appropriate to look at comparative data on life expectancy (table XXVII) [95]. Consider Greece and Italy as representative of the Mediterranean and Japan as representative of the Far Eastern diet for economically relatively advanced countries. Compare life expectancy at any age. Note that Italian males have better life expectancies than US males at all ages from birth through age 35.

Table XXVII. Life expectancy at selected ages, by sex and country, 1978

Country	Age, years									
	0	1	5	10	15	25	35	45	55	65
Men										
Finland[1]	67.4	67.2	63.4	58.6	53.7	44.4	35.2	26.4	18.6	12.2
USA[2]	69.4	69.5	65.7	60.9	56.0	46.9	37.7	28.8	20.7	14.1
England, Wales[3]	70.2	70.3	66.5	61.6	56.7	47.1	37.5	28.2	19.7	12.7
Switzerland	78.9	78.4	74.6	69.7	64.8	55.0	45.3	35.8	26.7	18.2
Sweden	72.5	72.2	68.3	63.4	58.5	49.0	39.6	30.4	21.8	14.3
Italy[1]	69.8	70.4	66.6	61.7	56.8	47.4	37.8	28.6	20.3	13.3
Greece	72.9	73.5	69.7	64.8	59.9	50.5	41.0	31.6	22.9	15.3
Japan	73.2	72.9	69.1	64.3	59.3	49.8	40.3	31.0	22.4	14.7
Women										
Finland[1]	76.3	75.9	72.1	67.2	62.3	52.5	42.8	33.4	24.3	16.1
USA[2]	77.3	77.3	73.5	68.6	63.7	54.0	44.4	35.1	26.5	18.6
England, Wales	76.3	76.3	72.4	67.5	62.6	52.8	43.0	33.6	24.7	16.7
Switzerland	72.0	71.7	67.9	63.1	58.2	48.9	39.5	30.2	21.6	14.2
Sweden	79.0	78.5	74.6	69.7	64.8	55.0	45.3	35.8	26.7	18.1
Italy[1]	76.1	76.5	72.6	67.7	62.8	53.0	43.3	33.8	24.8	16.5
Greece	77.6	78.0	74.1	69.2	64.3	54.6	44.8	35.2	26.0	17.5
Japan	78.6	78.2	74.4	69.4	64.5	54.7	45.0	35.5	26.3	17.8

[1] 1975 data.
[2] 1977 data.
[3] 1976 data.

Note that Greek males have much better life expectancies at all ages than US males, in fact, better also than all the other European populations, including the Swedish. Note further that Japanese males at most ages are second only to the Greeks in life expectancy; at birth they are the first in longevity. Of course, many factors influence life expectancy and dissecting their individual contributions is a most difficult task. But one thing is clear from these data: They lend no support to the thesis that the US diet is optimal health-wise, or to the thesis that progressive change in eating pattern in Mediterranean and Far Eastern directions has a potential for harm to health. Quite the contrary. For Greek and Japanese females too, life expectancy is better than for US females at every age from birth through 45.

Table XXVIII. Criteria for assessing the etiological significance of epidemiological associations

1 Strength of the association
2. Graded nature of the association
3 Temporal sequence
4 Consistency
5 Independence
6 Predictive capacity
7 Coherence: data from other research methods; reasonable pathogenetic mechanism

Finally, there is the recent assertion that the recommended diet may not be safe because of a possible inverse association between serum cholesterol and cancer, particularly colon cancer. Space does not permit a review of the literature available to date on this possible association [41]. Suffice it to note the following: such an association has been reported for men (but not for women) in some, but not in other, within-population prospective epidemiologic studies, i.e. the association has not been consistently recorded. When found, it is a weak one. The data available in no way fulfill the criteria (table XXVIII) [72, 74, 75, 78, 94], for inferring that the relationship is etiologic, i.e. there is no evidence that low serum cholesterol causes cancer. On the contrary, evidence from the largest study on this matter indicates that low serum cholesterol may be a consequence of occult cancer not manifest clinically on baseline examination of populations [23]. Moreover, all the studies were observational, i.e. they dealt with the question of relationship between serum cholesterol and cancer in populations eating their usual diets, frequently – as with the US samples – diets high in saturated fat and cholesterol. There is no evidence whatsoever relating *reduction* of serum cholesterol by diets low in saturated fat and cholesterol to cancer [41]. In fact, cross-population studies indicate that the *lower* the saturated fat and cholesterol intake, the *lower* the risk of major non-smoking-related cancers, particularly colon and breast cancer (table XXIX) [44]. Thus, it is possible that adoption of Mediterranean and Far Eastern type eating styles is protective against these cancers.

Finally, a methodological note is in order. However important it may be to pose possibly relevant questions about safety, these are hypotheses by definition since they have little or no foundation in fact. Hypotheses cannot be sound bases for failing to implement nutritional policy for CHD prevention. To propose not to act in the face of massive evidence because

Table XXIX. Nutrients available for consumption per capita, 1954–1965, and mortality from specified cancers, age 35–74, 1967–1973, 20-country study

Nutrient, per capita	Simple correlation coefficient	
	breast cancer women	colon cancer, men and women
Total calories	0.818**	0.811**
Total protein, g	0.620*	0.797**
Animal protein, g	0.664**	0.830**
Total fat, g	0.826**	0.779**
Saturated fat, g	0.743**	0.778**
Cholesterol, mg	0.747**	0.893**
Oleic acid, g	0.814**	0.854**
Polyunsaturated fat, g	0.318	0.075
Unsaturated fat, g	0.817**	0.708**
Carbohydrate, g	0.001	–0.037

* $p \leq 0.01$; ** $p \leq 0.001$.

hypothetical risks have been set forth is neither sound biomedical science nor wise public health policy.

Conclusion. Extensive evidence is available indicating both the efficacy and safety of the nutritional recommendations for prevention of the atherosclerotic diseases. Assertions about possible dangers of these recommendations are not buttressed by data, but rather are hypothetical, hence are not a sound basis for judgment about applying available knowledge to control epidemic disease and improve life expectancy with better health.

General Comment and Conclusion

In the foregoing pages, eight assertions casting doubt on the diet-CHD relationship, its etiological significance, and its implications for policy and practice in medicine and public health were assessed in detail. Based on the considerable amount of research evidence available, it was concluded that each of them is invalid. There are others that have currency. Space limitations here do not permit further detailed exposition. Moreover, they have been critically considered at length elsewhere over the years, item-by-item; the reader is referred to extant monographs and reviews [68–72, 74–80]. They too are without sound significant foundation in fact or theory.

Of the several aspects of modern life-style and life-style-related risk factors implicated in the etiology of epidemic atherosclerotic disease in 'western' industrialized countries ('rich' diet, cigarette smoking, sedentary habit, incongruent behavior patterns), 'rich' diet and diet-dependent risk factors play key, primary, and essential roles. Improvement in eating habits of the whole population, and especially its higher risk strata, with consequent prevention and control of diet-dependent risk factors, is a decisive component of a sound long-term strategy for conquering the epidemic of atherosclerotic disease.

References

1 Allen, A.; Stamler, J.; Stamler, R.; Gosch, F.; Cooper, R.; Trevisan, M.; Persky, V.: Observational and interventional experiences on dietary sodium intake and blood pressure. Proc. on the Role of Salt in Cardiovascular Hypertension, Philadelphia 1981.
2 American Heart Association, Central Committee for Medical and Community Program: Dietary fat and its relation to heart attacks and strokes (Am. Heart Ass., New York 1961).
3 American Heart Association, Nutrition Committee of the Steering Committee for Medical and Community Program: Diet and coronary heart disease (Am. Heart Ass., Dallas 1978).
4 Anitschkow, N.: Experimental arteriosclerosis in animals; in Cowdry, Arteriosclerosis, pp. 271–322 (Macmillan, New York 1933).
5 Anonymous: Recent trends in mortality from cardiovascular diseases. Stat. Bull. Metropol. Life Ins. Co. *60:* 3–8 (1980).
6 Armstrong, M. L.: Regression of atherosclerosis; in Paoletti, Gotto, Atherosclerosis reviews, vol. 1, pp. 137–182 (Raven Press, New York 1976).
7 Armstrong, M L.; Megan, M. B.; Warner, E. D.: Intimal thickening in normocholesterolemic rhesus monkeys fed low supplements of dietary cholesterol. Circulation Res. *34:* 447–454 (1974).
8 Beaglehole, R.; LaRosa, J. C.; Heiss, G. E.; Davis, C. E.; Rifkind, B. M.; Muesing, R. M.; Williams, O. D.: Secular changes in blood cholesterol and their contribution to the decline in coronary mortality; in Havlik, Feinleib, Proc. Conf. on the Decline in Coronary Heart Disease Mortality. NIH Publ. No. 79–1610, pp. 282–295 (US Department of Health, Education, and Welfare, Public Health Service, National Institutes of Health, Washington 1979).
9 Berkson, D. M.; Stamler, J.: Epidemiology of the killer chronic diseases; in Winick, Nutrition and the killer diseases, pp. 17–55 (Wiley & Sons, New York 1981).

10   Brown, H.B.: Diet and serum lipids-lipoproteins: controlled studies in the USA; in Farinaro, Stamler, Mancini, Proc. of the Int. Symp. on Epidemiology and Prevention of Atherosclerotic Disease, Anacapri 1981. Prev. Med. (in press).
11   Brown, M.S.; Kovanen, P.T.; Goldstein, J.L.: Regulation of plasma cholesterol by lipoprotein receptors. Science 212: 628–635 (1981).
12   Burke, B.S.: The dietary history as a tool in research. J. Am. diet. Ass. 23: 1041–1046 (1947).
13   Byington, R.; Dyer, A.R.; Garside, D.; Liu, K.; Moss, D.; Stamler, J.; Tsong, Y.: Recent trends of major coronary risk factors and CHD mortality in the United States and other industrialized countries; in Havlik, Feinleib, Proc. Conf. on the Decline in Coronary Heart Disease Mortality. NIH Publ. No. 79–1610, pp. 340–379 (US Department of Health, Education, and Welfare, Public Health Service, National Institutes of Health, Washington 1979).
14   Caggiula, A.W.; Christakis, G.; Farrand, M.; Hulley, S.B.; Johnson, R.; Lasser, N.L.; Stamler, J.; Widdowson, G.: The multiple risk factor intervention trial (MRFIT). IV. Intervention on blood lipids. Prev. Med. 10: 443–475 (1981).
15   Connor, W.E.; Cerqueira, M.T.; Rodney, M.S.; Wallace, R.B.; Malinow, M.R.; Casdorph, H.R.: The plasma lipids, lipoproteins, and diet of the Tarahumara Indians of Mexico. Am. J. clin. Nutr. 31: 1131–1142 (1978).
16   Cornfield, J.: Joint dependence of risk of coronary heart disease on serum cholesterol and systolic blood pressure: a discriminant function analysis. Fed. Proc. 21: suppl. II, 58–61 (1962).
17   Dyer, A.; Stamler, J.; Paul, O.; Berkson, D.M.; Lepper, M.H.; McKean, H.; Shekelle, R.B.; Lindberg, H.A.; Garside, D.: Alcohol consumption, cardiovascular risk factors, and mortality in two Chicago epidemiologic studies. Circulation 56: 1067–1074 (1977).
18   Dyer, A.R.; Stamler, J.; Paul, O.; Berkson, D.M.; Shekelle, R.B.; Lepper, M.H.; McKean, H.; Lindberg, H.A.; Garside, D.; Tokich, T.: Alcohol, cardiovascular risk factors and mortality: the Chicago experience. Circulation 64: suppl. III, 20–27 (1981).
19   Elveback, L.R.; Connolly, D.C.; Kurland, L.T.: Coronary heart disease in residents in Rochester, Minnesota. II. Mortality, incidence, and survivorship, 1950–1975. Mayo clin. Proc. 56: 665–672 (1981).
20   Goldstein, J.L.; Brown, M.S.: Lipoprotein receptors: from basic biology to clinical cardiology. George Lyman Duff Memorial Lecture 54th Scientific Sessions (Am. Heart Ass., Dallas 1981).
21   Gordon, T.: The Framingham diet study: diet and the regulation of serum cholesterol. Section 24; in Kannel, Gordon, The Framingham study; an epidemiological investigation of cardiovascular disease (US Government Printing Office, Washington 1970).
22   Hegsted, D.M.; McGandy, R.B.; Myers, M.L.; Stare, F.J.: Quantitative effects of dietary fat on serum cholesterol in man. Am. J. clin. Nutr. 17: 281–295 (1965).
23   International Collaborative Group: Serum cholesterol and risk of death from cancer: experience of an International Collaborative Group. J. Am. med. Ass. (in press).
24   Inter-Society Commission for Heart Disease Resources. Atherosclerosis Study Group and Epidemiology Study Group: Primary prevention of the atherosclerotic diseases; in Wright, Fredrickson, Cardiovascular diseases – guidelines for prevention and care, pp. 15–58 (US Government Printing Office, Washington 1974).

25 Jacobs, D.R., Jr.; Anderson, J.T.; Blackburn, H.: Do zero correlations negate the relationship? Am. J. Epidem. *110:* 77–87 (1979).

26 Jones, J.L.: Are health concerns changing the American diet? National Food Situation, NFS-159, pp. 27–28 (US Department of Agriculture, Washington 1977).

27 Jones, J.; Weimer, J.: A survey of health-related food choices. National Food Review, NFR-12, pp. 16–18 (US Department of Agriculture, Washington 1980).

28 Kagan, A.; Harris, B.R.; Winkelstein, W., Jr.; Johnson, K.G.; Kato, H.; Syme, S.L.; Rhoads, G.G.; Gay, M.L.; Nicham, M.Z.; Hamilton, H.B.; Tillotson, J.: Epidemiologic studies of coronary heart disease and stroke in Japanese men living in Japan, Hawaii and California: demographic, physical, dietary, and biochemical characteristics. J. chron. Dis. *27:* 345–364 (1974).

29 Kahn, H.A.; Medalie, J.H.; Neufeld, H.N.; Riss, E.; Balogh, M.; Groen, J.J.: Serum cholesterol: its distribution and association with dietary and other variables in a survey of 10,000 men. Israel J. med. Sci. *5:* 1117–1127 (1969).

30 Kato, H.; Tillotson, J.; Nicham, M.Z.; Rhoads, G.G.; Hamilton, H.B.: Epidemiologic studies of coronary heart disease and stroke in Japanese men living in Japan, Hawaii and California. Serum lipids and diet. Am. J. Epidem. *97:* 372–385 (1973).

31 Katz, L.N.; Stamler, J.; Experimental atherosclerosis (Thomas, Springfield 1953).

32 Katz, L.N.; Stamler, J.; Pick, R.: Nutrition and atherosclerosis (Lea & Febiger, Philadelphia 1958).

33 Keys, A. (ed.): Coronary heart disease in seven countries. Circulation *41:* suppl., 1–211 (1970).

34 Keys, A.: Mortality and coronary heart disease in the Mediterranean area. Proc. 2nd Int. Congr. on the Biological Value of Olive Oil, Torremolinos 1976, pp. 281–286.

35 Keys, A.: Serum cholesterol and the question of 'normal'; in Benson, Strandjord, Multiple laboratory screening, pp. 147–170 (Academic Press, New York 1969).

36 Keys, A. (ed.): Seven countries – a multivariate analysis of death and coronary heart disease (Harvard University Press, Cambridge 1980).

37 Keys, A.: WO Atwater Memorial Lecture: overweight, obesity, coronary heart disease, and mortality. Nutr. Rev. *38:* 297–307 (1980).

38 Keys, A.; Anderson, J.T.; Grande, F.: Serum cholesterol response to changes in the diet. I. Iodine value of dietary fat versus 2S-P. II. The effect of cholesterol in the diet. III. Differences among individuals. IV. Particular saturated fatty acids in the diet. Metabolism *14:* 747–787 (1965).

39 Krehl, W.A.; Lopez-S, A.; Good, E.I.; Hodges, R.E.: Some metabolic changes induced by low carbohydrate diets. Am. J. clin. Nutr. *20:* 139–148 (1967).

40 Levy, R.I.: Testimony before the Subcommittee on Nutrition of the Committee on Agriculture, Nutrition, and Forestry, United States Senate, 96th Congr. 1979, pp. 30–59 (US Government Printing Office, Washington 1979).

41 Levy, R.I.: Cholesterol and non cardiovascular mortality. Proc. 54th Annual Scientific Sessions (Am. Heart Ass., Dallas 1981).

42 Lewis, B.; Hammett, F.; Katan, M.; Kay, R.M.; Kerkz, I.; Nobels, A.; Miller, N.E.; Swan, A.V.: Towards an improved lipid-lowering diet: additive effects of changes in nutrient intake. Lancet *ii:* 1310–1313 (1981).

43 Liu, K.; Stamler, J.; Dyer, A.; McKeever, J.; McKeever, P.: Statistical methods to assess and minimize the role of intra-individual variability in obscuring the relationship between dietary lipids and serum cholesterol. J. chron. Dis. *31:* 399–418 (1978).

44 Liu, K.; Stamler, J.; Moss, D.; Garside, D.; Persky, V.; Soltero, L.: Dietary cholesterol, fat, and fibre, and colon-cancer mortality – an analysis of international data. Lancet *ii:* 782–785 (1979).
45 Liu, K.; Stamler, R.; Stamler, J.; Cooper, R.; Shekelle, R. B.; Schoenberger, J. A.; Berkson, D. M.; Lindberg, H. A.; Marquardt, J.; Stevens, E.; Tokich, T.: Methodological problems in characterizing an individual's plasma glucose level. J. chron. Dis. *35:* 475–485 (1982).
46 Liu, K.; Stamler, J.; Trevisan, M.; Moss, D.: Dietary lipids, sugar, fiber and mortality from coronary heart disease: a bivariate analysis of international data. Arteriosclerosis *2:* 221–227 (1982).
47 Mahley, R. W.: The role of dietary fat and cholesterol in atherosclerosis and lipoprotein metabolism. West. J. Med. *134:* 34–42 (1981).
48 Mahley, R. W.; Innerarity, T. L.; Bersot, T. P.; Lipson, A.; Margolis, S.: Alterations in human high-density lipoproteins, with or without increased plasma-cholesterol, induced by diets high in cholesterol. Lancet *ii:* 807–809 (1978).
49 Mahley, R. W.; Weisgraber, K. H.; Bersot, T. P.; Innerarity, T. L.: Effects of cholesterol feeding on human and animal high density lipoproteins; in Gotto, Miller, Oliver, High density lipoproteins and atherosclerosis, pp. 149–176 (Elsevier/North Holland, Amsterdam 1978).
50 Marmot, M. G.; Syme, S. L.; Kagan, A.; Kato, H.; Cohen, J. B.; Belsky, J.: Epidemiologic studies of coronary heart disease and stroke in Japanese men living in Japan, Hawaii and California. Prevalence of coronary and hypertensive disease and associated risk factors. Am. J. Epidem. *102:* 514–525 (1975).
51 Mattson, F. H.; Erickson, B. A.; Kligman, A. M.: Effect of dietary cholesterol on serum cholesterol in man. Am. J. clin. Nutr. *25:* 589–594 (1972).
52 McGill, H. C., Jr. (ed.): Geographic pathology of atherosclerosis (Williams & Wilkins, Baltimore 1968).
53 McGill, H. C., Jr.: The relationship of dietary cholesterol to serum cholesterol concentration and to atherosclerosis in man. Am. J. clin. Nutr. *32:* 2664–2702 (1979).
54 Mistry, P.; Miller, N. E.; Laker, M.; Hazzard, W. R.; Lewis, B.: Individual variation in the effects of dietary cholesterol on plasma lipoproteins and cellular homeostasis in man. J. clin. Invest. *67:* 493–502 (1981).
55 National Diet-Heart Study Research Group: The National Diet-Heart Study final report. Circulation *37:* suppl. 1, 1–428 (1968).
56 Nichols, A. B.; Ravenscroft, C.; Lamphier, D. E.; Ostrander, L. D.: Independence of serum lipid levels and dietary habits. J. Am. med. Ass. *236:* 1948–1953 (1976).
57 Pooling Project Research Group: Relationship of blood pressure, serum cholesterol, smoking habit, relative weight and ECG abnormalities to incidence to major coronary events: Final report of the Pooling Project. J. chron. Dis. *31:* 201–306 (1978).
58 Putnam, J. J.; Weimer, J.: Nutrition information – consumer's views. National Food Review, NFR-14, pp. 18–20 (US Department of Agriculture, Washington 1981).
59 Rhomberg, H. P.; Braunsteiner, H.: Excessive egg consumption, xanthomatosis, and hypercholesterolaemia. Br. med. J. *i:* 1188–1189 (1976).
60 Rickman, F.; Mitchell, N.; Dingman, J.; Dalen, J. E.: Changes in serum cholesterol during the Stillman diet. J. Am. med. Ass. *228:* 54–58 (1974).
61 Rizek, R. L.; Jackson, E. M.: Current food consumption practices and nutrient sources

in the American diet (Consumer Nutrition Center-Human Nutrition Science and Education Administration, US Department of Agriculture, Hyattsville 1980).

62 Robertson, T. L.; Kato, H.; Gordon, T.; Kagan, A.; Rhoads, G. G.; Land, G. E.; Worth, R. M.; Belsky, J. L.; Dock, D. S.; Miyanishi, M.; Kawamoto, S.: Epidemiologic studies of coronary heart disease and stroke in Japanese men living in Japan, Hawaii and California. Coronary heart disease risk factors in Japan and Hawaii. Am. J. Cardiol. *39:* 244–249 (1977).

63 Robertson, T. L.; Kato, H.; Rhoads, G. G.; Kagan, A.; Marmot, M. G.; Syme, S. L.; Gordon, T.; Worth, R. M.; Belsky, J. L.; Dock, D. S.; Miyanishi, M.; Kawamoto, S.: Epidemiologic studies of coronary heart disease and stroke in Japanese men living in Japan, Hawaii and California. Incidence of myocardial infarction and death from coronary heart disease. Am. J. Cardiol. *39:* 239–243 (1977).

64 Sacks, F. M.; Castelli, W. P.; Donner, A.; Kass, E. H.: Plasma lipids and lipoproteins in vegetarians and controls. New Engl. J. Med. *292:* 1148–1151 (1975).

65 Saunders, J. B.; Beevers, D. G.; Paton, A.: Alcohol-induced hypertension. Lancet *ii:* 653–656 (1981).

66 Shekelle, R. B.; Shryock, A. M.; Paul, O.; Lepper, M.; Stamler, J.; Liu, S.; Raynor, W. J., Jr.: Diet, serum cholesterol, and death from coronary heart disease – The Western Electric Study. New Engl. J. Med. *304:* 65–70 (1981).

67 Shekelle, R. B.; Stamler, J.; Paul, O.; Shryock, A. M.; Liu, S.; Lepper, M.: Dietary lipids and serum cholesterol level: change in diet confounds the cross-sectional association. Am. J. Epidem. *115:* 506–514 (1982)

68 Stamler, J.: Nutrition, metabolism and atherosclerosis – a review of data and theories, and a discussion of controversial questions; in Ingelfinger, Relman, Finland, Controversy in internal medicine, pp. 27–59 (Saunders, Philadelphia 1966).

69 Stamler, J.: Lectures on preventive cardiology (Grune & Stratton, New York 1967).

70 Stamler, J.: Practical aspects of the control of the coronary epidemic. Panel on diet; in Hansen, Schnohr, Rose, Ischaemic heart disease – the strategy of postponement, pp. 132–162 (FADL Publishing, Copenhagen 1977).

71 Stamler, J.: Introduction to risk factors in coronary artery disease; in McIntosh, Baylor College of Medicine Cardiology Series, vol. 1, part 3 (Medical Communications, Northfield 1978).

72 Stamler, J.: Lifestyles, major risk factors, proof and public policy. George Lyman Duff Memorial Lecture. Circulation *58:* 3–19 (1978).

73 Stamler, J.: Overweight, hypertension, hypercholesterolemia and coronary heart disease; in Mancini, Lewis, Contaldo, Medical complications of obesity, pp. 191–216 (Academic Press, London 1979).

74 Stamler, J.: Population studies; in Levy, Rifkind, Dennis, Ernst, Nutrition, lipids and coronary heart disease – a global view, pp. 25–88 (Raven Press, New York 1979).

75 Stamler, J.: Improved life styles: their potential for the primary prevention of atherosclerosis and hypertension in childhood; in Lauer, Shekelle, Childhood prevention of atherosclerosis and hypertension, pp. 3–36 (Raven Press, New York 1980).

76 Stamler, J.: The fat modified diet: Its nature, effectiveness, and safety; in Lauer, Shekelle, Childhood prevention of atherosclerosis and hypertension, pp. 387–403 (Raven Press, New York 1980).

77 Stamler, J.: Can an effective fat-modified diet be safely recommended after weaning for

infants and children in general? D. General discussion; in Lauer, Shekelle, Childhood prevention of atherosclerosis and hypertension, pp. 407–410 (Raven Press, New York 1980).

78　Stamler, J.: The established relationship among diet, serum cholesterol and coronary heart disease. Acta med. scand. *207:* 433–446 (1980).

79　Stamler, J.: Primary prevention of coronary heart disease: the last 20 years. Keynote Address, Eleventh Bethesda Conference. Am. J. Cardiol. *47:* 722–735 (1981).

80　Stamler, J.: Primary prevention of epidemic premature atherosclerotic coronary heart disease; in Yu, Goodwin, Progress in cardiology, vol. 10, pp. 63–100 (Lea & Febiger, Philadelphia 1981).

81　Stamler, J.; Farinaro, E.; Mojonnier, L. M.; Hall, Y.; Moss, D.; Stamler, R.: Prevention and control of hypertension by nutritional-hygienic means. Long-term experience in the Chicago Coronary Prevention Evaluation Program. J. Am. med. Ass. *243:* 1819–1823 (1980).

82　Stamler, J.; Katz, L. N.; Pick, R.; Rodbard, S.: Dietary and hormonal factors in experimental atherogenesis and blood pressure regulation; in Pincus, Recent progress in hormone research, vol. 11, pp. 401–452 (Academic Press, New York 1955).

83　Stamler, J.; Rhomberg, P.: Relationship of egg ingestion to serum lipids and atherosclerosis in experimental animals and man – a review of the literature (in press).

84　Stamler, R.; Stamler, J. (guest eds): Asymptomatic hyperglycemia and coronary heart disease – a series of papers by the International Collaborative Group, based on studies in fifteen populations. J. chron. Dis. *32:* 683–837 (1979).

85　Tillotson, J.; Kato, H.; Nichaman, M. Z.; Miller, D. C.; Gay, M. L.; Johnson, K. G.; Rhoads, G. G.: Epidemiology of coronary heart disease and stroke in Japanese men living in Japan, Hawaii and California. Methodology for comparison of diet. Am. J. clin. Nutr. *26:* 177–184 (1973).

86　US Department of Agriculture and US Department of Health, Education, and Welfare: Nutrition and your health – dietary guidelines for Americans (US Department of Agriculture, US Department of Health, Education, and Welfare, Washington 1980).

87　White, P. D.; Sprague, H. B.; Stamler, J.; Stare, F. J.; Wright, I. S.; Katz, L. N.; Levine, S. L.; Page, I. H.: A statement on arteriosclerosis, main cause of 'heart attacks' and 'strokes' (National Health Education Committee, New York 1959).

88　Wissler, R. W.: Principles of the pathogenesis of atherosclerosis; in Braunwald, Heart disease – a textbook of cardiovascular medicine, pp. 1121–1245 (Saunders, Philadelphia 1980).

89　Wissler, R. W.: Animal experimental findings: a key foundation of the preventive effort; in Farinaro, Stamler, Mancini, Proc. Int. Symp. on Epidemiology and Prevention of Atherosclerotic Disease, Anacapri 1981. Prev. Med. (in press).

90　Working Group on Arteriosclerosis of the National Heart, Lung, and Blood Institute: Report of the Working Group on Arteriosclerosis of the National Heart, Lung, and Blood Institute – summary, conclusions, and recommendations. NIH Publ. No. 81-2034, vol. 1 (US Department of Health and Human Services, Public Health Service, National Institutes of Health, Bethesda 1981).

91　Working Group on Arteriosclerosis of the National Heart, Lung, and Blood Institute: Report of the Working Group on Arteriosclerosis of the National Heart, Lung, and Blood Institute. NIH Publ. No. 81-2035, vol. 2 (US Department of Health and Human Services, Public Health Service, National Institutes of Health, Bethesda 1981).

92 Working Group on Arteriosclerosis of the National Heart, Lung, and Blood Institute: Population based research. Report of the Working Group on Arteriosclerosis of the National Heart, Lung, and Blood Institute. NIH Publ. No. 81–2035, vol. 2, pp. 157–260 (US Department of Health and Human Services, Public Health Service, National Institutes of Health, Bethesda 1981).

93 Working Group on Arteriosclerosis of the National, Heart, Lung, and Blood Institute: Prevention. Report of the Working Group on Arteriosclerosis of the National Heart, Lung, and Blood Institute. NIH Publ. No. 81–2035, vol. 2, pp. 261–422 (US Department of Health and Human Services, Public Health Service, National Institutes of Health, Bethesda 1981)

94 Working Group on Arteriosclerosis of the National Heart, Lung, and Blood Institute: Risk factors for the arteriosclerotic diseases. Report of the Working Group on Arteriosclerosis of the National Heart, Lung, and Blood Institute. NIH Publ. No. 81–2035, vol. 2, pp. 423–443 (US Department of Health and Human Services, Public Health Service, National Institutes of Health, Bethesda 1981).

95 World Health Organization: World Health Statistics Annual, 1980. Vital statistics and causes of death (WHO, Genève 1980).

96 Worth, R.M.; Kato, H.; Rhoads, G.G.; Kagan, A.; Syme, S.L.: Epidemiologic studies of coronary heart disease and stroke in Japanese men living in Japan, Hawaii and California. Mortality. Am. J. Epidem. *102:* 481–490 (1975).

97 Zilversmit, D.B.: Atherogenesis: a postprandial phenomenon. George Lyman Duff Memorial Lecture. Circulation *60:* 473–485 (1979).

J. Stamler, MD, Professor and Chairman, Department of Community Health and Preventive Medicine, and Dingman Professor of Cardiology, Northwestern University Medical School, Chicago, IL 60611 (USA)

Subject Index

Age trends, blood pressure 53
Alcohol
   correlation with HDL 92
   influence in hypertension 32, 57–59, 251
   in LRC study 98, 103–106
Apolipoproteins 111–118, 129, 130, 196–199, see also Lipoproteins
   dietary influence 161
   obese subjects 143
      fed a hypocaloric diet 144–148
   relation with body weight, fatness 141–150
Atherogenesis, lipid levels 148
Atherosclerosis, see also Coronary heart disease
   dietary influences 151–161, 166, 167, 247
   influence of hormones 167
   nutrition-related risk factors 245–302
   peanut oil effect 152
   reversible disease 247

Body weight, see also Obesity
   blood pressure 180–182
   HDL-C 142
   index 142–144
   lipoprotein components 141–148
      following dietary treatment 146, 147

Carbohydrates
   dextrin 157
   effects on cholesterolemia, atherosclerosis 157–159
   effects on HDL cholesterol 127
   glucose, fructose, sucrose, lactose, starch 157
   influence on lipids 159
Cardiovascular diseases, see Coronary heart disease
   diet 166, 230
   effect of nutrients 192
Central nervous system (CNS), factors in NaCl hypertension 216–218
Cholesterol
   dietary
      capacity for absorption 151
      composition of fats 288–293
      coronary heart disease (CHD) 248, 259
      effects 110, 115–117
         on CHD mortality 280–283
         on lipoproteins 195
      form, as factor 151
      intra- and interindividual variances 252
      plasma cholesterol 251
      recommendations 265, 293, 297
      transport 111
      US goal 294
      Western Electric Study, findings 254–258
   in blood
      age 16

Subject Index

dietary cholesterol 251
effect of fava beans 174
effect on CHD incidence 167, 246, 264–266, 271
low levels, Mediterranean 123
postprandial phase 131–135
trend, US men 276
variations induced by diet 117, 141–148
with fat intake 196
Cholesterolemia
effect of dietary components 161
effect, unsaturation of dietary fat 152
peanut oil effect 152
trans fatty acids 153
Chylomicrons 130, 139, 197
in obese subjects 142
in postprandial phase 132
relation with VLDL and HDL 131
Coronary heart disease (CHD), *see also* Atherosclerosis, Cardiovascular diseases
age, duration of follow-up 18
association
with hyperlipidemia 246
with hyperlipoproteinemia 89
cholesterol 230
dietary fat hypothesis 81
dietary intakes 167–173
effect of fava beans 174
epidemiological observations
China 1
Eskimo 20
Finland 6, 7
Germany 1, 2
Italy 2–5
Japan 5, 6
Japanese in Hawaii, California 5, 6
Java and Netherlands 1–3
Madrid 3
Minnesota 3
Naples 2, 3
Navajos 20
Rosetta 20
Scandinavia 1, 2
South Africa 4
etiology and pathogenesis 246, 247, 251
experimental models 174–176

mathematical models 14–18
minerals in water and diet 80–88
mortality rates
decline, US 277, 278
trends, specific countries 168
obesity 141
prevention, Rome Project 230–243
prospective studies 13–19
relation with dietary lipid, serum cholesterol 248, 249
risk factors 247
in childhood 123
role of dietary fats 170–174
socioeconomic factors 247
within culturally homogeneous groups 167

Diet, *see also* Hypertension
apolipoproteins, influence 161
atherosclerosis 166, 167, 247
new experimental models 166
multidisciplinary approach 167
cardiovascular diseases 166, 230
dietary goals, US 294
hypertension 31–71, 192–195
age trends 53
anthropological observations 31, 33–39
conclusion 62
dietary factors 31
discrepancies 63
epidemiological observations 44–64
experimental change in sodium intake 53
genetic susceptibility 49
physiological and clinical observations 32
population trends 47–50
preventive practice, public health 64–71
racial differences 50, 51, 63
regional differences 51
social and historical observations 39–44
within populations 51
hypocaloric, variations in lipoproteins 141–148

lipids 195–202
  in Naples 123–128
  low-fat, response 1–7, 117
  nature versus nurture 32
  relation to heart, controversial nature 248
  therapeutic, effect on lipoproteins 118

Elea 24

FAO 2, 124, 248
Fat intake 196
Fatness 141–148, *see also* Obesity
  high density lipoprotein 142
Fatty acids 59, 60, 98, 112
Fiber
  bran 160
  effect on hyperlipidemia 198
  in diet 126–128
  interaction with protein 160
  major classes 159
  pectin 159, 160
  prevents elevated cholesterol 250
  protective effect 159
  saponin components 198
Framingham 13–15, 51, 52, 59, 142

Genetic aspects, hypertension, susceptibility 49, 60–62, 177, 193, 223

High density lipoproteins (HDL) 7, 18, 91–92, 155
  changes with dietary cholesterol 195, 197, 198, 202
  cholesterol
    affected by diet 127, 167
    age, weight, height, ponderosity, correlation 93–95
    associated with carbohydrates 106
    correlation with selected variables 99
    distribution in North American populations 92
    percentile distribution of Quetelet index 92, 93
    relation to dietary cholesterol and fat 106
    relation to Quetelet index 95, 96
  correlation with alcohol 92
  correlation with lipoprotein lipase 138
  formation and catabolism 111–119
  HDL_2, HDL_3 in males, females 133–135, 138
  HDL_2/HDL_3 ratio 135–139
  HDL_2 in females 130
  increase following weight loss 144–146
    mobilization of cholesterol 146–148
  nutrients 91–92
  origin 130
  protective factors 129
  reduced levels with obesity 92, 141–143
  relation with chylomicrons and VLDL 131
  risk factor for CHD 91
  special role in atherogensis 148
  subfractions 130
Hyperlipidemia 89, 110
  associated with CHD 246
  dietary induction, models 192, 195–200
    fat intake 196
    fibers and saponins 198
    treatment models, polyunsaturated fatty acids 197
    vegetable proteins 198
    with dietary cholesterol 195
  spontaneous models 200–202
Hyperlipoproteinemias, associated with CHD 89–90
Hypertension, *see also* Diet
  Dahl rat strains 213
  diet 31–71, 192–195
  effect of proteins 193
  genetic susceptibility 49, 60–62, 177, 193, 223
  influence of kidney, rats 208
  mineralocorticoid 212
  obesity, risk factor 177–183, 194
  polyunsaturated fatty acids 194
  potassium intake 177, 186
  prevention and control 64–71
    government action 70
    industry action 69
    legislation 68
    multicenter trial 67
    national education programs 67

professional attitude change 66
public health recommendations 70
technology 69
voluntary agencies 70
relation to calorie imbalance 249
role of nutrition, experimental aspects 177–187
salt 32–71, 177, 183–186, 192–194, 214–223, 250
sodium/potassium ratio 177
treatment 224–227
  beta-blockers 225–227
  diuretics 224–227
  low-NaCl diet 224–227
  reducing cardiac output 226, 227
Hypertension Detection and Follow-up Programm 177
Hypertriglyceridemia, rats 118, 202
Hypocholesterolemia 167
  changes in VLDL, HDL, LDL 195–197
  dietary induction, animals 195
    rabbits 201, 202
  effect of fava beans 175
  familial 1
  involvement of hypothalamus 167

Intermediate density lipoproteins (IDL)
  effects of arginine 156, 157
  formation 130
  postprandial phase 133
Italian diet
  changes with time 169, 170
  effect on lipoprotein components 144
    metabolism 132–139
  southern Italy 123

Kidney 34, 35
  evolution 37–39
  influence on hypertension 208, 209

Lecithin-cholesterol acyltransferase (LCAT) 112–116, 129, 130
Lipemia in males, after fat meal 137
Lipid Research Clinics (LRC) Program 89–109
  alcohol intake data 98
  carbohydrate intake 106
  coronary primary prevention trial 91
  dietary cholesterol and fat 106
  dietary recall 96
  HDL cholesterol 92–96
  nutrient intake data 96
  objectives 90
  population studies 90
  prevalence study 89, 90, 91
    findings 92–95
  statistical analyses 98
Lipids
  atherogenicity of saturated fat 159
  body weight, fatness 141–150
  diet 195–202
    Naples 123–128
  diet effect 261
    animals 151–161
  dietary recommendations 265
  difference in sexes 138
  effect of cholesterol, animals 151
  following fat load 131–135
  independent relation to CHD 250
  newborns 123
  serum cholesterol, CHD 124, 248
Lipoprotein lipase 112, 117, 129, 130
  correlation with HDL 138
  reduced in obesity 142
Lipoproteins, *see also* Apolipoproteins, High density lipoproteins, Intermediate density lipoproteins, Low density lipoproteins, Very low density lipoproteins
  body weight, fatness 141–148
  dietary influences 151–161
  difference in sexes 138
  effects of dietary lipids 115–119
  factors affecting 90
  LDL/HDL ratio 148
  metabolism, dietary regulation 110–119
    endogenous cholesterol transport 113
    endogenous triglyceride transport 112
    transport of dietary fat and cholesterol 111
  nutrients 141, 166–176
  obese subjects 143
    following hypocaloric diet 144–148
  postprandial phase 129–139

response to dietary models 195-200
role in atherogenesis 129, 148
Low density lipoproteins (LDL) 112-119, 123, 132, 155, 196, 202, 260
cholesterol 127, 197
effect of fava beans 174
following weight loss 145
formation 130
LDL/HDL ratio 151
risk factor for cardiovascular diseases 129

Minerals
relation to CHD 32, 80-88
in drinking water 87
in food sources 87
Minnelea 23
Multifactorial preventive trials 230
Multiple Risk Factor Intervention Trial 262, 263, 275

Naples
diet and CHD, study 2-4
diet and lipids 123-128
National Diet-Heart Study 262, 264, 275
Ni-Hon-San Study, findings 267, 268
Nutrients, lipoproteins 166-176
Nutrition
impact on atherogenesis 148
role in hypertension
and cardiovascular diseases 230
experimental aspects 177-191
Nutrition-related risk factors for atherosclerosis 245-302

Obesity, *see also* Body weight
coronary heart disease 141
risk factor 283-288
correlation with
blood pressure 180
lipoproteins 141
non-insulin-dependent diabetes 249, 284
hypertension
adolescent, potent risk factor 178
anthropological observations 39
conclusions 62, 63

cross-cultural comparison 54
discrepant data for race 50, 54
major risk factor 194
mechanisms 182
migrant studies 54
social aspects 44
sociocultural influences 32
weight gain 54, 55
weight loss 55, 56
therapeutic effects 32
within cultures 32
within populations 54
inverse correlation with HDL 92

Phospholipids, postprandial phase 131, 134, 135
Polyunsaturated fatty acids 293-297
hypocholesterolemic effect 118, 119
prostaglandin metabolites 194
protective effect in hypertension 194
treatment for hyperlipoproteinemia 197, 198
Potassium
effect in hypertension 218
intake 177, 186
sodium/potassium ratio 186
Prostaglandins, NaCl hypertension, rats 221-223
Proteins
animal 197
and vegetable 32, 59, 154, 157
casein and soy 154
dietary, quantity 156
effect on hypertension 193, 194
methionine, lysine, proline 194
tryptophan deficiency 193
soybeans 119, 194, 199
effect on lipid metabolism 154-157
in diets 60
postprandial phase 134
vegetable, hypocholesterolemic activity 198-200, 250
Public health, aspects, recommendations 19-23, 64-71, 271-277, 300-302

Rome Project of Coronary Heart Disease Prevention (PPCC) 230-243

Subject Index

changes of risk factors 237
dietary behavior 240
entry examination data 236
hypertensive subjects, incidence data 242
mass education procedures 235
measurements 231
structure 231
treatment guidelines 234

Salt
  hypertension and 32–71, 192–194, 210–215, 250
    availability 41
    change in intake 53
    conclusions 62, 63
    cultural variations in use 35–38
    development of taste for 39
    dietary sources 42
    discrepant data for race 50
    epidemiological observations 32, 44–64
    genetic resistance 211, 218
    intake and population blood pressure 45
    intake within populations 51
    population trends 47
    public health recommendations 70, 71
  restriction, therapeutic effects 32
    social currency 40
    unacculturated villages 212
  intake 177, 183–186
    experimental models 184
    hypertension in man 185
      mechanisms 186
    population studies 183
Seven Countries Study 249, 265, 270
  findings 14, 17, 39, 54
  methodology 11, 15
  organization 7–9
Stroke
  rats 193
  with salt intake 47
Sucrose theory 20

Triglycerides 117, 118, 197
  associated with weight gain, loss 144–148, 249
  correlation with HDL_2 136–138
  endogenous transport 112
  following fat ingestion 130–133, 137
  formation 111
  Italian diet 127
  LRC prevalence study 93
  obese subjects 141
  variations induced by diet 141

USDA, food surveys 273

Very low density lipoproteins (VLDL) 123, 155
  changes with dietary cholesterol 195–197, 202
  effect of fava beans 174
  following weight loss 145–158
  obese subjects 141
  postprandial accumulation 132
  relation with chylomicrons and HDL 131
  risk factor for cardiovascular diseases 129, 130
  secretion 112–119

Water
  association with cardiovascular diseases 56, 57
    hardness, inverse relation 80, 81, 85
  mineral content 32, 87
  relation to CHD 80–88
  Zn/Cu ratio 80, 85
Western Electric Study 17, 254–258, 262, 280
World Health Organization (WHO) 2, 248
  European Multifactor Preventive Trial of CHD 230

Youth, blood pressure 61, 62

197003